EVOLUTION OF SUPPLY CHAIN MANAGEMENT

Symbiosis of Adaptive Value Networks and ICT

EVOLUTION OF SUPPLY CHAIN MANAGEMENT

Symbiosis of Adaptive Value Networks and ICT

edited by

Yoon S. Chang
KAIST, Republic of Korea

Harris C. Makatsoris
Orion Logic Ltd., United Kingdom

Howard D. Richards
MAPS, United Kingdom

KLUWER ACADEMIC PUBLISHERS
Boston / Dordrecht / New York / London

Distributors for North, Central and South America:
Kluwer Academic Publishers
101 Philip Drive
Assinippi Park
Norwell, Massachusetts 02061 USA
Telephone (781) 871-6600
Fax (781) 681-9045
E-Mail <kluwer@wkap.com>

Distributors for all other countries:
Kluwer Academic Publishers Group
Post Office Box 322
3300 AH Dordrecht, THE NETHERLANDS
Telephone 31 78 6576 000
Fax 31 78 6576 254
E-Mail <services@wkap.nl>

Electronic Services <http://www.wkap.nl>

Library of Congress Cataloging-in-Publication Data

EVOLUTION OF SUPPLY CHAIN MANAGEMENT: Symbiosis of Adaptive Value Networks and ICT/ Edited by Yoon S. Chang, Harris C. Makatsoris, Howard D. Richards.
p.cm.
Includes bibliographical references and Index.
ISBN: 1-4020-7812-9 (alk. paper)

1. Production, Operation, & Manufacturing Management. 2. Manufacturing Technology. 3. Operations Research/Management Science. I. Chang, Yoon S. II. Makatsoris, Harris C. III. Richards, Howard D.

Printed on acid-free paper.

Printed in the United States of America.

Dedication

Trust in the Lord with all your heart and lean not on your own understanding; in all your ways acknowledge him, and he will make your path straight (Proverbs 3:5-6)

To God, BokYoon Kim, YoonJoo Chang, ChangNam Hong, YoonSun Chang, Ha Young Moon, Pastor Jeon ByungWook--thanks for your love and prayer-Yoon S. Chang"

" To Katerina, Helen, George and Kostas – thank you for your support and understanding. – Harris Makatsoris, PhD"

" To Cicely – Howard D. Richards"

Contents

Preface

In the last half of the twentieth century industry encountered a revolutionary change brought about by the harnessed power of seemingly ever-increasing capacity, speed and functionality of computers and microprocessors. This strength provided management and workers within industries with new capabilities for management, planning and control, design, quality assurance and customer support. Organized information flow became the mainstay of industrial companies. New tools and information technology systems emerged and evolved to enable companies to integrate the various departments (Design, Procurement, Manufacturing, Sales and Finance) within companies, particularly the lager ones, including international corporations. This was to give them a chance to meet new demands for product time to market, just in time supply of orders, and customer support. To the smaller company these changes were not so apparent. Neither the tools nor systems nor indeed their economic value seemed appropriate to them except for special cases.

While all this was happening the structure of the larger companies began to disintegrate. Strong competitive pressures and globalization of the market place brought this about. Shedding unwanted competence and subcontracting it to others became common practice. Regional market pressures triggered companies to reorganize to create, produce, and distribute goods and services. Greater dependency on chains of supply from external companies became the norm. Medium and smaller sized companies began to gain some advantage and at the same time some were sucked into management and control systems governed by the larger companies. This was not so easy for them because they serviced more than one large company. As a result manufacturing service support and handling fluctuations in demand became difficult to achieve.

Improvements to communication technology such as the revolutionary Internet and the World Wide Web eased and helped the communication links between companies but it was manifestly obvious that something more had to be done. The competitive advantage of companies was for the main part dependent upon how well supply chains could be organized, managed and controlled to make them more effective and efficient and importantly fast acting. Moreover, every company has continued to strive to gain access to new markets and a wealth of new opportunities. As a consequence, solutions for new organizations and the way Information and Communication Technology (ICT) is used to support them to flourish is one of the most important challenges today. In addition Governments need to continue to sponsor research and developments to boost industrial research and to help SMEs. They also need to take action to cut local regulation and improve some of the legal aspects of electronic commerce and business.

This book focuses to a large extent upon supply chains that include manufacturing units and the ICT, tools and systems that will help them to move toward or create adaptable value networks. But is also applicable to other types of supply chain organization and contains ideas and concepts to meet the industrial challenge facing us all today and includes a number of case studies to illustrate our current understanding and to highlight the difficulties facing industry everywhere, large company or small. The volatility of demand (local and global), consumerism, and insatiable customer desires all contribute to the highly dynamic environment for business and supply chain alike.

The contents is intended to inspire the reader and lift the shades from their eyes to visualize new possibilities, which allow visualization of how ICT together with new gambits and research results create new possibilities for supply chains and those businesses wishing to collaborate with others in enterprise. The book seeks to provide a better understanding of the change process and the options available for the conception and formation of new organizations or the restructuring of existing ones to empower them as adaptable value networks. It deals with the adaptable and the collaborative aspects within organizations and seeks to give substance to value networks. Value networks are not someone's great idea but a way of categorizing a particular organization to serve new levels of effectiveness. New catalysts, which may be new technologies, components, tools, or system approaches for ICT, are also brought out in the book. These will be part of the crucible mix transforming ICT to permit new levels of efficiency in both supply networks and electronic marketplaces.

The book is structured in four parts with a final section for concluding remarks. Part one deals with the scale and scope of the process of metamorphosis. Part two deals with case studies. Part three deals with confronting new ways of working together. Part four deals with systems and tools. The front piece to each part contains a summary of what is to be found in the chapters contained within it. The book is aimed at managers and leaders in industry to provide an insight into current thinking and also to specialists who in part help to carry out the transformation process and also not forgetting the people who work within organizations who will make the transformation succeed.

PART 1. SCALE AND SCOPE

"Daring ideas are like chessmen moved forward, they may be beaten, but they may also start a winning game." Goethe

"The real voyage of discovery consists not in seeing new lands but seeing with new eyes" Proust

In the last half of the twentieth century new information and communication technology (ICT) has pervaded every business and every home. And we have seen, through a revolution brought about by microelectronics, cheaper and faster computers and networking technologies for industry that transformed the way business was done. There have been benefits to consumers with many new and affordable products and services to change the way we live. Television and newer forms of interactive communication as the Internet and the mobile phone accelerated this rise in consumerism. Expectations were raised: no longer were we prepared to put up with unreliable products, delays in getting new products or services, or delays to our orders or indeed the price we had to pay. Pressure was reflected back on industry, public services and governments at large.

How to make business processes more effective and efficient throughout the multitude of chains of supply and how to provide customers with the choice that they wished to receive had to be earnestly tackled, and in the end would have a lasting effect on society.

During the last decade considerable change has taken place to transform the global market beyond the recognition of the previous generation. But the change process is not a short overnight fix, it is a metamorphosis that will take a number of years before we will all know that a significant change has happened. There is no lack of technology, research and development, vendors offering a quick fix, consultants bombarding corporate offices to sell their services to guide change. But at the same time there is a general lack of understanding of what needs to be done and how to accomplish it. In this unsettled environment of confusion and anxiety, chaos is a source of creative experimentation. Out of this are emerging a surge of new ideas for new business organizational structures, as adaptive value networks, and new forms of market, as the e-Market, that will embrace new concepts for virtual working environments, new technologies, agents and tools that are the result of continuous human endeavor. These will in turn act as catalysts to the creation of further new ideas.

It is these ideas and the consequent implementations that will go some way to offsetting the soaring economic tide in developed countries racing in the from former underdeveloped countries, destabilizing the global economy, by creating and opening up many new good opportunities for business and the citizen alike. In part one of the book there are three chapters that will stir the imagination that will help to visualize the future,

Chapter 1, written by Shoumen Datta and his colleagues, sets out some of the challenges before adaptive value networks become a reality. The article ranges widely over methodologies and new practices, as game theory, autonomous agents, automated object identification, sensor networks and the semantic web, and throughout it provides new insights illustrated with many asides. Shoumen argues that managers must think differently for the support from ICT. They should take into account static systems reflecting a stable environment versus the alternative of adaptable systems ready to respond to any event in real time. They should also take into account the likely benefits that may be gained for decision making from an approach that is not centralized. The article presents the foundations of a possible route to the future with a convergence of feasible concepts and technologies to support it.

Chapter 2, written by Florent Frederix, presents a vision for manufacturing in the Western World and particularly in Europe as a post mass production renewal of adjusting products and services to need. Three value stages are brought out, pre-production, production and post-production, and arguing that the first and the latter of these will gain prominence in Europe. Each of these is amplified and examples cited on a roadmap to the future. Customer Relationship Management with two-way dialog, simultaneous engineering and concurrent manufacturing with distributed Tele-working, product life cycle planning and wireless tagging and control are all brought into the equation.

Chapter 3, written by Isidro Laso, presents a vision to turn the Internet into one single electronic market place in which trusted and secure collaborative business will take place, value networks created and the participation cost will be affordable to SMEs and Micro-Businesses as well as large companies. The chapter reviews the goals and aspirations and provides the focus of work in Europe as the Single Electronic European Market (SEEM). It is strewn with ideas for utilizing significant emerging technologies and also sets out what needs to be done through new research and development activities. It poses the challenges for the researcher and IT provider drawing amongst others upon previous research dedicated to isolated emarkets in Europe. Network centricity, virtualization, people orientation, company registries, personal assistant mobile electronic proxies, semantic web and web services are all dealt with amongst many others. And the chapter is illustrated with working scenarios to explain some of the future possibilities.

Chapter 1

ADAPTIVE VALUE NETWORKS

*Convergence of Emerging Tools, Technologies and Standards as Catalytic Drivers**

Shoumen Datta[1], Bob Betts[2], Mark Dinning[3], Feryal Erhun[4], Tom Gibbs[5], Pinar Keskinocak[6], Hui Li[1], Mike Li[1], Micah Samuels[7]
Massachusetts Institute of Technology[1], Timogen Inc.[2], Dell Corporation[3], Stanford University[4], Intel Corporation[5], Georgia Institute of Technology[6], Amazon.com[7]

Abstract: If a typhoon in the South China Sea impacts the shipment and delivery of memory chips to an assembly plant in Mexico City, you can count on the ripple effect to impact financial service providers, manufacturers and suppliers, shippers in charge of logistics and of course, the end-consumer. Can we plan to reduce the risk arising from such uncertainties? Can businesses (semiconductor plants, banks, logistics providers) cooperate to minimize uncertainties? Conventional wisdom states that uncertainties are equivalent to accidents and hence by nature remain unpredictable. However, application of tools and technologies based on emerging standards may partially disprove such wisdom. Focus on demand management may be the guiding light for supply chain practitioners. Can we collapse information asymmetries (between manufacturers and their lending institutions, for example) and add far more value to networks or demand webs? Real-time operational adaptability is key, especially in fast 'clockspeed' industries. Confluence of emerging tools, technologies and standards are required to converge to catalyze the evolution of such adaptable enterprise. Can real-time distributed data, in-network processing, Agent-based autonomy, taken together, tame the Bullwhip Effect? Can the (semantic) web catalyze the "Nash Equilibrium" of people (games) and information (theory) in our quest for real time "predictive" decision support systems? We will explore a few of these issues and how they may coalesce to enable the adaptive value network of the future.

Key words: Adaptive value networks, Game theory, Bullwhip effect, Agent, Automatic Identification Technologies

---

New technologies for supply chain management and flexible manufacturing imply that businesses can perceive imbalances in inventories at an early stage — virtually in real time — and can cut production promptly in response to the developing signs of unintended inventory build up.

Alan Greenspan
Testimony to the U.S. Senate Committee on Banking, Housing and Urban Affairs (13 February 2001)

---

*Disclaimer: This article is an over-simplified and incomplete exploration of a few tools and technologies that may converge to influence decision support systems in a manner that may catalyze the transformation of current-day supply chains toward an adaptive value network. In addition to named contributions, the corresponding author has used several sources of information in an effort to 'connect the dots' to suggest how apparently distant disciplines, if coalesced, may be complementary. Although the list of references is seriously incomplete, it should be amply clear that the original research is not due to the corresponding author. However, quotes, opinions, comments expressed in this article are solely attributable to the corresponding author and do not represent the views of MIT as an institution or the co-authors or their institutions or organizations. Some terms, for example, information asymmetry, may be used in a generic sense to imply lack of information or information visibility. This article is not an original research document but rather a synthesis of a few ideas, which, if taken together, may be catalytic in the transformation of supply chain management to become adaptive or perhaps even predictive. In this chapter, we suggest that adaptive may morph into 'predictive' through a confluence of principles. We advocate inclusion of ARCH (autoregressive conditional heteroskedasticity) and GARCH (generalized ARCH) in the context of supply chains to model high frequency (volume) real-time data (from RFID tags) that may also exhibit volatility (http://pages.stern.nyu.edu/~rengle/). All errors of content or coherence are solely due to the corresponding author.

1. INTRODUCTION

"At the science museum in Barcelona, I saw an exhibit that beautifully illustrated 'chaos.' A nonlinear version of a pendulum was set up so that a visitor could hold the bob and start out in a chosen position with a chosen velocity. One could then watch the subsequent motion, which was also recorded with a pen on a sheet of paper. The visitor was then invited to seize the bob again and try to imitate exactly the previous initial position and velocity. No matter how carefully it was done, the subsequent motion was quite different from what it was the first time. I asked the museum director what the two men were doing who were standing in a corner, watching us. He replied, "Oh, those are two Dutchmen waiting to take away the "chaos." Apparently, the exhibit was about to be dismantled and taken to Amsterdam. I have wondered ever since whether the services of those two Dutchmen would not be in great demand across the globe, by organizations that wanted their chaos taken away." (Gell-Mann 1994).

The holy grail of industry is to remove "chaos" from the supply chain in order to better adapt to demand fluctuations. Managing uncertainty is compounded by the increasing degree of information asymmetry[1] between the supply "chain" (value network[2]) partners (designers, suppliers, distributors, retailers, consumers) who have different and often conflicting objectives, that threaten to create barriers on the road to adaptive business networks of the future (Heinrich and Betts, 2003).

Ampex pioneered the video recorder market in 1956. Each unit was priced at $50,000 and the only competitors, RCA and Toshiba, were way behind. Sony, JVC and Matsushita were mere observers. Masaru Ibuka, co-founder of Sony and Yuma Shiraishi at JVC, issued directives for their respective engineers to produce an unit that would cost $500, a mere 1% of Ampex's price. In the 1980's, video recorder sales went from $17 million to $2 billion at Sony, $2 million to $2 billion at JVC, $6 million to $3 billion at Matsushita and $296 million to $480 million at Ampex. Failure to adapt eclipsed Ampex. (Tellis and Golder, 1996).

One business objective of suppliers is to secure large volume purchase commitments (with delivery flexibility) from manufacturers. It conflicts with the manufacturer's objective that must include rapid response to demand fluctuation yet the manufacturer must mass produce (to take advantage of economies of scale) yet production runs must adapt to fluctuations even though a certain run may have been planned based on demand forecast. Thus, manufacturers may need more or less raw materials and therefore seek flexibility in purchasing raw materials, which conflicts with the supplier's objective. Manufacturer's desire to run long production batches are in conflict with the warehouse and distribution centers that aim to reduce inventory due to storage capacity constraints. The latter increases cost of transportation for all the players.

During 2000, supply chain related costs in USA alone exceeded $1 trillion (10% of GDP), which is close to the entire GDP of Russia, is more than the GDP of Canada or Spain or the *combined* GDP of all the 22 nations who are members of the League of Arab Nations. The combined GDP of all 22 Arab nations is less than that of Spain (www.wrmea.com/archives/sept-oct02/0209044-2.html; www.bea.doc.gov; www.cia.gov). A mere 10% savings of supply chain costs in USA is nearly equal to the GDP of Ireland. Therefore, tools and processes that may reduce supply chain inefficiencies and help it better adapt to demand changes, are valuable. We will briefly explore some of the tools that may catalyze the adaptive value network.

Some emerging technologies may take leading catalytic roles but technology is *not* the solution. Ability to adapt supply chains will depend on continuous business process innovation led by management capable of envisioning use of technology as a tool to reduce (1) inefficiencies, (2) uncertainties and (3) information asymmetry within the value network. In essence, decision making processes should respond to (dynamic) information such that the system (enterprise) is able to rapidly adapt and/or respond.

One driver of this transformation (from 'push' based supply chain management to 'pull' based adaptive value networks) is the potential use of real-time information to catalyze or trigger autonomous decision steps capable of re-planning and execution. By some estimates, business in 2003 generated more than 1 terabyte of data per second (excludes data gathered by automatic identification technologies). Is this equivalent to information? It is unlikely that this data, as is, can be considered as information. Even when we extract the information, will it offer a "transactional" value? The ability to extract intelligence from data to manage information may be the differentiator between companies who will profit from data (such as automatic identification or sensors) versus those who will not. Data that is stored in business systems (such as ERP) may suffer from problems that reduce the value of their information. ERP systems may also compromise the efficacy of dynamic data if the data feeds static systems unable to respond in near real-time. When such ERP data and/or information sources are used by strategic planners for forecasting and optimization, it leaves room for speculation about the validity of the outcome since the process may have been optimized, or forecast delivered, based on "noise" rather than robust dynamic data. Stemming from poor data quality and information asymmetry between supply chain partners, these errors (of optimization, forecasting) accumulate at successive stages of a supply chain and manifests itself as the generic Bullwhip Effect (Forrester 1961, Sterman 1989, Lee et al., 1997) (Figure 1-1 and Figure 1-2).

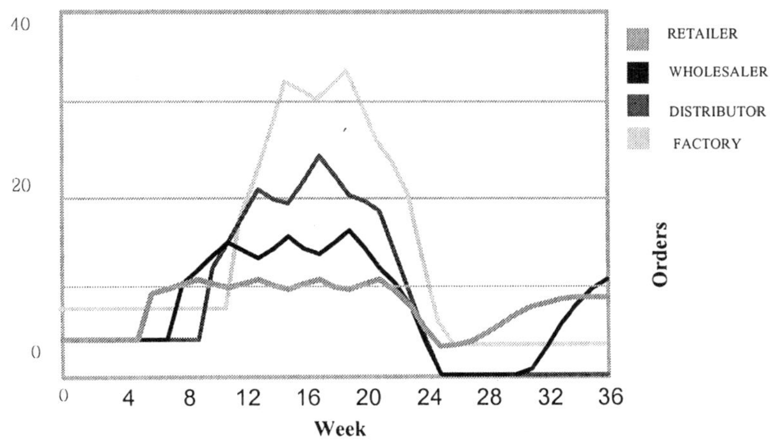

Figure 1-1. The Bullwhip Effect (Source: Joshi 2000)

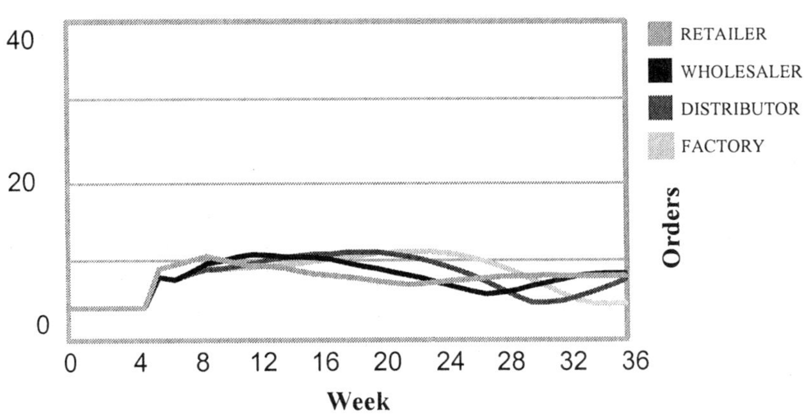

Figure 1-2. How do we Taming the Bullwhip Effect? (Source: Joshi 2000)

Example from the semiconductor equipment supply chain shows demand forecast versus actual purchase of equipment by Intel Corporation (Figure 1-3).

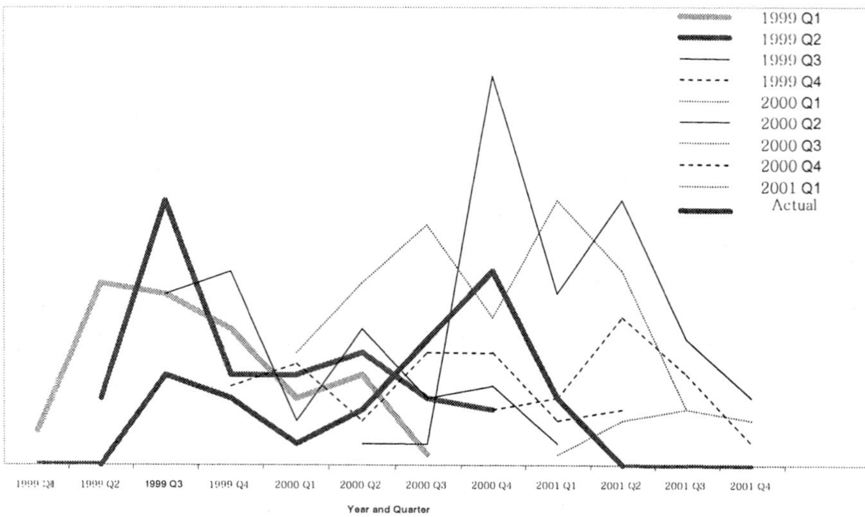

Figure 1-3. Intel Tool Order Data 1999-2001 (Source: Cohen, Ho, Ren and Terwiesch, 2003)

2. TOWARD ADAPTIVE VALUE NETWORKS: INFORMATION VISIBILITY (TRANSPARENCY)

Tools and technologies that may be catalytic in taming the Bullwhip Effect may also have an impact on making supply chains more adaptive by ushering in adaptive decision support systems. However, both assume the success of business process innovation to improve intra- and inter-enterprise information exchange as well as efforts to break down data silos as a segue to distributed data infrastructure. In thinking about adaptive supply chain management, it is helpful to analyze how the tools and technology catalysts may help connect ***objects to processes*** and ***processes to decision systems***. Some of these "catalysts" may be classified into two broad (albeit arbitrary) categories. We will make an attempt to show how a few of these catalysts may (converge to) transform current supply chains to become more adaptive (Table 1-1).

Table1-1. Tools and Concepts versus Data Mobility

Tools and Concepts	Data Mobility
Operations Research and Game Theory	Automatic identification technologies (RFID, UWB, GPS)
Distributed Artificial Intelligence and Agents	Wireless protocols (802.11, 802.16)
'Clockspeed' as defined by Charles Fine, MIT	Sensor networks
Semantic Web	
Simulation	

3. OPERATIONS RESEARCH AND GAME THEORY[3]

The workhorse of optimization (algorithms) is based on operations research. It is an area of intense research and innumerable sources of information are available. However, Game Theory was not such a "household" name until 1994 when John Nash, and later the movie about him, changed the public's perception so much so that generic business journals began touting the virtues of Game Theory.

> *Having game theory in your corporate 'bag of tricks' can mean the difference between success and failure. (Investor's Business Daily, 25 January 1996).*

> *Managers have much to learn from game theory - provided they use it to clarify their thinking, not as a substitute for business experience (The Economist, 15 June 1996).*

Game theory helps us model, analyze and understand the behavior of multiple self-interested parties who interact while making decisions. As such, Game Theory deals with interactive optimization problems. In particular, it is a tool for analyzing situations where the parties strive to maximize their (expected) pay-offs while choosing their strategies. Each party's final pay-off depends on the profile of strategies chosen by all parties. Most business situations can be modeled by a "game" since in any business interaction, involving two or more parties, the pay-off of each party depends on the other party's actions.

For centuries economists have worked on various game-theoretic models but Neumann and Morgenstern (1944) are credited as the fathers of modern Game Theory. Game Theory has since enjoyed an explosion of developments, including the concept of equilibrium (Nash, 1950), games with imperfect information (Kuhn, 1953), cooperative games (Aumann, 1959; Shubik, 1962) and auctions (Vickrey, 1961).

3.1 Let the Game Begin

The overarching theme in Game Theory is "interactions." In business, each decision maker is a player making a decision or choosing a strategy that will be impacted by the competitor.

"A chip manufacturer slashed prices of its desktop and mobile processors just days after a similar move by a rival. We're going to do what it takes to stay competitive' on prices, said representative. The company's aggressive price-chopping means the company doesn't want to give up market share gains, even at the cost of losses on the bottom line (CNet News.com, May 30, 2002)"

In this example, companies compete on price to gain market share. During Q1 of 2002, this semiconductor company increased processor shipments (compared to Q4 of 2001) worth $8 million but processor revenue declined by 3% (sold more chips for less money). Companies engaged in price wars rarely, if ever, benefit from such competition. Reducing prices slightly might increase the overall market potential but decreasing prices beyond a certain limit has a diminishing impact. Eventually the size of the "pie" does not increase anymore and firms fight harder to get a bigger "pie" by slashing prices and profits. Why do firms behave this way? In this situation and in some others, firms are caught in what is known in Game Theory as the "Prisoner's Dilemma" where the rational response may not be the optimal.

3.1.1 Prisoner's Dilemma

Two burglars, Alice and Bob, are arrested near the scene of a burglary and interrogated separately. Each suspect can either confess with a hope of a lighter sentence or may refuse to talk (does not confess). The police do not have sufficient information to convict the suspects, unless at least one of them confesses. Each must choose without knowing what the other will do. In other words, each has to choose whether or not to confess and implicate the other. If neither confesses, then both will serve one year on a charge of carrying a concealed weapon (Table 1-2). If both confess and implicate each other, both will go to prison for 10 years. However, if one burglar confesses and implicates the other but the other burglar does not confess, then the one who cooperates with the police will go free, while the other burglar will go to prison for 20 years on the maximum charge. The "strategy space" in this case is simple: confess or don't confess (each chooses one of the two strategies). The payoffs (penalties) are the sentences served. In effect, Alice chooses a column and Bob chooses a row.

Table 1-2. Prisoner's Dilemma

		Alice	Alice
		Confess	*Does not*
Bob	*Confess*	10, 10	0, 20
Bob	*Does not*	20, 0	1, 1

The two numbers in each cell show the outcomes for the two prisoners when the corresponding pair of strategies are chosen. The number to the left shows the payoff to the person who chooses the rows (Bob) while the number to the right tells the payoff to the person who chooses the columns (Alice). Thus (reading down the first column) if they both confess, each gets 10 years, but if Alice confesses and Bob does not, Bob gets 20 and Alice goes free. Therefore, what strategies are "rational" in this game if both of them want to minimize their sentences? Alice might reason: *"Two things can happen: Bob can confess or Bob can keep quiet. If Bob confesses, I get 20 years (if I don't confess) and 10 years if I do confess (cooperate), so in that case it is better to confess. On the other hand, if Bob doesn't confess and I don't either, I get a year but in that case, if I confess I can go free. Either way, it is better if I confess. Therefore, I will confess."*

But Bob can and presumably will reason in the same way. So they both reason rationally to confess and go to prison for 10 years each. But, if they had acted "irrationally" and did not confess, they each could have gotten off with a year (http://william-king.www.drexel.edu/top/eco/game/ game.html).

Prisoner's Dilemma[4] is an example of a non-cooperative static game where the players choose strategies simultaneously and are thereafter committed to their chosen strategies. The main issue of such games is the existence and uniqueness of Nash equilibrium (NE). NE is the point where no player has incentive to change her strategy since each player has chosen a strategy that maximizes his or her own payoff given the strategies of the other players. It may be prudent to point out that the fundamental distinction between cooperative and non-cooperative games is that cooperative games allow binding agreements while non-cooperative games do not. Study of cooperative games focuses on the outcome of the game in terms of the value created through cooperation of players, but does not specify the actions that each player will take. The study of non-cooperative games is more concerned with the specific actions of the players. Hence the former allows us to model outcomes of more complex business processes.

3.1.2 Dilemma in Prisoner's Dilemma

A key concept not captured in "Prisoner's Dilemma" is the repetition of interactions. In business, most players know they will be in the "game" for awhile. Hence, they may choose to cooperate, especially if they deem that cooperation today may increase the chances of cooperation, or even collaboration, in the future. With repeated actions, companies build a reputation, which influences the actions of others. For example, a restaurant might make a higher profit today by selling slightly less fresh food (prepared yesterday), but will it be worth the possible consequence of losing customers in the future? Thus, rationally speaking, companies aim to act strategically with competitors and partners. In practice the behemoths continually try to squeeze their suppliers (blinded by "cost" reduction) and could lose critical partners.

Intel uses its much envied supplier ranking and rating program - which tracks a supplier's total cost, availability, service, supports responsiveness and quality – to keep its top suppliers on a course for better quality. 'We reward suppliers who have the best rankings and ratings with more business,' says Keith Erickson, Director of Purchasing.

As an added incentive, Intel occasionally plugs high-quality suppliers in magazine and newspaper advertisements. The company even lets its top performing suppliers publicize their relationship with Intel. That's a major marketing coup, considering that Intel supplies 80% of chips used in PCs today and is one of the most recognized brand names in the world.

Given that each party in a supply chain acts entirely on self interest, individual choices collectively do not lead to an "optimal" outcome for the supply chain. Thus, supply chain profit of a "decentralized" supply chain composed of multiple, independently managed companies, is less than the total supply chain profit of the "centralized" version of the same chain where the partner interactions (suppliers, manufacturers, retailers) are managed by a single decision-maker (information symmetry) to optimize total supply chain profits. Sharing of information in centralized supply chains reduces inefficiencies that are obvious in decentralized supply chains due to "double marginalization" stemming from self-centered decision making.

3.1.3 Optimal Profit is Higher in Centralized Supply Chains with Information Sharing (Symmetry)

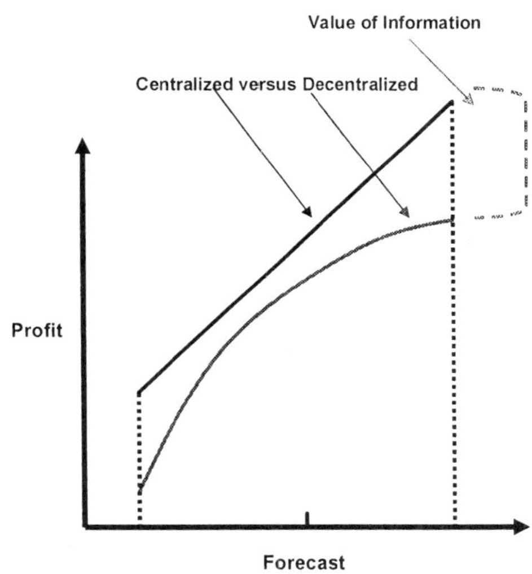

Figure 1-4. Optimal Profit versus Forecast (personal communication; Ozalp Ozer, Stanford University)

One strategy for reducing inefficiencies in decentralized supply chain (and consequent loss of profit) is "vertical integration" where a company owns every part of its supply chain. An example of vertical integration is Ford Motor Company. Earlier in the 20th century, in addition to automobile factories, Henry Ford owned a steel mill, glass factory, rubber plantation and iron mines. Ford's focus was on (mass production) making the same car, the Model T, cheaper. This approach initially worked well. The price of a Ford Model T fell from $825 in 1908 to $290 in 1924. By 1914, Ford had a 48% share of the US market. By 1920, Ford was manufacturing half the cars made worldwide. Vertical integration allows a company to obtain materials at a low cost and control the entire supply chain.

In today's economy, customer demand and preferences change rapidly. Companies that focus on core competencies are more likely to be nimble in order to stay ahead of competition and succeed. Hence, we see a trend towards "virtual integration" where supply chains are composed of independently managed but tightly partnered companies. Information sharing and vendor managed inventory (VMI) are strategies successfully used by some companies (such as Dell Corporation) for higher degree of virtual integration. However, most companies still find it difficult or face internal reluctance to usher changes in their supply chain practices. Similar issues apply to independently managed intra-company divisions, such as marketing, production, and sales.

Game Theory makes some assumptions in its study of the impact of interactions of multiple players and the resulting dynamics in a market environment. Two key assumptions are:

- Each player in the market acts on self-interest, but they pursue well-defined exogenous objectives, that is, they are rational;
- In choosing a strategy, a player considers the potential reaction of other players and takes into account his or her knowledge of other decision makers' behavior, that is, he or she reasons strategically.

These assumptions rule out games of pure chance, such as lotteries and slot machines, where strategies do not matter and games without strategic interaction between players, such as Solitaire. Credibility is a central issue in games.

Coca-Cola is developing a vanilla-flavored version of its best-selling flagship cola, a report says, extending the company's palette of flavorings from Cherry Coke and Diet Coke with lemon. But don't expect to see a vanilla-flavored Pepsi anytime soon. 'It's not something we're looking at,' said spokesman Larry Jabbonsky of Pepsi. 'We think it's a bit vanilla.' (USA Today, 1 April 2002).

PepsiCo is launching Pepsi Vanilla and its diet version in stores across the country this weekend. Coke came out with Vanilla Coke in May 2002 and it was a resounding success, selling 90 million cases. 'We're a little surprised that Pepsi decided to enter the vanilla segment,' said Mart Martin of Coca-Cola. 'When we came out with Vanilla Coke, Pepsi originally said the idea sounded 'a bit vanilla.' (CNN/Money August 8, 2003).

3.2 Game Theory in Quantity and Price Competition

Business decisions include what to produce/procure, sell, how much, and at what price. Study of competitive interactions around these issues can be addressed using game theoretic models that focus on price and quantity decisions (several excellent papers including that of Albeniz and Simchi-Levi, 2003).

Quantity competition (Cournot Game) is especially important for commodities where there is an inverse relationship between quantity and market price. Price competition (Bertrand Game), on the other hand, occurs in every market, as competing companies try to maintain or increase market share.

OPEC decided to slash its crude oil production by 1.5 million barrels a day (6%). The issue came to a head this autumn with weakening world economy, together with the uncertainty caused by the Sep 11 attacks on the US, dragged down prices some 30%. The cut is expected to lift OPEC's benchmark price to $22 a barrel, the group's minimum target price (CBS News, December 28, 2001).

Burger King will put its Whopper on sale for 99 cents. The move is likely to intensify and prolong the burger price wars that have been roiling the US fast-food industry in recent months. Burger King Officials had said earlier they had little choice given a $1 menu at McDonald's that included a Whopper-like hamburger called the Big 'N Tasty." (Chicago Sun-Times, January 3, 2003).

Tesco announced plans to slash £80 million from prices of more than 1,000 products, with some prices falling by more than 30%. The cuts came as rival Asda also said it was slashing selected prices. The cuts echo memories of the supermarket price wars in 1999 as stores fought to capture more customers and increased market share (Sunday Telegraph, January 5, 2003).

Cournot Game

A market with two competing firms, selling homogeneous goods, where the two firms choose production quantities simultaneously, is known as a Cournot Game. It is a static game where the player's action sets are continuous (each player can produce any non-negative quantity). This is a tacit collusion to raise prices to a jointly optimal level and thus is a "cartel." A cartel is defined as a combination of producers of any product joined together to control its production, sale and price, so as to obtain a monopoly and restrict competition in any particular industry or commodity (www.legal-database.com). Cartels can be quite unstable. At each stage, the players have a huge incentive to cheat.

On Tuesday, 23 September 2003, an agreement was submitted to the US District Court in Philadelphia for an out-of-court settlement for a suit filed by industrial purchasers in 1999. According to this agreement, International Paper, Weyerhaeuser and Georgia-Pacific will pay US$68 million to avoid litigation related to class-action lawsuits that accused them of conspiring to fix prices for container-board (packaging material).

The oil market is notoriously difficult to balance - demonstrated by the rollercoaster of prices over the last few years. Member states of OPEC do not have identical interests and find it difficult to reach consensus on strategy. Countries with relatively small oil reserves are often seen as 'hawks' pushing for higher prices. Meanwhile, producers with massive reserves and small populations fear that high prices will accelerate technological change and the development of new deposits, reducing the value of their oil in the ground (BBC News, February 12, 2003).

Bertrand Game

Models situations where firms choose prices rather than quantities. Assume two firms produce identical goods which are perfect substitutes from the consumers' perspective (consumers will buy from the producer who charges the lowest price). If the firms charge the same price, they will split the market evenly. There are no fixed costs of production and the marginal costs are constant. As in the Cournot Game, the firms act simultaneously. Therefore, when the costs and the products are identical, there exists a unique equilibrium in which all output is sold at the price equal to the marginal cost. Thus, the Bertrand Game suggests that when firms compete on price, and the costs are symmetric, we obtain a perfectly competitive market even in a duopoly situation. However, in real life, customers do not choose based on price alone. For example, Wendy's fast food chain decided to stay out of the Burger King and McDonald's price war (for a while) by aiming to gain market share by offering high quality food.

Stackelberg Game

In most business situations, firms choose their actions sequentially rather than simultaneously. In price wars, one firm responds after it observes another firm's actions. For our discussion, consider that firm 1 moves first and firm 2 responds. We call firm 1 the Stackelberg "leader," and the "follower" is firm 2.

3.3 Game Theory in Inventory Optimization

In time-dependent multi-period games, the players' payoff in each period depends on the actions in the previous, as well as, current periods. The payoff structure may not change from period to period (so called stationary payoffs). This resembles multi-period inventory models in which time periods are connected through the transfer of inventories and backlogs. Due to this similarity, time-dependent games have applications in supply chain management, for example, Stochastic Games. (For detailed mathematical review, see Cachon and Netessine, 2003).

Stochastic Games may help in analyzing:

- two-echelon game with the wholesaler and retailer making stocking decisions;
- price and service competition;
- game with the retailer exerting sales effort and wholesaler stocking inventory and van;
- two-period game with capacity choice in 1st period and production decision under capacity constraint in 2nd period.

These games involve a sequence of decisions that are separated in time, but many supply chain models rely on continuous-time processes. Such applications of Differential Games are especially valuable in the area of dynamic pricing and in marketing-production games with manufacturer and distributor.

Biform Games have been successfully adopted in supply chain management (Anupindi et al., 2001). Consider a game where multiple retailers stock at their own locations, as well as, at several centralized warehouses. In the first (non-cooperative) stage, retailers make stocking decisions. In the second (cooperative) stage, retailers observe demand and decide how much inventory to trans-ship (cross-dock) among locations to better match supply and demand and how to appropriate the resulting additional profits. Variations on this theme are:

- allow retailers to hold back residual inventory. This model has three stages: inventory procurement, decision about how much inventory to share with others and, finally, the trans-shipment stage;
- option of pooling their capacity and investments to maximize the total value. In the first stage, firms choose investment into effort that affects market size. In the second stage, firms bargain over division of market and profits.

3.4 Game Theory in Contracts (Revenue Sharing)

This model is motivated by revenue sharing contracts implemented in practice. Blockbuster purchases movies from studios (suppliers) and rents them to customers. The supplier's wholesale price impacts how many videos Blockbuster orders and hence, how many units are available for rent. Before 1998, purchase price of a video tape from the studio was around \$65. Given that video rental fees are \$3-\$4 per tape, Blockbuster could purchase only a limited number of videos and suffered lost demand during the initial release period (peak demand <10 weeks). About 20% of customers could not find the desired tape to rent. The studio's high wholesale price impacted on the quantity purchased by Blockbuster and in turn, revenues and profitability of both firms. Thus, Blockbuster and the studios crafted a revenue sharing agreement such that Blockbuster pays only \$8 per tape initially but then gives a portion (30 to 45%) of rental revenues to the studio (supplier). Since this agreement reduced Blockbuster's initial investment, it could order more tapes to meet peak demand and generate more revenues even with contracted revenue sharing with the studio (supplier). Blockbuster increased its overall market share from 25% to 31% and improved its cash flow by 61% (CNet News.com, October 18, 2000).

3.5 Games with Incomplete Information (Game Theory and Information Asymmetry)

Ubiquitous knowledge about players and decisions or payoffs is rarely a reality in real world supply chains. It is common that one firm may have a better demand forecast than another or a firm may possess superior information regarding its own costs and operating procedures. If a firm knows that another firm may have better information, it may choose actions that take this into account. Game Theory provides tools to study cases with information asymmetry with increasing analytical complexity.

Signaling Game

In its simplest form, a Signaling Game has two players, one of which has better information than the other. The player with the better information makes the first move. For example, a supplier must build capacity for a key component for a manufacturer's product. The manufacturer has a better demand forecast than the supplier. In an ideal world, the manufacturer may truthfully share his or her demand forecast with the supplier so that the supplier could build the appropriate capacity. But the manufacturer benefits from a larger capacity at the supplier in case of higher demand. Hence, the manufacturer has an incentive to inflate his or her forecast. However, the supplier will bear the cost of building capacity if it believes the manufacturer's (inflated) forecast. The manufacturer hopes the supplier believes the (inflated) forecast and builds capacity. Fortunately, the supplier is aware of the manufacturer's "game" to inflate (distort) forecast. What move (signal) from the manufacturer may induce the supplier to believe the forecast is credible? Consider the example below (from Ozalp Ozer of Stanford University).

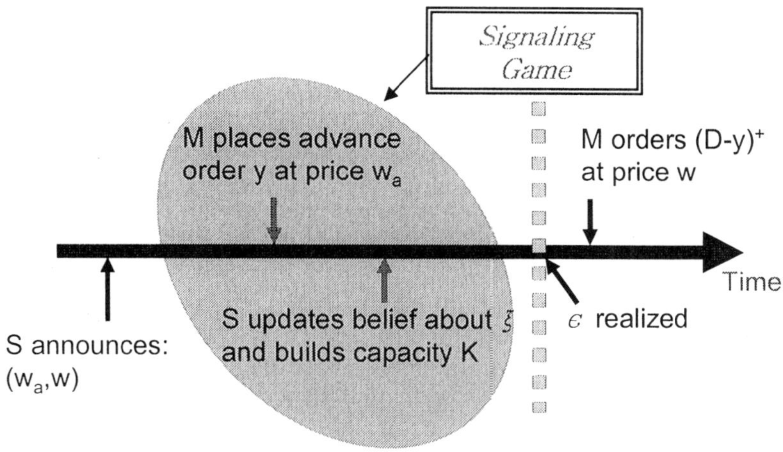

Where, $$D = \mu + \xi + \varepsilon$$

Figure 1-5. Signaling Game

In its simplest form, in this example, Demand (D) is represented as a sum of three forecasts (Figure 1-5). A market forecast mu (μ) is common information and published by commercial analysts. The manufacturer has sources and/or experience to derive private forecast information xi (ξ) which is unknown to the supplier in a decentralized system (information asymmetry). However, the supplier can categorize the manufacturer into certain "types" based on prior actions or credibility of the manufacturer. Thus, the supplier updates its "belief" about the "type" of the manufacturer's forecast information and may select some value of ξ that is spread over a distribution (function). This introduces a random variable. The general market uncertainty is given by epsilon (ε) and neither the manufacturer nor the supplier can control its value, although using appropriate tools, a closer to reality approximation of ε is possible. This introduces another random variable which is also spread over a distribution (function).

The Signaling Game, shown here, commences with a price announcement by the supplier: w (regular) and w_a (advance purchase) price. The manufacturer creates a demand (D) forecast and based on the strength of forecast, reacts to the supplier's price package by placing an advanced order (y) to be purchased at w_a. The volume of y sends a "signal" to the supplier. The "signal" is used to update the supplier's "belief" about the credibility of manufacturer's forecast (D). Based on this, the supplier can determine how much capacity to build (K) to optimize his or her profits (inventory risk). Down the timeline, the market uncertainty is realized and using this value of ε the manufacturer may update its forecast. The volume of the D minus y is the remaining volume the manufacturer orders from the supplier at a price w. While optimization based on Signaling Games may increase profits for manufacturer and supplier, it will still remain vulnerable to the value chosen for the variables ξ and ε. But, this may be further reduced using near real-time data (from automatic identification technologies, as we shall discuss in a later section), which offers greater adaptability to demand.

If signaling favors optimization of the supplier's capacity planning, then what is the manufacturer's incentive to signal? Does the manufacturer incur a cost to signal? Is the manufacturer's expected profit in the signaling equilibrium lower than what it would be if the manufacturer's type were known to the supplier to update his or her "belief" with certainty?

An ideal action for a high demand manufacturer is one that sends the signal of his or her high demand forecast at no cost. If a costless signal does not exist, then the goal is to seek the lowest cost to signal. Whether or not a costless signal exists depends upon what commitments the manufacturer can impose on the supplier. Suppose the manufacturer dictates (contractually) to the supplier a particular capacity level and the supplier accepts the terms. By accepting the contract, the supplier has essentially no choice but to build that level of capacity (severe penalty for non-compliance). This is referred to as "forced compliance" and in this case many costless signals exist for the manufacturer. However, if the supplier could potentially deviate without penalty, referred to as voluntary compliance, then the manufacturer's signaling task becomes more complex. One solution for a high demand manufacturer is to give a sufficiently large advance payment to the supplier. Since the high demand manufacturer's profit is higher than the low demand manufacturer's profit, only a high demand manufacturer could offer such an advance payment. This is referred to as signaling by "burning money" (who can afford to burn money?). A better signal is a contract that is costless to a high demand manufacturer, but expensive to a low demand manufacturer. An example of such a signal is a minimum commitment. The latter is costly only if realized demand is lower than the commitment and the manufacturer is forced by contract to purchase excess inventory. That scenario is less likely for a high demand manufacturer but a minimum commitment may be costly for a low demand manufacturer. Should a manufacturer agree to a minimum commitment if it possesses perfect information? Likely, because these contracts could be used solely for the purpose of signaling information.

Screening Game

In this game the player who lacks information makes the first move. For example, a supplier offers a menu of contracts with the intention of getting the manufacturer to reveal his or her type via the contract selected (in economics this is referred to as mechanism design). The supplier is in charge of designing a mechanism to extrapolate the manufacturer's information.

The space of potential contract menus may be large. How many contracts should be offered and what form should they take? Furthermore, for any given menu, the supplier needs to infer for each manufacturer type which contract that manufacturer will choose. The revelation principle begins with the presumption that a set of optimal mechanisms exists. Associated with each of these mechanisms is a Nash Equilibrium (NE) that specifies which contract each manufacturer type chooses and the supplier's action given the chosen contract. (NE is the point where no player has incentive to change her strategy since each player has chosen a strategy that maximizes her own payoff given the strategies of the other players.) However, it is possible that some manufacturer type chooses a contract that is not designated for that type. For example, a high demand manufacturer chooses an option that the supplier had designed for the low demand manufacturer. Therefore, even though this game does not seem desirable, it is possible that this mechanism is still optimal in the sense that the supplier may not be able to do better on average because the supplier ultimately only cares about optimizing expected profit (not the means by which that profit is achieved). Auction design in the context of supplier procurement contracts and inventory contract design are two of the potential applications of the revelation principle in supply chain management.

Even though an optimal mechanism may exist for the supplier, this does not mean that the supplier earns as much profit as he or she would if he or she knew the manufacturer's type. The gap between what a manufacturer earns with the menu of contracts and what the same manufacturer would earn if the supplier knew her type is called an information rent. The separation of manufacturer types goes hand in hand with a positive information rent, that is, a manufacturer's private information allows the manufacturer to keep some rent that the manufacturer would not be able to keep if the supplier knew his or her type. Hence, even though there may not be any cost involved in information revelation with a Signaling Game, the same is not true with a Screening Game.

Bayesian Games

With a Signaling or Screening Game, actions occur sequentially, such that information may be revealed through observation of actions. There also exist games with private information that do not involve signaling or screening. Consider that a single supplier has a finite amount of capacity. There are multiple retailers and each knows his or her demand, but not the demand of other retailers. The supplier announces an allocation rule, the retailers submit their orders. Then, the supplier produces and allocates units. If the retailer's total order is less than the supplier's capacity, then each retailer receives his or her entire order. If the retailer's total order exceeds the supplier's capacity, the supplier's allocation rule is implemented to allocate the capacity. To what extent does the supplier's allocation rule influence the supplier's profit, retailer's profit and the supply chain profit? In this setting the firms (retailers) that have the private information choose their actions simultaneously (no information exchange among retailers). If the supplier's capacity is fixed before the game starts, the supplier is unable to use any information from retailers (demand) to adapt capacity planning. However, it is possible that correlation exists in the retailers demand information, that is, if a retailer observes his or her demand type to be high, then he or she might assess that other retailers may have high demand types as well (if there is a positive correlation). Thus, each player uses Bayes' rule to update his or her belief regarding the types of the other players in a Bayesian Game. Bayesian Equilibrium is a set of strategies for each type that is optimal given the updated beliefs with that type and the actions of all other types. If a player deceptively inflates demand (high type) and other players use this information to update their "beliefs" then this effect may contribute to the observed Bullwhip Effect.

3.6 Temporary Conclusion

God definitely plays dice! Combined GT/OR may offer approaches to use (data) dynamic information for continuous optimization in terms of location and real-time availability (improve from visibility to transparency, among players) as a step toward an adaptive value network.

4. AGENTS

Linearization of real world conditions to fit mathematical models, such as Game Theory, may stifle real-time adaptability of value networks. As an example (see preceding section), a Bayesian Game potentially could contribute to the Bullwhip Effect representing wide fluctuations in supply chain. The discrete, dynamic and distributed nature of data and applications require that supply chain solutions do not merely respond to requests for information but anticipate, adapt and (support users to) predict. In that vein, 'intelligent' autonomous Agents are an essential tool for adaptive value networks to emerge.

The idea of Agent originated with John McCarthy in the 1950's at MIT. The term "Agent" was coined by Oliver Selfridge, a colleague of McCarthy's at MIT. Recent trends, beginning 1977, in Agent systems are based on research in distributed artificial intelligence. Research from MIT, DARPA, Carnegie-Mellon University and University of Michigan at Ann Arbor has made significant contributions.

We define an autonomous Agent as a software entity that functions continuously in an environment, often inhabited by other Agents. Continuity and autonomy empower Agents to (plan) execute processes in response to changes in the environment without requiring constant human guidance, intervention or top-down control from a system operator. Thus, Agents offer the ability to rapidly adapt. An Agent that functions continuously in an environment over a period of time also learns from experience (patterns). In addition, Agents that inhabit an environment with other Agents in a Multi-Agent System (MAS) are able to communicate, cooperate and are mobile between environments. Agents work best for clearly discernible tasks or processes, such as, to monitor data from, for example, automatic identification technologies (radio frequency identification or RFID), ultrawideband (UWB) transponders, global positioning system (GPS), WiFi and sensors. Data Agents can share this data with high level Information Agents and offer real-time information to Process Agents (Inventory Agent, Purchasing Agent). The emergence of Multi-Agent Systems (MAS) may be slow to take-off unless the Semantic Web sufficiently permeates the environment for ubiquitous deployment of Agents.

Design of Agent-Based Modeling (ABM) draws clues from natural behavior of biological communities. Although it still remains a paradox, it is increasingly undeniable that simple individual behaviors of bugs like ants and wasps, *collectively*, may offer intelligent models of complicated overall behavior. In fact, this may have been known for centuries. One ancient observer, King Solomon, knew from his father, David, of the elaborate court organizations of oriental kings and preparations needed for military campaigns. He marveled that insects could accomplish both these tasks without any central control. Thinking of the complex systems needed to maintain the palace commissary, he wrote, "Go to the ant, consider her ways and be wise. Having no guide, overseer or ruler, she prepares her bread in summer and gathers her food at harvest time." He knew the complexity of a military organization and was impressed that "locusts have no king, yet all of them go forth by companies." Nearly 3000 years later, a participant in the NCMS Virtual Enterprise Workshop (1994) commented, "we used to think that bugs were the problem. Now we suspect they may be the solution!" (Parunak 1997)

Adaptability in biological systems is a fundamental characteristic of nature, and thus, models based on and inspired by such superior systems can contribute significantly to reduce key inefficiencies (and stem the loss of profit) between centralized and decentralized supply chains. Most software is based on equations that link rates and flows (consumption, production). Variables (cost, rebates, transportation time, and out-of-stock) evaluate or integrate sets of ordinary differential equations (ODE) or partial differential equations (PDE) relating these variables. Operations research provides the framework to optimize for the "best" result. What if the "best" result is not necessarily the optimal "best" for that situation? Shortest lead time could plan a route through an area with a high probability of flash flood due to a brewing storm or threat of sniper attack on a portion of the highway. Planning software (today) fails to, or is incapable of, modeling such random events that may have profound implications for business, at that time. Thus, the "best" solution may not be adaptive to supply chain events at hand.

Even excluding random events or decisions that require integration with other models (weather, road construction), what is the half-life of 'best' solution in a fickle economy or high "clockspeed" industry? Compared to ABMs, a significant shortcoming of such Equation-based (ODE, PDE) models (EBM) is that EBM based software processes assume that these parameters are linear in nature and relevant data is available (for optimization). In the real world, events are non-linear, actions are discrete, information about data is distributed (CRM, PLM, SCM data silos) and data is corrupted with "noise" (according to a study by Ananth Raman of Harvard Business School and Nicole DeHoratius of the University of Chicago, for a global retailer, in some cases, 65% of SKUs (bar coded) were found to be inaccurately represented between system data, back-store and availability on store shelf, see Dehoratius, 2002).

Virtually all computer-based modeling, up to this point has used system dynamics, an EBM approach. But the struggle to adapt and respond in real-time will eventually and collectively fuel a paradigm shift that will make it imperative to model business software based *both* with Agents and equations. The question is no longer whether to select one or the other approach, but to establish a business-wise mix of both and develop criteria for selecting composition of software-based on one or the other approach that can offer valuable combinatorial solutions. The "balance" is subject to dynamic change (seek analogy with Screening Games). For traditionalists in supply chain management, the situation is analogous to a "push-pull" strategy where the dynamic push-pull boundary shifts with changing demand (pull).

ABM and EBM, both simulate the system by constructing a model and executing it on a computer. The differences are in the form of the model and how it is executed. In ABM, the model consists of a set of Agents that encapsulate the behaviors of the various individuals that make up the system, and execution consists of emulating these behaviors, which is essentially dynamic. In Equation-Based Modeling (EBM), the model is a set of equations (pre-determined, static) and execution consists of evaluating them. Thus "simulation" is a generic term that applies to both methods, which are distinguished as Agent-based *emulation* and equation-based *evaluation*.

Thus, the need for supply chains to be adaptive should rapidly trigger demand for Agent integration with existing EBM systems. But the demand for Agents software is slow to materialize. One reason may be gleaned from the observation by Norman Poire, an economist (Figure 1-6, blue lines, http://www.smalltimes.com/document_display.cfm?document_id=2141). As shown in figure 1-6, it takes about a quarter of a century for a technology to gain acceptance. Then it fuels a period of rapid growth lasting an additional half a century. Almost after a century since "invention" or introduction, the innovation may become a commodity and grows in line with fluctuations in macroeconomic forces. We propose that Agents, in principle linked to some of the fundamentals from distributed artificial intelligence (DAI), may follow a similar trajectory which suggests increasing adoption beginning about 2005 (Figure 1-6, red line). These Agents are the types that are capable of machine learning and utilize learning algorithms, such as (ant-based) swarm intelligence, genetic algorithms, and neural networks (single and multilayer perceptions, Hopfield networks, Kohonen networks, radial basis function networks).

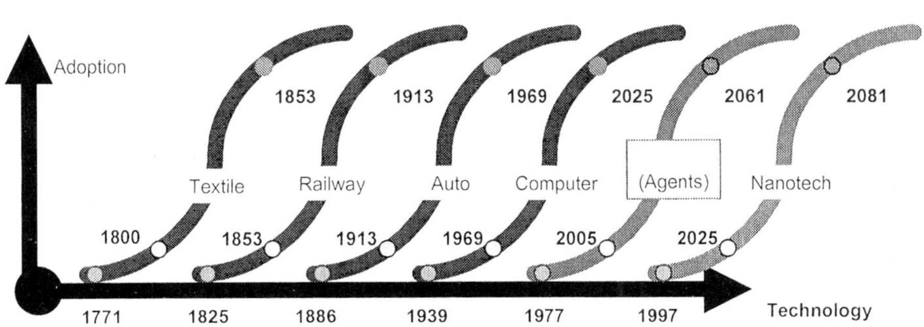

Figure 1-6. How Conceptual Advances Lead to the Wealth of Nations

Continuity and autonomy of biology offer behavior patterns that are flexible, adaptive and responsive to change. Thus, the mobile, networked, autonomous, self-learning, adaptive Agent may have different principles compared to those that were developed for monolithic systems. Examination of naturally occurring Agent-based systems suggests design principles for Agents. While some circumstances may warrant deliberate exceptions, in general, Agents are aligned with the concepts listed below from Parunak (1997) and Parunak et al., (1998):

1. Agents should correspond to "things" in the problem domain rather than to abstract functions;
2. Agents should be small in mass, time (able to forget), and scope (avoid global knowledge action);
3. Multi-Agent Systems should be decentralized (no single point of control/failure);
4. Agents should be neither homogeneous nor incompatible but diverse;
5. Agent communities should include a dissipative mechanism (entropy leak);
6. Agents should have ways of caching and sharing what they learn about their environment;
7. Agents should plan and execute concurrently rather than sequentially.

4.1 Agents versus Equations: Conceptual and Practical Considerations

The difference in representational focus between ABM and EBM has consequences for how models are modularized. EBM represents the system as a set of equations that relate observables to one another. The basic unit of the model, the equation, typically relates observables whose values are affected by the actions of multiple individuals. ABM represents the internal behavior of each individual. An Agent's behavior may depend on observables generated by other (Agents) individuals, but does not directly ***access*** the representation of those individual behaviors, thus, maintains boundaries among individuals. This fundamental difference in model structure gives ABM a key advantage in commercial applications such as an adaptable value network where partners may interact over an e-marketplace.

First, in an ABM, each firm has its own set of Agents. An Agent's internal behaviors are not required to be visible to the rest of the system, so firms can *maintain proprietary information* about their internal operations. Groups of firms can conduct joint modeling exercises (Public MarketPlaces) while keeping their individual Agents on their own computers, maintaining whatever controls are needed. Construction of EBM requires disclosure of the relationships that each firm maintains on observables so that the equations can be formulated and evaluated. Distributed execution of EBM is not impossible, but does not naturally respect commercially important boundaries (why the early wave of e-MarketPlaces failed to survive).

Second, in many cases, simulation of a system is part of a larger project whose desired outcome is a control scheme that more or less automatically regulates the behavior of the entire system. Agent systems may correspond 1-to-1 with the individuals (firms or divisions) in the system being modeled, and the behaviors are analogs of real behaviors. These characteristics make Agents a natural locus for the application of adaptive techniques that can modify their behaviors as the Agents execute, so as to control the *emergent behavior* of the system. Migration from simulation model to adaptive control model is more straightforward in ABM than in EBM. One can imagine a member of adaptable business network using its simulation Agent as the basis for an automated control Agent that handles routine interactions with trading partners. It is unlikely that such a firm would submit aspects of its operation to an external "equation manager" that maintains specified relationships among observables from several firms.

EBM most naturally represents the process being analyzed as a set of flow rates and levels. ABM most naturally represents the process as a set of behaviors, which may include features difficult to represent as rates and levels, such as step-by-step processes and conditional decisions. ODEs are well-suited to represent purely physical processes. However, business processes are dominated by non-linear, discrete decision-making.

Both ABMs and EBMs can be validated at the system level by comparing model output with real system behavior. In addition, ABM's can be validated at the individual level, since the behaviors encoded for each Agent can be compared with local observations on the actual behavior of the domain individuals. ABMs support direct experimentation. Managers playing 'what-if' games with the model can think directly in terms of business processes, rather than translate them into equations relating observables. A purpose of what-if experiments is to identify improved business practices that can be implemented. If the model is expressed and modified in terms of behaviors, implementation of its recommendations is a matter of transcribing the modified behaviors of Agents into task descriptions for the underlying physical entities in the real world.

In many domains, ABM gives more realistic results than EBM, for manageable levels of representational detail. The qualification about the level of detail is important. Since PDEs are computationally complete, in principle, one can construct a set of PDEs that completely mimics the behavior of any ABM (thus produce the same results). However, the PDE model may be much too complex for reasonable manipulation and comprehension (for example what we observe in repetitive Stochastic Games with incomplete information). EBMs (like system dynamics) based on simpler formalisms than PDEs may yield less realistic results regardless of the level of detail in the representation. For example, the dynamics of traffic networks achieved more realistic results from traffic models that emulate the behaviors of individual drivers and vehicles, compared with the previous generation of models that simulate traffic as flow of a fluid through a network. The latter example bears strong similarities to the flow-and-stock approach to supply chain simulation.

The disadvantages of EBM in this and other examples result largely from the use of averages of critical system variables over time and space. EBM assumes homogeneity among individuals but individuals in real systems are often highly heterogeneous. When the dynamics are non-linear, local variations from the averages can lead to significant deviations in overall system behavior (outcome). Refer back to the section on Game Theory and in light of ABM vs. EBM, re-consider the example of the Signaling Game: the choice of values (of ξ and ε from the distribution) can significantly impact capacity planning (inventory risk) and profit optimization (price risk). In such business applications, driven by "if-then" decisions, non-linearity is the rule. Because ABM's are inherently local and can adapt to changes, it is beneficial to let each Agent monitor the value of system variables locally (for example, real-time data for ε, in the Signaling Game), without averaging over time and space.

Ant-based algorithms based on naturally occurring systems, enables the Agent to forget (ant pheromones evaporate and obsolete paths leading to depleted food sources disappear rather than misleading members of the colony). The mechanism of forgetting is an important supplement to the emphasis in conventional artificial intelligence (AI) systems on mechanisms for learning. In a discrete-event system, forgetting can be as complex as learning since both represent discrete state transitions. In a time-based system, forgetting can take place "automatically" through the attenuation of a state variable that is not explicitly reinforced. The Agents ability to "forget" is a boon to real-world adaptable business networks. EBM based demand forecasting generally uses a weighted-average of past consumption data. If there was a marked variation (for example, spike in sales, 20 weeks ago) the planning algorithm continues to consider that value because equation-based modeling cannot "forget" facts, although the weight will decrease successively in each planning cycle unless manual intervention or program insertion specifies a "forget" rule. The forecasting engine, therefore, may continue to reflect the effect in its subsequent forecast for weeks or months. Consider the cumulative error from such events, if aggregated over geographies prior to generating a global forecast that may guide procurement or production. Such events produce the Bullwhip Effect. Agents can improve forecasting and with real-time data, accuracy may be further enhanced. As a result, for example, the manufacturer may adjust production to manage inventory better and reduce waste. Reduced inventory decreases working capital charges which improves return on assets because manufacturing the cash cycle gets shorter.

In a traditional system, forecast determines production planning and subsequently, execution of the plan. Some manufacturers develop a schedule each night that optimizes manufacturing the next day, a process not much different from grocery chains that order perishables the day before it is displayed in stores. Engineers in industries as diverse as auto, semiconductors, aerospace, and agricultural equipment will agree that a daily schedule is obsolete less than an hour after the day begins. But Agents seek to avoid the "plan ***then*** execute" mode of operation and instead responds dynamically to changes in the environment. In concurrent planning and execution, the actual time at which a job will execute may not be known until the job starts. The resource does not schedule a newly-arrived job at a fixed point in time but estimates probabilistically the job's impact on its utilization over time, based on information from the customer about acceptable delivery times. The width of the window within which the job can be executed is incrementally reduced over time, as needed, to add other jobs (may be rated by priority, at that time) to the resource's list of tasks. If the resource is heavily loaded, the jobs organize themselves into a linear sequence but if it is lightly loaded, the actual order in which jobs are executed is decided at the moment the resource becomes available, depending on the circumstances that exist at that time. Figure 1-7 shows simplified view of agent in system architecture.

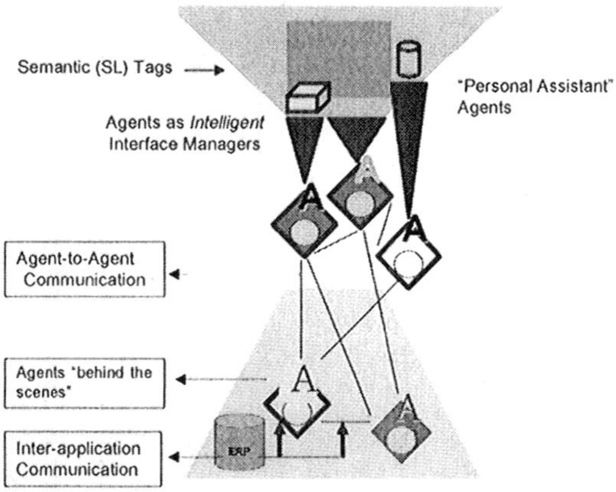

Figure 1-7. A Simplified View of Agents within the System Architecture

4.2 Agents in Maintenance (US Air Force Case Study)

This example of a Multi-Agent framework (and this case study) was developed by Shehory, Sycara and Sukthankar in 1999 (Agent aided aircraft maintenance) at the Carnegie-Mellon University, Pittsburgh (Shehory et al., 1999). It provides information retrieval and analysis in support of decision making for aircraft maintenance and repair for the US Air Force (USAF). Although the solution was developed for a specific type of aircraft, the Agents and interactions were designed to apply to a range of similar maintenance scenarios.

Aircraft maintenance in the USAF is performed at different levels. Basic and intermediate levels are usually performed at the base where the aircraft is deployed, whereas periodic, comprehensive maintenance is performed at special depots. Initially, mechanics inspect the aircraft for discrepancies (and may also receive such information from pilots). For each discrepancy, the mechanic consults the technical manuals for a standard repair procedure. In case such a repair procedure is found and the resources (parts) are available, the mechanic proceeds with the repair. In cases where parts are not available or they are too expensive or require too much time and additional machinery for replacement or in cases where a procedure is not provided in the technical manual, a mechanic needs to consult an expert engineer. The engineer, in turn, may consult external sources of information. These include manuals, historical maintenance data and may even include consultation with experts.

Inventory of parts is based on traditional data input from goods received. Locating spares, therefore, could be a time consuming and arduous undertaking that can be automated to a significant extent by use of automatic identification technologies (UWB, RFID) and to link inventory object related data with service/maintenance processes to offer transparency of the spares supply chain.

Until recently, no automation was introduced to the consultation processes, either, of this information-rich environment. Hard-copy repair manuals are used by mechanics and engineers. Search for relevant information may be time consuming and incomplete. Historical data (records of previous similar repairs) is scarcely used, since it is stored in paper format with no search mechanisms and usually only kept for short periods (distributed along remotely located service centers). Expert engineers may be located remotely and their advice is available by voice or fax messages, usually delayed for hours or days. All of these factors contribute to a slow, inefficient maintenance that compromises readiness.

The inspection, consultation and repair process consists of the following steps:

1. Aircraft arrives at a maintenance center, either at its home base or depot (depending on the type of maintenance required). In both cases, the maintenance procedures must be completed within a limited time period. This period varies. Basic and intermediate maintenance must be completed within hours or a few days, whereas depot maintenance may be scheduled for several weeks (depends on aircraft).
2. Mechanics inspect the aircraft and locate discrepancies. For each discrepancy a mechanic performs the following:

a) browse the technical manual for repair procedures;
b) in case an appropriate procedure is located, mechanic needs to verify whether it can be completed given limitations on repair time and parts availability. Mechanic may also need to consider the price of repair. For example, the technical manual may require replacing a whole wing if a crack in the wing is greater than some given threshold. This may take too long and become too expensive thereby causing delay or compromise operational activity or readiness;
c) if the procedure found can be performed, the mechanic performs it. If not, mechanic proceeds to fill out form 202a, standard USAF form for reporting aircraft discrepancies and requesting advice. The mechanic may attach supporting information. The mechanic may consult Illustrated Part Breakdown (IPB) technical manuals and possibly other experienced mechanics. Form 202a is sent for advice and authorization for non-standard repair.

3. An engineer, upon receipt of a Form 202a, proceeds to:

a) use experience, historical repair information and manuals to design appropriate repair;
b) fill in a Form 202b, standard US Air Force form for discrepancy repair instructions. To this form the engineer may attach graphical illustration to clarify required repair procedure;
c) file 202a and 202b for future use as historical repair information.

4. When a standard repair procedure is found or on receipt of Form 202b from engineer, the mechanic performs the repair as instructed. The current inspection, consultation and repair processes, as described above, have several problems. The Multi-Agent System (MAS) implementation reported here attempts to address these problems. The majority of the information, both historical repair information and technical manuals, is found in hard-copy format as well as hand-written pieces. Mechanics and engineers spend precious time on:

a) Browsing manuals and searching for historical repair information;
b) Drawing graphical discrepancy and repair illustrations;
c) Mechanics are idle, waiting for Form 202b to arrive from engineers in reply to their Form 202a;

d) Historical information is unused when stored remotely or local hard-copy is difficult to browse.

For information needs of mechanics, using manuals during inspection for diagnosis is inefficient and at times impossible due to the physical constraints of the inspection environment. Scribbled information both from historical forms and the current Form 202 may have limited comprehensibility. The problem intensifies due to deterioration in the quality of such information when it is transmitted via fax or photo-copied. Historical forms are kept only for two years. Time and effort spent on paperwork and filing should be used instead for diagnosis and repair. Technical manuals (IPB) are not consistently updated.

The problem consists of decision support in a physically distributed, dynamically changing environment, rich in multi-modal information, where users have diverse (varying over time) information needs. This is the type of problem for which RETSINA (**RE**usable **T**ask-based **S**ystem of **I**ntelligent **N**etworked **A**gents) is a solution. It is a Multi-Agent infrastructure that was developed for information gathering and integration from web-based sources and decision support tasks. It includes a distributed MAS organization, protocols for inter-Agent interaction as well as collaboration and a reusable set of software components for constructing Agents. Each Agent in RETSINA specializes in a special class of tasks. When Agents execute tasks or plan for task execution, they organize themselves to avoid processing bottlenecks and form teams to deal with dynamic changes in information, tasks, number of Agents and their capabilities.

In RETSINA, the Agents are distributed and execute on different machines. Based on models of users, an Agents and tasks, the Agents decide how to decompose tasks, whether to pass them to others, what information is needed at each decision point, and when to cooperate with other Agents. The Agents communicate with each other to delegate tasks, request or provide information, find information sources, filter or integrate information, negotiate to resolve inconsistencies in information and task models. The RETSINA infrastructure consists, by convention, of 3 broad types of Agents:

- Interface Agents;
- Task Agents;
- Information Agents.

In the RETSINA Multi-Agent infrastructure, Interface Agents interact with users receiving specifications and delivering results. They acquire, model and utilize user preferences. The Interface Agents hide the underlying structural complexity of the Agent system. Main functions of an Interface Agent include:

- collecting relevant information from the user to initiate a task;
- presenting relevant intermediate and final results;
- requesting additional information during task execution;

Task Agents formulate and execute plans. They have knowledge of the task domain and which other Task Agents or Information Agents are relevant for performing various parts of a task. Task Agents have strategies for resolving conflicts and fusing information retrieved by Information Agents. A Task Agent:

- receives user delegated task specifications from an Interface Agent;
- interprets the specifications and extracts problem-solving goals;
- forms plans to satisfy these goals;
- identifies information-seeking sub-goals that are present in its plans;
- decomposes plans and cooperates with appropriate Task Agents or Information Agents for planning, execution, monitoring and results compilation.

Information agents provide intelligent access to heterogeneous collection of information sources. They have models of information resources and strategies for source selection, information access, conflict resolution, and information fusion. Information Agents can actively monitor information sources.

Middle agents collect and provide information about the location, availability and capabilities of other Agents (possibly additional information about them). They may also serve as mediators, hiding the identities of either service requester Agents or service provider Agents or both. Middle Agents (Matchmakers) provide RETSINA-based MAS with openness. That is, Agents may leave and enter the system dynamically. When an Agent appears, it advertises itself with a Middle Agent. When it leaves, it informs the Middle Agent, as well. Agent disappearance as a result of Agent or network failure is detected by a Middle Agent via a pinging mechanism. The RETSINA internal Agent architecture is based on a multi-module, multi-thread design. It consists of two component types: functional units and data stores. Given its properties, we found the RETSINA infrastructure appropriate to solve the USAF maintenance problem. By developing and using Agent architecture, we gain the following advantages:

1. The RETSINA architecture can be used to wrap legacy software systems by equipping them with a Communicator module. Thus the resulting system remains backwardly compatible with the older systems, without restricting future software development to an obsolete model. For instance, in 1999 the Warner Robins Air Force Base (AFB) engineers were experimenting with entering some of the data into an Access

database format, as a temporary measure while waiting for (the ITL-ALC) another system to become available. With this design, separate Information Agents can easily be designed to accommodate both data sources. Since the maintenance personnel only interact with Interface Agents, they are shielded from internal data discontinuities;

2. The information required by the maintenance engineers is likely to be distributed among several computers in different geographic locations. RETSINA architecture provides built-in networking support useful for developing distributed systems, in the form of the Communicator. The Agent Name Sever (Matchmaker) allows service requesters to locate service providers. Although the current focus is on handling the repair operations described in Form 202A, which are performed locally in Warner Robins AFB, additional Agents can be added to the system to access collections of Form 00-107 (immediate repair requests), which can be filed from multiple locations. These Agents would be located on computers at the local Air Force base performing the repair and would communicate to agents at the central F-15 repair location (Warner Robins AFB);
3. The Warner-Robins Air Force Base is in a transitional phase of reorganizing their data and also training personnel. Rapid prototyping of a group of Agents are underway to address the current situation and slowly add to the "Agent population" as new information sources become available electronically. Since the Interface Agent is decoupled from the Information Agents, it is possible to replace older Information Agents without disruptions or disturbance to the users.

4.3 Agents in Manufacturing

Commercial aerospace industry makes fewer products and sells to a different set of customers than the retail industry (Figure 1-8 shows typical aerospace supply chains). Some (modular) parts and components are shared between different models (variants) of aircraft. Significant profit in this (and the automobile industry) is derived from the aftermarket sale of parts and service. The companies therefore have access to a large amount of usage data. Premature failure of two hydraulic pumps in different corners of the world prompts an Agent to explore the pattern. Both pumps came from the same manufacturing lot. The Agent prompts maintenance technicians to perform non-routine vibration analysis. Results indicated that the manufacturing lot had a defect. If vibration analyses data from manufacturer's test results were available to the Agent in this value network, a pattern may have emerged even before a single pump failed. Comparative analysis involves access to massive data processing that is beyond human reach in a reasonable time frame. Agents could accomplish such tasks rapidly and be able to predict, thereby avert, a potential catastrophe. The information required for such Agent operations to recognize a pattern from manufacturer data, lot information, date of installation and hours of usage are possible in value networks with integrated points of access to distributed data, but impossible in silos of supply "chains" which are common today.

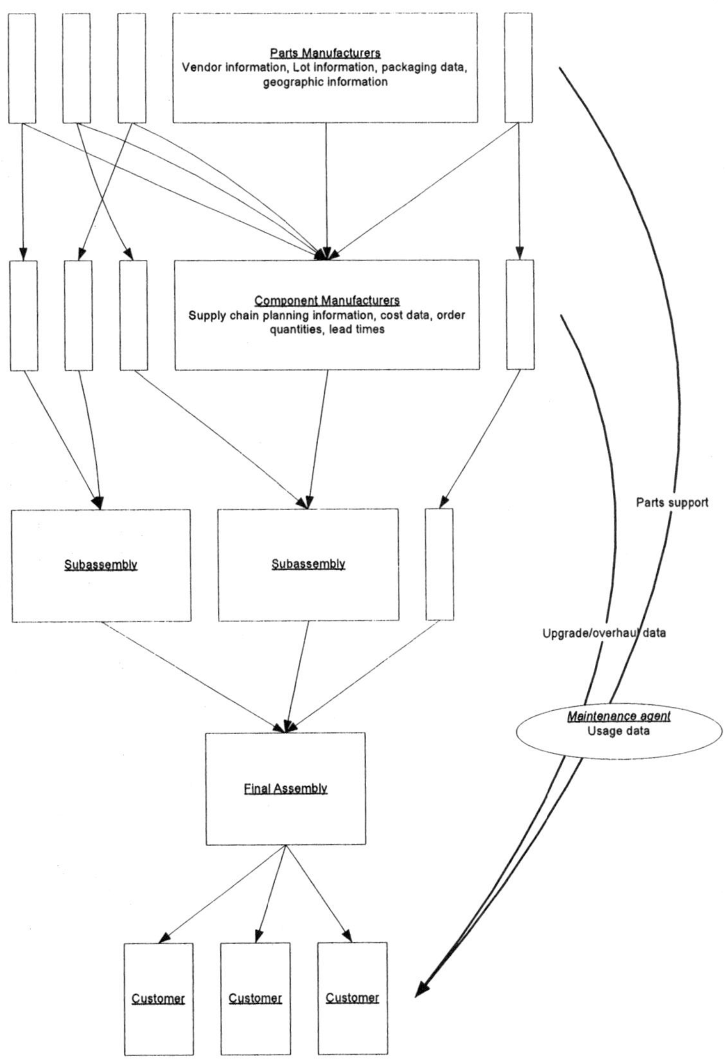

Figure 1-8. Commercial Aerospace Industry Supply Chain: Information Collection

4.4 Future Agents at Work?

Transistor Titikaka Promethium (TTP), a small electronics retailer, starts selling a digital camera (named, CELC) and soon runs out of inventory due to the popularity of the new product. TTP places another order. A week later some customers returned the cameras and others call with questions. TTP is unable to determine the cause and loses time and revenues. Think different.

You are Must-See-Borgium Corporation, the bleeding-edge retailing behemoth. You started selling CELC and soon your return center in Moose Jaw is flooded with CELC from unhappy customers. Fortunately, your ex-VP (exiled to Timbuktu) had created a liaison with a tiny institute around Boston. She quietly integrated a system called MY-CAH that offered no satisfactory ROI to your bean counters. Within a week of mounting CELC returns to Moose Jaw, Must-See-Borgium's MY-CAH Agent sends an e-mail alert (cc you) to N. E. Shee in Urawa (manufacturer's headquarter) that many US customers who returned CELC to Moose Jaw also bought a certain brand of BELL notebooks with Dumb-Bell Mobile Bambino. In your in-box you also find a response from Shee-san that the camera's software is incompatible with systems installed with Dumb-Bell Mobile Bambino without a special patch from MacroHard that can be downloaded from www.bosonic-hadrons.net and the CELC website will soon upload the link for customers. MY-CAH Agents already posted an update on the corporate website, informed Moose Jaw Center, CELC customers who registered or returned their products, sent e-mail to only those customers who bought CELC with Must-Have-Borgium credit-loyalty card and printed out an exact number of stickers (per inventory) with instructions to be affixed to CELC boxes in all your local mega-stores. You find a note of gratitude from Miss Fermionic Baryons at TTP who saw the notice about CELC on your website and could inform TTP's customers by phone. You had no problem getting out of a mess and a bad PR wrap because MY-CAH actually works! Didn't you vociferously object to the VP's proposal to sponsor research at that tiny institute around Boston? Anyway, you solved the problem.

What really happened? Your store was running an Agent system that analyses data for trends. The Agent was able to identify this trend in minimal time. The missing patch could have been identified without the use of an Agent, but it would have taken much longer and resulted in many more unhappy customers, which would generate bad publicity. Why did an Agent work in this situation? Data and information derived from data is the key enabler for decision systems to be agile. In this example, the Agents were autonomously collecting product, customer, and service related information. Customer purchases were compared for people who bought and returned this new product (SKU). How does a company know what information to collect? Easily enough, companies should collect the same information that was needed to find previous patterns if the company had data mining capabilities. In this case, real-time data over short time windows were constantly under analysis and random associations were easier to track by Multi-Agent Systems monitoring multiple operations both within the company and its interactions in the value network. Concurrently, it was analyzing legacy data (ERP) to ***learn*** or create analytic parameters from past data patterns.

In another scenario, consider an Agent system that operates in a services business area (only) charged with the analysis of returns. The Agent spots that the rate of return for a certain manufacturer's products has risen above a certain level in recent weeks. Why? They are a relatively high value product, which weighs more than 15 pounds and the majority was shipped 500 miles or more. An alert from the Agent reaches the manager and she intuitively inspects the packaging and... Voila! It is different than the packaging for products that have a lower return rate. A phone call confirms the hunch that the manufacturer recently switched to a different packaging vendor in an effort to conserve costs. The Agent succeeded in creating the alert because the Agent system collects, processes, correlates and cross-references vendor data, shipping method data, shipping distance information data and other cradle-to-grave stages and any related ePC data that it can extract from the local data store connected with goods movement (RFID/UWB tags attached to this item). SKU information (only) still exists as a barcode on the outer packaging. The Agent also extracts the UPC code from the store master data (redundant information). If packaging type information was stored on RFID/UWB tags for each SKU sold, the Agent system may have been able to spot the trend without the aid of a human, the manager (Figure 1-9).

Agents can also help with marketing. Dell sells computers that consumers can configure. Bundling is a marketing technique that pairs two products together to sell at a single price, which is lower than the normal price of the two if sold individually. Single price gives a greater revenue and profit than if either item were sold alone. Dell stores exabytes of information on customer buying patterns. An "analytic" Agent is able to spot a pattern where 40% of customers who buy extra memory also buy a certain high-speed processor. A "marketing" Agent can "talk" to the "pricing" Agent to offer discounts if memory is bought together with the processor. As the trend of choices for combinations (memory vs. processor speed) changes or differs in demographics or geographies, the data from "analytic" Agent can be used for the "marketing" and "pricing" agent to adapt and offer new bundling options (dynamic pricing). This can augment demand for the memory and increase total revenue and profit. Customers who are likely to buy a product may be targeted for marketing (may not buy without bundle discount).

The number of potential product combinations increases if three or more options are thrown into the mix, not to mention accessories like cameras, MP3 players or printers. It is simple for Agents to analyze gargantuan amounts of data and spot potential (multiple) bundling opportunities and adapt to the fluctuations in demand in near real-time much faster (by several orders of magnitude) than a human or software based on equation (EBM). Bundling strategies can be catalytic to sell slow moving inventory or end of life (EOL) product prior to new product introduction.

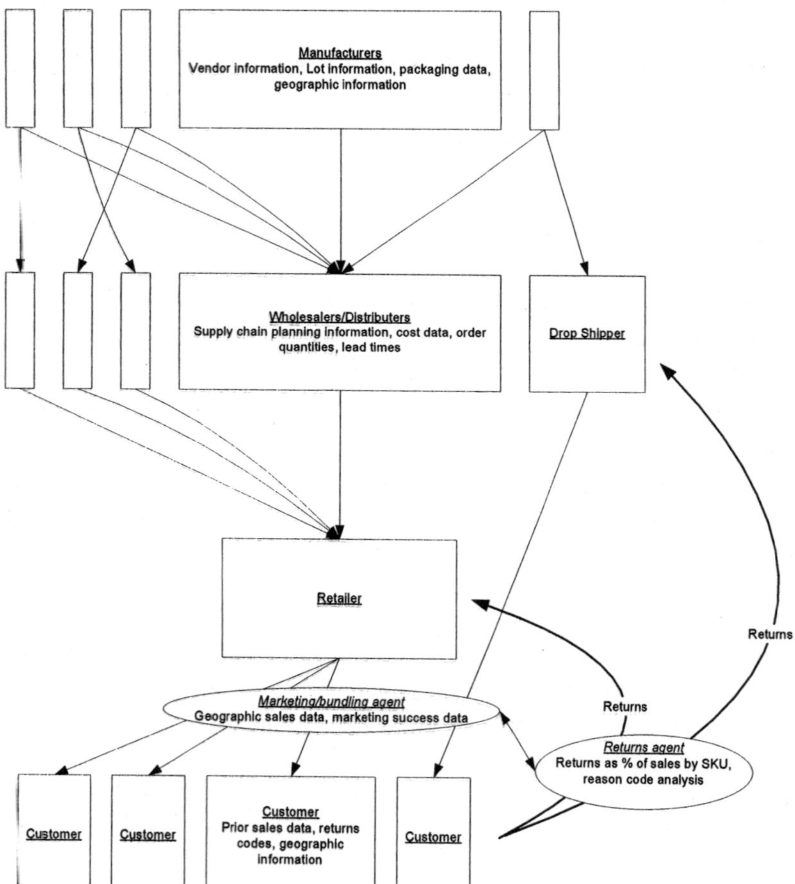

Figure 1-9. Agents in Retail Industry (also shows where "returns" and "bundling" Agents may integrate)

When I want to go out to the movies, rather than read reviews, I ask my sister-in-law. We all have an equivalent who is both an expert on movies and an expert on us. What we need to build is a digital sister-in-law ("Less Is More: Interface Agents as Digital Butlers" by Nicholas Negroponte, 1994).

4.5 WHY THINK DIFFERENTLY?

The approach to system design and management with Agents in the software landscape is at odds with the centralized top-down tradition in systems engineering. The question usually arises in terms of the contrast between local and global optimization. Decision-makers fear that by turning control of a system over to locally autonomous Agents without a central decision-making body, they will lose value that could have been captured by an integrated (enterprise) global approach.

Benefits of Agent-based architecture versus centralized ones are conditional, not absolute. In a stable environment, a centralized approach can be optimized to out-perform the efforts of an opportunistic distributed system of Agents. If the system has appropriate learning capabilities, it will eventually become as efficient. The appropriate comparison for systems designers of enterprise software is not between local and global optima but between static versus adaptable systems. Thus, evaluate the competing options (in any particular case) theoretically, strategically, tactically and practically.

Theoretically, there are decentralized mechanisms that can achieve global coordination. For example, economists have long studied how local decisions can yield globally reasonable effects. Recently these insights have been applied to a number of domains that were not traditionally considered as economic, such as network management, manufacturing scheduling and pollution control.

Strategically, managers must weigh the value of a system that is robust under continual change against one that can achieve a theoretical optimum in a steady-state equilibrium (that may never be realized). A company that anticipates a stable environment may well choose centralized optimization. One that *also* incorporates Agent-based software does so because it cannot afford to be taken by surprise.

Tactically, the life-cycle software costs may be lower for Agent-based systems than for centralized enterprise software. Agents can be modified and maintained individually at a fraction of the cost of opening up a complex enterprise software system. In systems that must be modified frequently, losses due to sub-optimal performance can be recovered in reduced system maintenance expenses.

Practically, Agent-based systems that follow these principles have been piloted or deployed operation (US Air Force case study by CMU). The Agents reflect the principles outlined rather than those of centralized systems. Growing acceptance of Agents in competitive business environments may be evidence of the benefit they bring to their adopters (Figure 1-10)

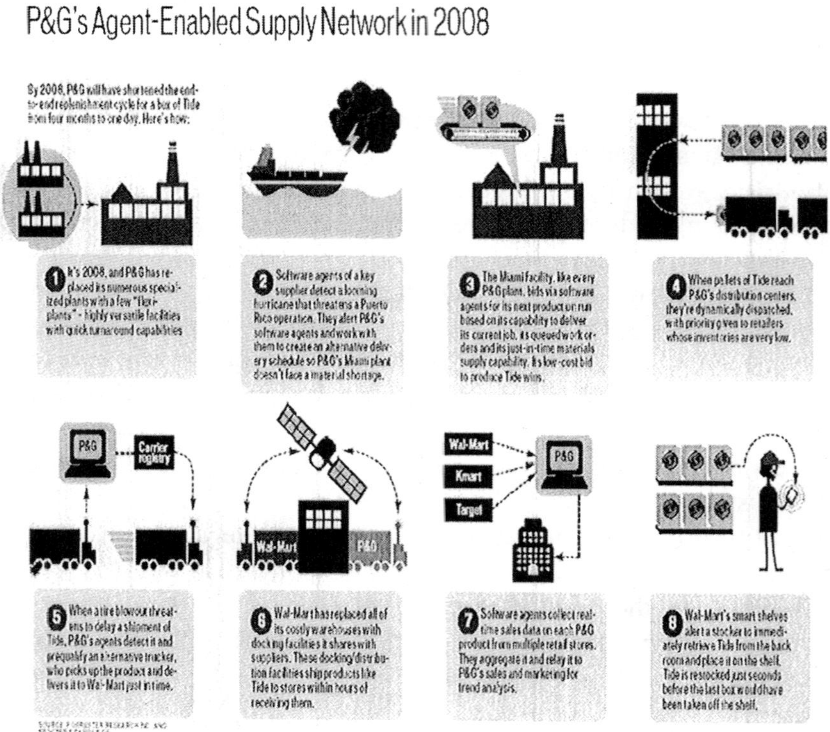

Figure 1-10. P&G's Agent-enabled Supply Network in 2008

5. AUTOMATIC IDENTIFICATION TECHNOLOGIES

Automatic identification technologies offer tools to acquire data about ***objects*** (e.g. IV pumps, toothpaste, and ammunition). Innovation and leadership lies in the effective use of the data, not in its acquisition.

In 1894, Oliver Lodge demonstrated how to communicate (data) using radio waves. Half a century later, with the discovery of the RADAR at MIT, it was likely that the natural frequency spectrum was going to "make waves" for quite some time. Near the tail end of the last century, with the establishment of the MIT Auto-ID Center (which morphed into Auto-ID Center), once again, more than a century later, a radio frequency-based identification (RFID) and communication protocol created waves whose impact will be inescapable in the future and for the future of most businesses that were present in the past.

Neither the technology nor concepts are new but the two thinker-founders of the MIT Auto-ID Center (Sarma and Brock, 1998) created a "storm in a tea cup" by reversing the conventional thinking (kilobytes of data on RFID tags) in their proposed standardization of a format for ***minimal*** data on RFID tags, referred to as electronic product code (ePC) that will serve only as a ***reference*** to a physical object, data about which may be stored on the internet (Figure 1-11). The generic organization of ePC was to extend the Universal Product Code (UPC) format currently used in bar codes. Thus, ePC was re-using an 'old bag of tricks' yet 'disruptive' to the *status quo* since the business of RFID usage had been around for half of the 20th century. A 'killer' ePC application may be a simple way to connect bits (information) with atoms (physical objects) in a manner that may make it feasible for ***widespread*** business adoption by offering low cost tags and use of the internet as the 'data' store. Low cost passive tags suffer from some limitations (signals absorbed by metal, such as beverage cans) which can be circumvented by a combinatorial approach to include emerging technologies, such as the active ultra wideband (UWB) tags. UWB tags transmit data at distances 30-300 meters using low power levels and the signals can penetrate metal barriers as well as concrete walls.

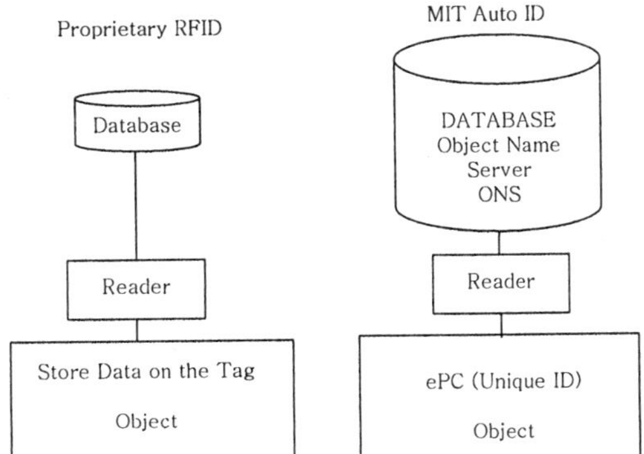

Figure 1-11. Evolution of ePC

The 96 bit electronic Product Code (ePC) as proposed by the Auto-ID Center, is made up of Header, ePC manager (manufacturer's information, also in bar code), object class (product category similar to bar code) and serial number space that is expected to be unique for each unit, such as an individual can of Coke. The ePC manager is defined by 28 bits that can uniquely represent more than 268 million companies. Similarly 16 million different product classes (object) can be defined by 24 bits. Coke Classic and Diet Coke belong to 2 different object classes. The 36 bit serial number space refers to the maximum number of individual items in a specific product class that may be assigned a unique number. Thus, more than 68 billion individual Coke Classic cans may be individually identified if each can had a RFID tag encoded with ePC (Figure 1-12). In 2000, The Coca Cola Corporation, the largest bottler, sold 3.8 billion 'unit cases' each containing 192 ounces. About 42%, or 1.6 billion, 'unit cases' were Coke Classic (19.2 billion individual cans, assuming that all Coke was sold in 16 ounce cans). If each 16 oz. can had a unique identifier (19.2 billion cans per year), even then the ePC serial number space, as defined by the Auto-ID Center, will accommodate individual numbering of Coke Classic cans for many years!

If the company made a sensible business process decision that the granularity of information at the level of each can was unnecessary, it could still track and trace 'unit cases' affixed with RFID tags encoded with ePC. If we use 2000 sales figures for Coke Classic (1.6 billion unit cases), the ePC serial number space will accommodate unique numbering of each Coke Classic 'unit case' for about 40 years. The 96 bit ePC serial number space will be sufficient for nearly a century for Perrier, the French bottling plant that produces 3 million bottles of Perrier per day.

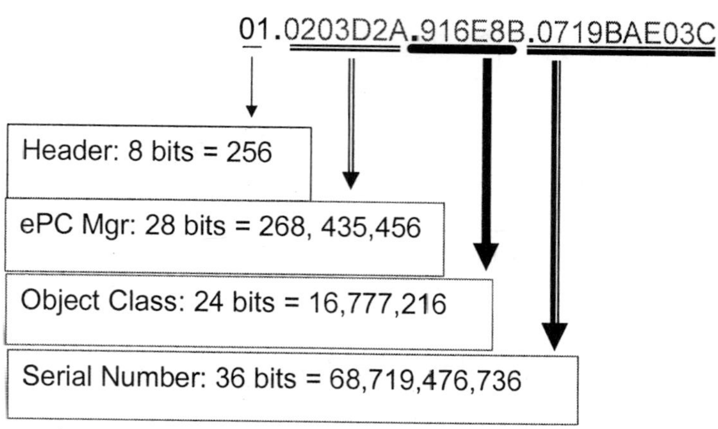

Figure 1-12. 96 bit Electronic Product Code (ePC) proposed by the Auto-ID Center

But, these are only tools which may act as catalysts if thought leaders develop a vision to use this rich yet raw data. Businesses may manage uncertainty, reduce inefficiencies, and information asymmetry if corporate leaders are capable of utilizing real-time data to stimulate business process innovation. Can real-time information compress time between supply and demand? Auto identification technologies, as enablers of data acquisition about objects, to be valuable, must feed real-time information to update processes (maintenance, cross-docking) or decision systems (planning, execution) to trigger adaptive response(s). As an extension of adaptive decision support capabilities, real-time data can offer 'transparency' if pervasive and accessible via distributed data infrastructure among the value network. Transparency may be the key to further catalyze the practice of supply chain management to evolve toward an adaptive value network. Point A to B data visibility may augment a few operations and offer savings, but is far from the supply chain profits from real-time adaptability through pervasive real-time data (RFID) usage.

The impact of pervasive RFID (or UWB) deployment will create an avalanche of data, but can we extract the information from this data that will be valuable for business transactions? (Figure 1-13) In US alone, there are 1.5 million retail outlets, 160,000 grocery store chains, 400,000 factories and 115 million homes. The US consumer packaged goods (CPG) industry produces 1 billion items per year. If we read each item 10 times (in the supply chain) it translates to 300,000 "reads" per second. At 100 bytes to store each 'read/event' data, we will be faced with 1000 terabytes of static data storage each year, from CPG operation alone. The road to ubiquitous tagging of objects will dwarf the current internet that now holds about 1 billion web pages with only 10 petabytes of data. Current (year 2003) estimates suggest that we generate about 1 terabyte of data per second. The future requires a radically different mechanism of data and information handling, through Agents and use of semantic tags, to make sense of it all.

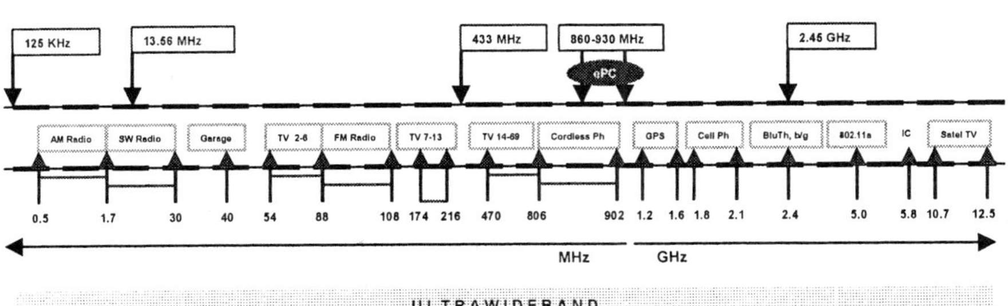

Figure 1-13. The Ultra High Frequency (UHF) Range for which an ePC Standard is Now Available

Given the potential impact, the 'RFID' market is, naturally, in quagmire, in part, spawned by unrealistic claims by some proponents of RFID who are focused on cost. Others discuss supply chain but understand less of its implication in terms of transparency in a value network. Still other contributors include vendors pushing products and consultants pushing services to offer you awe-inspiring ROI. Both groups want to "make hay while the sun shines." Nay-sayers (with other commercial interests) are eager to point out shaky ROI because ***their*** methods still cannot prove the value. Another component has emerged in the form of individuals or groups (in search of media attention) who are quite vociferous about privacy of information but offers little substance to explain what constitutes violation of privacy in the context of an ePC alphanumeric string serving as a reference for Jiffy Peanut Butter or Wrigley's Chewing Gum.

More than 5 billion bar codes are scanned each day, worldwide, but its potential for ubiquity may be cut short by the up-start ePC, but not anytime very soon. The inventors of the first linear bar code system were, naturally, decades ahead of their time. Bernard Silver and Norman Joseph Woodland applied to patent the system in 1949 and their patent was granted in 1952. Both were graduate students at the then Drexel Institute of Technology in Philadelphia and the idea was triggered by over-hearing a conversation in 1948 between the President of a grocery chain store imploring the Dean at DIT to develop an automated checkout system. Woodland took a job at IBM after graduation but IBM expressed limited interest in this work for bar codes. Disappointed, the duo sold their patent to Philco. Bernard Silver died in 1962. In the late 1960's when their patent expired, several new technologies converged to make product scanning commercially feasible. In 1970, ten grocery companies formed a committee to choose a standard for encoding product data (the present day universal product code, UPC, the predecessor to ePC). By then IBM wanted "in" on the action and brought in Norman Woodland, still an employee at IBM, to help launch the bar code research effort. In 1973, Woodland's leadership may have persuaded the standards committee to choose IBM's symbol over six other competitors. On 26 June 1974, in a Marsh Supermarket in Troy, Ohio (USA) a package of Wrigley's Chewing Gum was the first item scanned using the (universal product code) bar code (Scanlon, 2003).

" In contrast, at highly successful firms such as McKinsey and Company [...] Hundreds of new MBAs join the firm every year and almost as many leave. But the company is able to crank out high-quality work year after year because its core capabilities are rooted in its processes and values

rather than in its resources (vision). I sense, however, that these capabilities of McKinsey also constitute its disabilities. The rigorously analytical, data-driven processes that help it create value for its clients in existing, relatively stable markets render it much less capable [...] in technology markets." (Christensen, 2000).

Given the volume (some of dubious quality) of information already available on every facet of RFID and its applications in various industries, it is not necessary to add any technology or application review in this article. Our view of RFID deployment from a process perspective includes, albeit in stages, gradual integration with Agents in the system, for possible transition from real-time to adaptive to predictive states within the value network. The following figure outlines this convergence (Figure 1-14).

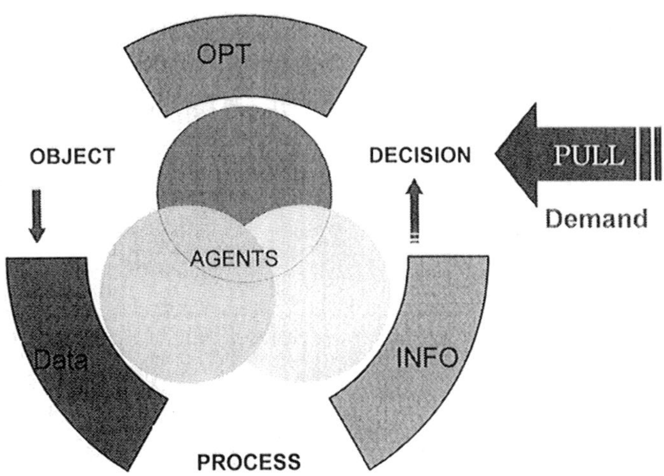

Figure 1-14. Convergence: Near Real-time Predictive Model

The version of the above illustration that may be forwarded as the "real-time model" can be viewed by removing the Agents and the 'Pull' signal from the above figure. Similarly, the "real-time adaptive model" may be visualized by excluding the "Pull" signal but including Agents. In general, the ability to identify any object in real-time (without errors and manual data entry), offers data that may be sieved through "intelligent" middleware to improve or adapt processes. High level or aggregated information and/or learnings may enable precision planning in the decision layers. Prior to execution, decision systems will be able to optimize how many objects may be distributed, displayed or destroyed. The ability of Agents to monitor and process vast surges of data in near real-time will enhance the adaptive abilities implied in the model. However, what the customer "wants" to buy still remains the predominant market uncertainty, ε in the Signaling Game (see section on Game Theory and Operations Research). Are there mechanisms or innovative strategies that can "extract" this future demand signal to move the push-pull boundary? Actual "pull" data that is verifiably robust (value of ε) is at the core of the 'predictive' model since such customer "pull" data for future demand may be one pivotal factor in reducing supply chain inefficiencies by taming the Bullwhip Effect (Figure 1-15).

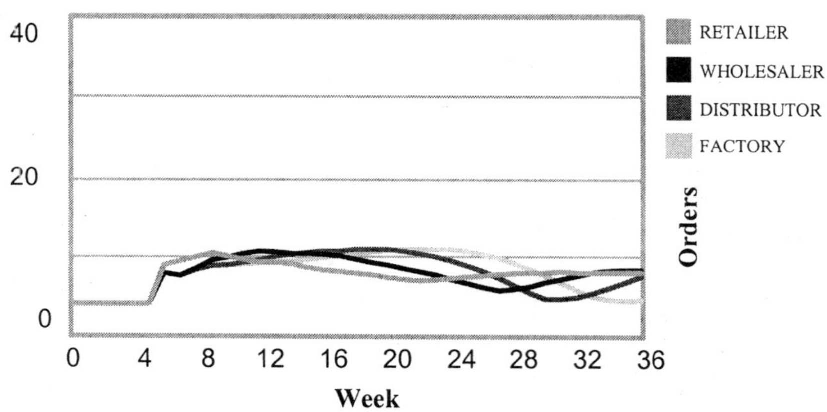

Figure 1-15. Bullwhip Effect after RFID? (Source: Yogesh V. Joshi, 2000. MIT Thesis)

The immense diversity of the "end" consumer makes it impossible to suggest any general mechanism. As an example, consider super-market type retailers who sell both dry goods and perishables. Retailers operating in single digit profit margins dream about improving accuracy of demand forecasting. Consider a down-to-earth scenario where a family of four living on San Silvestro in Venezia does ***not*** own an internet linked, Agent impregnated, ePC reader enabled refrigerator (from Being Digital Inc).

Instead, this family has a note pad on the refrigerator door. If Kathleen is using all the pesto, she writes Pesto (Butoni) on the super market shopping list, which keeps growing since the last shopping trip to Tesco. Charles wants fresh bananas and adds it to the list. Colin, manager of the Albertson's Super Store, due to open next week near the Rialto Bridge, visits you. He is engaging and talks about his last job in Garden City. As a part of Albertson's marketing campaign, Colin offers you a sleek tablet PC-like personal digital assistant (PDA). You are struck by the logo of Carleton urging all of us to "invent" and it inspires you to think different. Colin explains that Albertson's has teamed up with Moore Inc who bought Boingo Wireless from Sky Dayton. Colin is very convincing and you realize that this is not "a pie in the sky" scheme. You just may be on the road when the future arrives. The PDA is wireless internet accessible. You can use it at a T-Mobile "Hot Spot" such as one in the McDonald's in San Marco. However, Colin would like you to use the magnetic holder of the PDA and slip it on the refrigerator door. Every time Matt is close to emptying the shampoo or Elaine finishes the Barilla tortellini, they should add these to the shopping list, as usual, but on the PDA with the sensor pen. What's that to Albertson's? Well, if you wrote down Barilla Pasta and bought Barilla Pasta the next time you shopped at Albertson's with your Club Card, you shall receive a 2% discount, which also applies to all the items you scribbled on the PDA, if you actually bought those items at the store. What happens if you shopped online at Albertson's virtual store, A_Pea_in_the_Pod.com? Colin explains that the PDA is still going to save you money. If you can plan ahead such that you can wait 24 hours before home delivery, then you get a discount. If you wait 48 hours, you receive 2.5% off your total bill. What if I can wait for 5 days? Colin explains that any wait longer than 48 hours is rewarded with a massive discount of 3%. But if you did go to the store with your PDA, it will wirelessly guide you to find things on your list and offer other tips or alert you to manufacturers or competitors e-coupons for things on your list. The first 100 people to sign up for Albertson's offer also gets an autographed copy of the book of poetry "Moy Sand and Gravel," by the Pulitzer Prize winning author Paul Muldoon of Princeton University. Kathleen loves "Daffodils" and you want "in" on the action. Does it matter if Albertson's gets to know today what I want tomorrow?

Convergence of falling prices on PDAs, low cost wireless/wired access and some "intelligent" software is the infrastructure a retailer may need to capture the "pull" demand directly from some customers, as illustrated in the near real-time predictive model. Can this data from customers reduce your waste of perishables by 10% or adapt forecasting to reduce your purchasing capital by 1%? Real-time POS data from RFID tagged (ePC encoded) objects and the data flow from customers' pre-shopping lists may be combined for accurate forecasting and planning, particularly in procurement of perishables with short half-lives. In case of the latter, a final purchase order is sent only 36-24 hours prior to expected store delivery from producers (farmers, poultry, dairy). You can model the metrics in this scenario and claim that there may not be sufficient ROI to justify investment in this "pull" signal. How do you model the behavior of customers, say, in an area where more than 50% of the adults are internet users?

In 1959, GE recruited the reputable consulting firm of Arthur D. Little Inc. in Boston to conduct a study to determine whether there was a market for portable TV sets that GE could now build using solid state transistors. Several months later in 1959, after spending a staggering amount of money (millions) in focus groups and discussions, Arthur D. Little Inc. sent their analysis to GE suggesting that they do not believe there is any market for such TV sets. GE management pushed aside the project proposed by its engineers. Just before Christmas in 1959, Sony introduced a small B&W television in the US market. Sony sold more than 4 million television sets within months (Tellis and Golder, 1996).

5.1 ULTRAWIDEBAND: THE NEXT GENERATION RFID?

Instead of the customer's pre-shopping list in the retail scenario, what if that list was for spare parts at the Warner Robbins US Air Force Base (Agents case study) or US Army Aviation and Missile Command in Huntsville, Alabama? Can MRO (maintenance, repair and overhaul) improve its efficiency if the mechanics had visibility of the inventories of approved spare parts? In these and several other scenarios, it is likely that the benefits of using active ultrawideband tags will exceed low cost RFID tag usage. Only a brief overview of UWB is provided below since there is a mountain of original work in this area, especially from Dr. Gerry Ross and Dr. Robert Fontana (www.aetherwire.com/CDROM/Welcome.html).

The origin of ultra wideband technology stems from work in time-domain electromagnetics that began in 1962. At the Sperry Research Center, then part of the Sperry Rand Corporation, Dr. Gerry Ross, the father of baseband technology, applied these techniques to various applications in radar and communications. The experimental phases of these studies were aided by the development of the sampling oscilloscope by Dr. Bernard Oliver of the Hewlett-Packard Corporation (1962). In April 1973, Sperry Research Center was awarded the first UWB communications patent, due to Dr. Gerry Ross. Through the late 1980's, this technology was alternately referred to as baseband, carrier-free or impulse. The term "ultra wideband" was first applied by the US Department of Defense in 1989. By 1989, Sperry Research Center had been awarded over 50 patents including UWB applications such as communications, radar, collision avoidance, positioning systems, liquid level sensing and altimetry.

One recent application of UWB communications technology is the development of highly mobile, multi-node, *ad hoc* wireless communications networks for the US Department of Defense. The system is designed to be secure with low probability of intercept and detection. UWB ad hoc wireless network supports encrypted voice/data (128 kbps) and high-speed video (1.544 mbps). A parallel effort, funded by the Office of Naval Research, under the Dual Use Science and Technology (DUST) program is developing a state-of-the-art, mobile *ad hoc* network (MANET) based upon Internet Protocol (IP) suite to provide a connectionless, multihop, packet switching solution for survivable communications in a high link failure environment. The thrust of DUST is toward commercialization of UWB technology for applications to high-speed (>20 mbps) wireless applications for the home office. The Hummingbird collision avoidance UWB sensor (originated from a US Marine Corps project) was created for an electronic license plate commissioned by the US National Academy of Science (Transportation Research Board). The UWB Electronic License Plate provides a dual function capability for both automobile collision avoidance and (RF) tagging for vehicle to roadside communications.

Therefore, UWB usage in tagging is a proven technology. A comparative analysis of RFID versus UWB shows that UHF RFID has a spatial capacity of 1 kbpspm^2 (grouper.ieee.org/groups/802). Spatial capacity of UWB is 1000 kbpspm^2 or 1000-fold more than RFID. Spatial capacity focuses not only on bit rates for data transfer but on bit rates available in confined spaces (retail stores) defined by short transmission ranges. Measured in bits per second per square meter, spatial capacity is a gauge of "data intensity" that is analogous to the way lumens per square meter determine illumination intensity. Growing demand for greater wireless data capacity and crowding of regulated radio frequency may increasingly favor usage of spectrum that will offer appreciable bit rates that will function despite noise, multi-path interference and corruption when concentrated in smaller physical areas (grocery stores and warehouses). Spatial capacity limits may clog (like cholesterol in arteriosclerosis) "interrogation" systems when and if item level tags are a reality and readers in smart shelves continually emit electromagnetic signals to solicit tag data from objects. Part of this reasoning is evident in independent efforts by Hitachi and Sony who are exploring BlueTooth options with spatial capacity of 30 kbpspm^2 and others in asset management (Rockwell Automation) are exploring 802.11a compliant 5GHz with spatial capacity of 55 kbpspm^2 (spare parts). Unfortunately, 802.11a is non-compliant with 802.11b but 802.11g is compliant with 802.11b.

In the past couple decades, several companies have engaged in commercializing UWB. As implied by its name, UWB spans several gigahertz of spectrum at low power levels below the noise floor of existing signaling environment. Conventional narrow band technology relies on a base "carrier" wave modulated to embody a coded bit stream. Carrier waves are modified to incorporate digital data through amplitude, frequency or phase shift key modulation. These mechanisms are, therefore, susceptible to interference and the coded bit stream (for example, electronic product code or ePC) may be decoded/intercepted. UWB wireless technology uses no underlying carrier wave (hence secure military use) but modulates individual pulses either as a bipolar modulation or amplitude modulation or pulse-position modulation, where it sends identical pulses but alters the transmission timing. UWB offers narrower pulse time (300 picoseconds) and covers a broader bandwidth extending up to several gigahertz. Because UWB operates in picosecond bursts, power requirement is as low as 200 mW (compare 802.11b at 500 mW or 802.11a at 2000 mW). High data rate (0.1 to 1.0 gbps[2]) for UWB compares poorly with 802.11b (0.01 gbps[2]) or 802.11g (0.05 gbps[2]).Thus, UWB is used for wireless transmission of data, video as well as networked games, toys and appliances. There are robotic vacuum cleaners (from iRobots) and lawn mowers that may clean the living room or manicure the garden without touching the sofa or grazing the rose bush. Universal appeal for UWB is enhanced by its capability to accommodate several standards (ePC, GTAG). Without spectrum restrictions specific to country or region, UWB may become a global wireless medium.

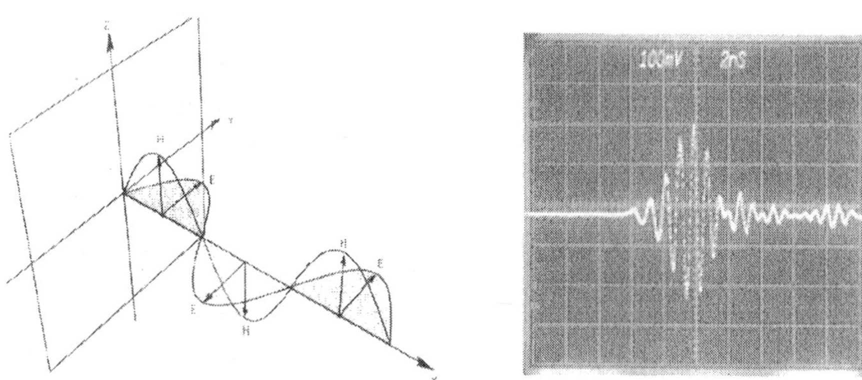

Figure 1-16. 860-930MHz RFID (left) and "Pulse" Transmission of UWB (right)

After the events of 11 September 2001, UWB transmitters (like RFID readers) were mounted on robots for search missions at the World Trade Center. UWB is not hindered by metal or layers of concrete. On 14 February 2002, the FCC gave qualified approval to use UWB (www.fcc.gov/e-file/ecfs.html) in the range >960 MHz, 3.1-10.6 GHz and 22-29 GHz. Active

UWB tags cost \$1-\$10 while the transmitters may be cheaper than RFID readers because they do not need many analog components to fix, send and receive on specific frequencies. However, software defined radio (SDR) based readers may soon arrive. UWB is not without its critics. Dispute stems from claims that UWB transmissions could interfere with spectrum used by GPS, cell phones and air traffic control. FCC is investigating, but plans to open up more spectrum for UWB commercial applications. Without the burden of fees for spectrum usage, the commercial floodgates for UWB usage may be unstoppable. Telecommunication giants who rushed to buy spectrum seduced by the future of 3G services are fighting to keep UWB off the news after investing billions in auctions to buy spectrum. Perhaps worse affected are the GSM sponsors in EU and USA. UWB is a tool for data acquisition (healthcare, hazardous chemicals) and thus a contributor to the future of adaptive value networks. An added value is its dual ability to provide data about objects when tagged to objects and form a wireless network to upload the data (over distances of 30-300 meters through metal and/or concrete) to the data infrastructure in much the same way that WiFi (802.11b) wireless networks may be used to upload RFID data in warehouses, stores or hospitals. Figure 1-17 shows plot of data rate and range capabilities of UWB.

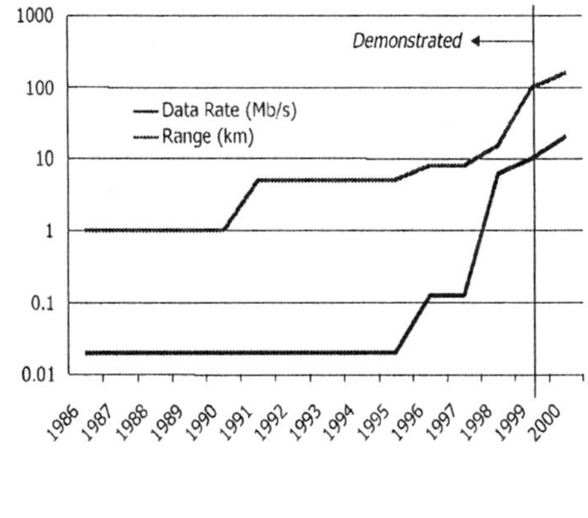

Figure 1-17. Data Rate and Range Capabilities of UWB (www.multispectral.com/pdf/APPsVGs.pdf)

6. SENSOR NETWORKS

Wireless sensor networks may be the first example of pervasive computing. Its applications extend from sensing blood pressure in arteries and transmitting them to a patient monitoring device to suggesting trends of warehouse shelf occupancy to a plethora of uses in the security industry. Sensors do not transmit identification data or ePC. Sensor data models cannot be used in the same manner as data models from UWB or RFID. Sensors are self-powered and form wireless *ad hoc* networks that upload through specific nodes which may be connected to data stores or the internet (Figure 1-18). However, each sensor has certain analytical abilities and due to in-network processing, the sensor network transmits analyses of the data rather than the raw bits of data to provide "answers" rather than "numbers" to the system. Embedded sensors are most likely to influence various forms of supply chain-related functions. For example, sensors attached to spindles in drilling machines may continually upload the status of the spindle such that it is serviced or replaced within a reasonable time to avoid breakdown of the machine and systemic downtime. Metrics like meantime between failure (MTBF) and other parameters may be helpful to determine when the service may be scheduled. Sensor data may require different thinking in terms of "adaptive flow" databases where the data (or analyses from sensor nodes) stream through the database where the query is stored. For example, embedded light emitting sensor network in a secure room sends positive light emission data on which the query (is anybody entering the room) need not act. Only when an obstruction causes a break in the *ad hoc* network or occludes the light signal from a sensor or group of sensors, then the query comes into effect. Service supply chains (such as heating, cooling companies) may benefit from sensor-based information to pre-dispatch technicians to stem problems before they reach break-points or require emergency attention. The key is to try to understand how to integrate sensor data to benefit supply chains functions. With the flood of nanosensors soon to arrive, the involvement of Agents may be absolutely imperative to harvest the benefits by extracting intelligence from such data.

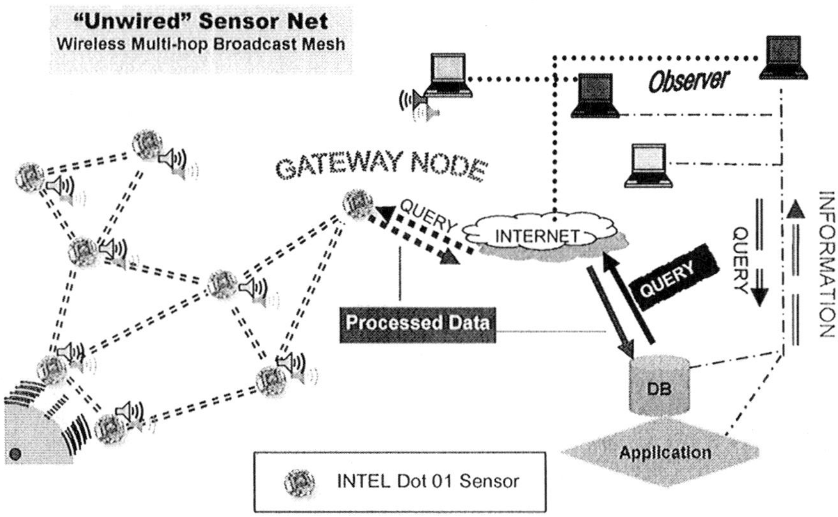

*Figure 1-18.*Unwired Sensor Net

7. THE SEMANTIC WEB (IS SPREADING)

The average user will never see this web but the buzz about the Semantic Web is as intense as the internet itself. Semantic metadata will let you do things with meaning. The massive amounts of data that we are likely to experience will be useless unless meaningful correlations and connections help us drive innovations, the profitable ones. But just because it is hidden from view does not mean that you can bypass the evolution of the Semantic Web, although it is intended for computers to improve searches, viewing data, interacting with services and sharing information. Taken together, it can offer process transparency across language and geographic boundaries to connect partners in a value network even if individual partners define or perform certain functions differently from others.

Tim Berners-Lee of MIT, the creator of the world wide web as we know it today (while at CERN, Geneva), had described the Semantic Web concepts perhaps as early as 1995 and certainly more clearly by 1998. Since that time, Tim Berners-Lee's vision has matured and significant progress has taken place in research communities around the world to demonstrate that semantic web can solve a variety of today's business problems. Semantics is a collection of Resource Description Framework (RDF) data (or any other semantic language) which describes the meaning of data through links to ontologies, which act as global decentralized vocabularies. In philosophy, ontology is a theory about the nature of existence (of what types of things exist). Ontology as a discipline studies such theories. Artificial intelligence and semantic web researchers have co-opted the term to indicate a document or file that formally defines the relations among terms. Computers, empowered with this metadata, will be far more "meaningful" in their understanding of the data without human intervention provided data is in machine readable format.

Michael Dertouzos of MIT and James Hendler of the University of Maryland have authored books and articles which are excellent resources to understanding the concepts of semantic web (Dertouzos 2002). Human language thrives when using the same term to mean somewhat different things, but automation does not. The authors provide this example: Imagine that I hire a clown messenger to deliver balloons to my customers on their birthdays. Unfortunately, the service transfers the addresses from my database to its database, not knowing that the "addresses" in mine are where bills are sent and that many of them are post office boxes. My hired clowns end up entertaining a number of postal workers, not necessarily a bad thing, but certainly not the intention. An address that is a mailing address can be distinguished from one that is a street address and both can be distinguished from an address that is a speech, with the tools from the Semantic Web.

This is not the end of the story, because two databases may use different identifiers for what is, in fact, the same concept, such as *zip code*. A program that wants to compare or combine information across the two databases has to know that these two terms are being used to mean the same thing. Ideally, the program needs to discover such common meanings for whatever databases it encounters. For example, an *address* may be defined as a type of *location* and *city codes* may be defined to apply only to *locations*. Classes, subclasses and relations among entities are a very powerful tool for web use. We can express a large number of relations among entities by assigning properties to classes and allowing subclasses to inherit such properties. If *city codes* must be of type *city* and cities generally have web sites, we can discuss the web site associated with a *city code* even if no database links a city code directly to a web site. Inference rules in ontologies supply further power. Ontology may express the rule "if a city code is associated with a state code, and an address uses that city code, then that address has the associated state code." A program could then readily deduce, for instance, that a Cornell University address, being in Ithaca, must be in New York State, which is in the US and therefore should be formatted to US standards. The computer doesn't truly "understand" any of this information, but it can now manipulate the terms much more effectively in ways that are useful and meaningful to the human user.

The real power of the Semantic Web will be realized when Agents collect web content from diverse sources (stock quotes from Bloomberg), process the information (in relation to your business) and exchange the results with other programs or data (demographic data from the US Census Bureau). The effectiveness of such Agents will increase exponentially as more machine-readable web content and automated information services (such as, real time-data) become available. The Semantic Web promotes the synergy between Agents that were not expressly designed to work together but can now transfer data among themselves if data comes with semantics (which levels the playing field in terms of the meaning of data, such as, your purchase order is the supplier's sales order). With ontology pages on the Web, solutions to terminology (and other) problems begin to emerge. The meaning of terms or XML codes used on a web page can be defined by pointers from the page to ontology. Of course, the same problems as before now arise if you point to an ontology that defines *addresses* as containing a *zip code* and one that uses *postal code*. This kind of confusion can be resolved if ontologies (or other web services) provide equivalence relations: one or both of our ontologies may contain the information that a zip code is equivalent to a postal code.

The scheme for sending in the clowns to entertain customers is partially solved when the two databases point to different definitions of *address*. The program, using distinct URIs (universal resource indicators) for different concepts of address, will not confuse them and in fact will need to discover that the concepts are related at all. The program could then use a service that takes a list of *postal addresses* (defined in the first ontology) and converts it into a list of physical *addresses* (the second ontology) by recognizing and removing post office boxes and other unsuitable addresses. The structure and semantics provided by ontologies makes it easier to provide such a service and can make its use completely transparent.

Ontologies can enhance the functioning of the web in many ways. They can be used in a simple fashion to improve the accuracy of web searches. Advanced applications will use ontologies to relate the information on a page to the associated knowledge structures and inference rules. An example of a page marked up for such use is www.cs.umd.edu/~hendler. If you send your Web browser to that page, you will see the normal web page entitled "Dr. James A. Hendler." As a human, you can readily find the link to a short biographical note and read there that James Hendler received his PhD from Brown University. A computer program trying to find such information, however, would have to be very complex to guess that this information might be in a biography and to understand the English. For computers, the page is linked to an ontology page that defines information about computer science departments. For instance, professors work at universities and they generally have doctorates. Further markup on the page (not displayed by the typical web browser) uses the ontology's concepts to specify that James Hendler received his PhD from the entity described at the URI http://www.brown.edu (the web page for Brown University in Rhode Island). Computers can also find that James Hendler is a member of a particular research project, has a particular e-mail address. All that information is readily processed by a computer and may be used to answer queries (where did Dr. Hendler receive his degree?) that currently would require a human to sift through the content turned up by a search engine. In addition, this markup makes it much easier to develop programs that can tackle complicated questions whose answers do not reside on a single Web page. Suppose you wish to find the Miss Cook you met at a trade conference last year. You do not remember her first name, but you remember that she worked for one of your clients and that her son was a student at your alma mater. An intelligent search program can sift through all the pages of people whose name is "Cook" (sidestepping all the pages relating to cooks, cooking, the Cook Islands and so forth), find the ones that mention working for a company that's on your list of clients and follow links to Web pages of their children to track down if any are in school at the right place.

An important facet of (Agent) functioning will be exchange of "proofs" written in the Semantic Web's unifying language using rules and information such as those specified by ontologies. For example, suppose Miss Cook's contact information was located by an online service which places her in Johannesburg. Naturally, you want to check this, so your computer asks the service for a proof of its answer, which it promptly provides by translating its internal reasoning into the Semantic Web's unifying language. An inference engine in your computer readily verifies that this Miss Cook indeed matches the one you were seeking and it can show you the relevant Web pages if you still have doubts. Although they are still far from plumbing the depths of the Semantic Web's potential, some programs can already exchange proofs in this way, using the preliminary versions of the unifying language. Figure 1-19 shows Tim Berners-Lee's Semantic Web layers. Many automated web services already exist commercially without semantics and their claims may be doubtful. Even if these services had Agents, at present Agents have no way to locate a service that will perform a specific function.

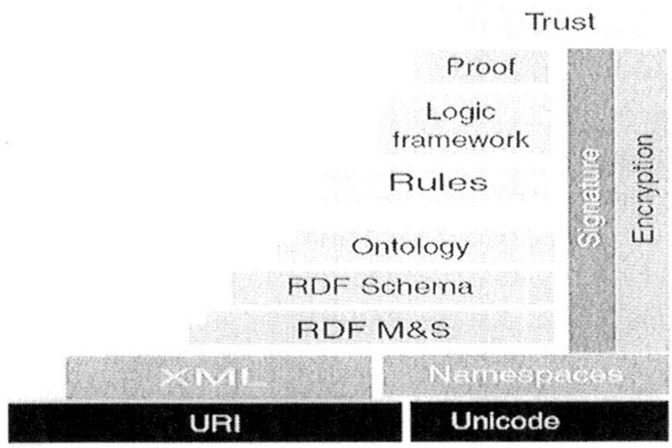

Figure 1-19. Semantic Web layers
(http://www.csail.mit.edu/research/abstracts/abstracts03/web/02berners-lee.pdf)

This process, called service discovery, can happen only when there is a common language to describe a service in a way that lets other Agents "understand" both the function offered and how to take advantage of it. Services and Agents can advertise their function by, for example, depositing such descriptions in directories analogous to the Yellow Pages. Some low-level service-discovery schemes are currently available, such as Microsoft's Universal Plug and Play, which focuses on connecting different types of devices. Sun Microsystems's Jini aims to connect services. These initiatives, however, attack the problem at a structural or syntactic level and rely heavily on standardization of a predetermined set of functionality descriptions. Standardization can only go so far because we cannot anticipate all possible future needs.

The Semantic Web, in contrast, is more flexible. The consumer and producer Agents can reach a shared understanding by exchanging ontologies, which provide the vocabulary needed for discussion. Agents can even "bootstrap" (learn) new reasoning capabilities when they discover new ontologies. Semantics also makes it easier to take advantage of a service that only partially matches a request. A typical process will involve the creation of a "value chain" in which sub-assemblies of information are passed from one Agent to another, each one "adding value" to construct the final product requested by the end user. To create complicated value chains automatically on demand, Agents will increasingly exploit more and more artificial intelligence technologies in addition to the Semantic Web. But the Semantic Web will provide the foundations and the framework to make such technologies more feasible.

8. CONCLUSION

Scientists use models to represent the basic nature of the universe. Businesses use models to optimize profits, products and services. Models may even predict future action. But, as ubiquitous as models are, they are, for the most part, isolated from one another. In other words, a model from one domain, such as weather forecasting, does not interact with another, such as purchasing or customer behavior. Can we harness the power of multiple individual data models into larger aggregates? What if we could make predictions based on not a few parameters in an equation based model but billions of diverse facts and functions that Agent based models might be able to accommodate? The latter may result in an unprecedented increase in productivity through the optimal use of resources, ability to adapt and prepare for change according to prediction. We may dramatically reduce the cost of goods and services through the elimination of inefficiencies.

To build these models, individually and then test them in combinations may be a worthy endeavor for generations of engineering and business students, supported by businesses. However, the business community may wish to embrace the key elements (tools and technologies) mentioned in this chapter and seek ways to bring about the convergence, repeatedly mentioned throughout this article. Principles from Game Theory empowered by real-time data from automatic id technologies may enhance your profit optimization. Reducing information asymmetry with partners in your value chain through secure Agents-based systems may exponentially eliminate inefficiencies. Deriving more meaning from data through the Semantic Web will allow you to enhance inter-operability between diverse environments of the partners in a value network. Convergence will determine, in part, the pace of your ability to adapt. Translating convergence to create a merger between bits and atoms is an evolution and is underway. The ability to use it in your business processes to innovate or invent is only limited by your imagination. You cannot visualize the future if your imagination is out of focus.

The payoff from information technology is in making transactions and processes more effective and efficient, it's not about creating a new economy or creating new models of industry. It is about taking a tool, powerful tool, and saying, "How can I make my supply chain more effective and efficient?" (Lou Gerstner, CEO, IBM, The New York Times, 10 March 2002).

ACKNOWLEDGEMENTS

For discussions, information and material in this article, the author wishes to gratefully acknowledge the contributors and help from:

Benson Adams, US Army
David Brock, MIT
Gerard Cachon, University of Pennsylvania
Nicole DeHoratius, University of Chicago
H. van Dyke Parunak, Altarum
Charles Fine, MIT
Gerry Frizelle, University of Cambridge
Neil Gershenfeld, MIT
Henry Lieberman, MIT
Grace Lin, IBM
Serguei Netessine, University of Pennsylvania
Ozalp Ozer, Stanford University
Greg Parlier, University of Alabama
Ananth Raman, Harvard Business School
Sanjay Sarma, MIT
David Simchi-Levi, MIT
Katia Sycara, Carnegie-Mellon University
Artificial Intelligence Laboratory, MIT

Support for the MIT Forum for Supply Chain Innovation provided by:

General Electric; General Motors; Intel Corporation; Michelin Corporation; MIT School of Engineering; MIT Sloan School of Management; SAP ; Timogen Systems Inc; US Department of Defense; Xerox Corporation

---

The physicist Leo Szilard once announced to his friend Hans Bethe that he was thinking of keeping a diary: "I don't intend to publish. I am merely going to record the facts for the information of God." "Don't you think God knows the facts?" Bethe asked. "Yes," said Szilard. "He knows the facts, but He does not know ***this version of the facts***."

Hans Christian von Baeyer in *Taming the Atom*

APPENDIX

CLUES FOR BUSINESS PROCESS RELATED INNOVATION TO USHER IN ADAPTIVE SCM:

This article deals with the ideas and concepts that may converge for the future of adaptive value networks. Often the key question is where to get started. We have made the point that process is the key and technology is a catalyst. Here are some clues with respect to processes that may offer room for innovation (Simchi-Levi et al, 2002).

Table 1-3. What's Good for Your Business: Optimize or Adapt?

	Optimize?	Adapt?
	Global Optimization	Managing Uncertainty
Distribution to Customer Assignment		
Distribution Logistics Strategies		
Distribution Network Configuration		
Production-Distribution Schedule		
Inventory Control by SKU and Nodes		
Vendor Managed Inventory		
Supply Contracts		
Outsourcing & Procurement		
Strategic Partnerships		
Product Design and Differentiation		
Plant-Product Assignment		
Customer Value/Profitability		
Decision Support Systems		
Information Technology		

Table 1-4. SCM Models and Parameters

SCM : Model Simple, Think Complex	Paradigm or Parameters
Deterministic analytical models	Variables are known and specified
Stochastic analytical models	At least one variable is unknown and follows a probability distribution
Economic models	Game Theory models of buyer-supplier
Cost-based simulation CBS for material	Order quantities, response times, cost data
CBS of production control	Lot size, lead time, material response time
CBS of finished goods stockpile	EOQ, demand data, production lead times
CBS of distribution	Ordering policies, transportation time requirements, demand & cost data, fill rate

Model sources of **uncertainty** (with certainty theory, Bayesian updating, fuzzy sets):

- Customer demand (Bullwhip Effect or "Hog-Cycle" Effect);
- Supply deliveries;
- External markets.

Quick Wins from Logistics Network Configuration:

- Storage at manufacturing plants (raw material, WIP, finished goods);
- Pick, load, ship to warehouse or DC;
- Unload and store at warehouse or DC;
- Pick, load and ship for delivery to next node (customer);

Savings may be possible from analytical optimization of (Table 1-5):

Table 1-5. Process and Impact of Real Time Data

Process	Impact of Real-Time Data	Value from RFID
Dock receiving capabilities		
Storage capabilities		
Receiving methodologies		
Order-generation capabilities		
Delivery time constraints		
Pricing and Promotions		
Merchandising requirements		

Data Collection based on model:

- Group products and/or product families (demand per product per customer);
- Group accounts by customer value and/or geography (zone) plus delivery frequency;
- Product shipment mapped by source warehouse vs. customer/zone;
- Demand by SKU per product (family) per zone;
- Production capacity (annual) at each plant;
- SKU storage capacity in warehouses (BOM for delayed differentiation products);
- Transportation mode, rates (TL/LTL by SKU) and cost (product/mile) between nodes;
- Service level (observed vs. expected vs. promised vs. industry best);
- Inventory carrying cost for safety stock to reach service level;
- Delivery time by customer/zone (map locations vs. transportation distances);
- Order processing cost (labor) and fixed operating cost (by nodes);
- Return and warranties (service cost);
- Wastage and shrinkage (costs).

Estimate/quantify time lag between processes *(consider total system benefit)*:

- point of origin of data/information and data upload/update/accessibility;
- systems visibility of data from any single point of contact;
- data/information usage in systems (disruption management delay);
- information application/use to improve (adapt) decision support system (DSS).

One Outcome:

Industry "clockspeed" vs. "lag" may suggest process innovations for real-time adaptive SCM.

Quick Wins from Inventory Management (Raw Material, WIP, Finished Product)

The source of system-wide savings forecast can be based on near-actual customer demand or "pull" strategies, such as buy-back or revenue-sharing contracts. Inventory carrying cost is about 20-40% of (turnover) value/year. Most (software) planners use Economic Lot Size Model (1915) to calculate economic order quantity (EOQ) that minimizes the cost function:

$$Q^* = \sqrt{(2KD)/h}$$

Parameters of this model may still be valid, but assumptions in this formula are likely targets for (real time data catalyzed) improvements toward "adaptive" supply chain management. Model assumptions (Table 1-6):

Table 1-6. Assumptions

Assumptions: subject to change if process modified/benefits from→	Real Time Data (RFID)
Demand is constant at a rate of D items per day	
Order quantities fixed at Q items per order (safety stock)	
Fixed cost K is incurred each time warehouse places an order	
Inventory carrying/holding cost (h) accrues per unit per day	
Lead time is zero	
Initial inventory is zero (shift inventory cost to supplier)	
Cycle Count Frequency (labour)	
Infinite planning horizon (periodic review)	

If one can factor in the "improvements" from real-time data that may help reduce variability (lead time heterogeneity, demand fluctuations) then, this formulation may still remain an effective model to indicate when orders should be received at warehouses (precisely when the inventory level drops to zero). Implementing ZIOP (zero inventory ordering property) involves precision real-time data synchronization across value chain partners that may make it possible to delay orders until inventory is zero (for whom?). In a centralized system, practice of ZIOP may approach near-reality but in a decentralized system concepts like CPFR along with real-time data sharing may be required as a precursor to practice of ZIOP.

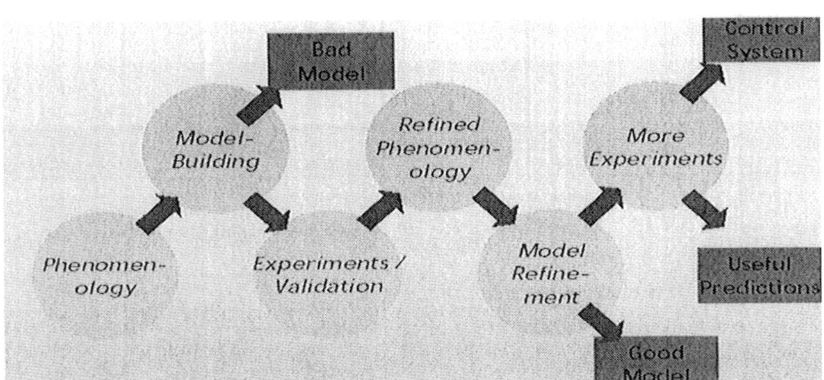

Figure 1-20. Steps in Model Building

Traditional SCOR Model based on "Push" System: Is it still relevant for your need? (Table 1-7)

Table 1-7. Perspectives, Metrics and Measures

Perspectives	Metrics	Measure	Real-Time Data
Supply Chain Reliability	On-time delivery	Percentage	
	Order fulfillment lead time	Days	
	Fill Rate	Percentage	
	Perfect order fulfillment	Percentage	
Flexibility & Responsiveness	Supply chain response time	Days	
	Upside production flexibility	Days	
Expenses	SCM cost	Percentage	
	Warranty as % of revenue	Percentage	
	Value added per employee	USD/EUR	
Asset / Utilization	Total inventory days of supply	Days	
	Cash-to-cash cycle time	Days	
	Net asset turns	Turns	

(One solution fits all?) SCM Software: Can it help with strategic effects?

Top Line Revenue Growth:

- Reduced time from concept to production;
- Minimize engineering change orders after production release;
- Increased rate of innovation;
- Better on-time delivery (fewer canceled orders; fewer late penalties);
- Higher quality (fewer returns).

Reduced Requirements for Working Capital

- Raw material, WIP and finished goods inventory;
- Inventory obsolescence;

Higher Return on Fixed Assets
Higher Margins
- Lower shipping cost;
- Lower manufacturing costs;
- Lower wastage;
- Improved product mix;
- Reduced inventory carrying cost.

Above measures/metrics may be driven by the following applications:
- Collaborative Product Design;
- Collaborative Planning and Forecasting;
- Optimized Manufacturing Planning;
- Inventory Planning and Optimization;
- Synchronized Planning;
- Detailed Scheduling;
- Accurate Order Promising;
- Optimized Transportation Routing.

Strategic "quick wins" categories likely to benefit from real-time information/data (RFID):
- Reduced requirements for working capital;
- Higher margins.

NOTES

1. Information Asymmetry is a concept borrowed from economics. In 1776, in The Wealth of Nations, Adam Smith put forward the idea that markets by themselves lead to efficient outcomes. The mathematical proof specifying the conditions under which it was true, was provided in 1954 by Gerard Debreu (Nobel Prize 1983) of the University of California at Berkeley and Kenneth Arrow (Nobel Prize 1972) of Stanford University (Arrow, K. and G. Debreu (Existence of an equilibrium for a competitive economy. Econometrica 3 265–290). However, the latter result showing that when information is imperfect (information asymmetry) or markets are incomplete, competitive equilibrium is not efficient is due to B. Greenwald and J. Stiglitz in 1986 (Globalization and Its Discontents by Joseph E. Stiglitz).
2. Value Networks refers to concepts by Clayton Christensen (Harvard Business School) which builds on the concepts of Giovanni Dosi and Richard Rosenbloom (The Innovator's Dilemma, 1997, Harvard Business School Press). We may often use supply chain management and value networks interchangeably.
3. It is beyond the scope of this chapter to delve into even a moderate level of discussion of Operations Research and Game Theory. Our intent is to offer some simple descriptions and indications about the possibilities of Game Theory applications in SCM. Game Theory applications, per se, are unlikely to make SCM more adaptive but these models can help the current processes by providing deeper insights. It is not uncommon to find businesses that are severely under-optimized in their current SCM practices. In such cases, it is speculative whether real-time information or efforts to be adaptive will meet with success.

Optimization, including game theoretic tools, may be necessary to "tune the engine" before adaptive SCM can offer value.

4. Prisoner's Dilemma was authored by A. W. Tucker of Princeton University [PhD advisor of John Nash]. Al Tucker was on leave at Stanford in the Spring of 1950 and, because of the shortage of offices, he was housed in the Psychology Department. One day a psychologist knocked on his door and asked what he was doing. Tucker replied: "I'm working on game theory," and the psychologist asked if he would give a seminar on his work. For that seminar, Al Tucker authored the Prisoner's Dilemma. (www.nobel.se/economics/laureates/1994/nash-lecture.pdf)

REFERENCES

1. Albeniz, V., Simchi-Levi, D. Competition in the supply option market. Working Paper, MIT, 2003
2. Anupindi, R. Bassok Y., Zemel. E. A general framework for the study of decentralized distribution systems. Manufacturing and Service Operations Management 2001; 3: 349-368
3. Aumann, R. "Acceptable Points in General Cooperative n-person games." In *Contributions to the Theory of Games, Volume IV*, Tucker, A.W. and Luce, R.D., ed. Princeton University Press, 1959.
4. Cachon, G. P., Netessine, S. "Game Theory in Supply Chain Analysis." to appear In *Supply Chain Analysis in the eBusiness Era*, Simchi-Levi, D., Wu D., and Shen, Z., to be published by Kluwer Academic Publisher, 2003.
5. Christensen, C. *The Innovators Dilemma*: *When New Technologies Cause Great Firms to Fail*, (Chapter 8, pp 168-169), Harvard Business School Press; June 1997
6. Cohen, M, Ho, T. Ren J., Terwiesch, C., Measuring imputed cost in the semiconductor equipment supply chain, Working Papers, Operations and Information Management Department, the Wharton School, University of Pennsylvania, 2003
7. DeHoratius, N.(2003), First Measure, Then Manage, *Capital Ideas*, 4: 1: summer, 2002. Retrieved October 2003 from: http://gsbwww.uchicago.edu/news/capideas/summer02/measuremanage.html
8. Dertouzos, M., *The Unfinished Revolution*, Harper Collins, 2002.
9. Gell-Mann, M., *The Quark and The Jaguar*, New York: W. H. Freeman and Company, 1994.
10. Heinrich, C. and Betts, B., *Adapt or Die: Transforming Your Supply Chain into an Adaptive Business Network*. John Wiley and Sons, 2003.
11. Forrester, J., *Industrial Dynamics*, MIT Press, 1961.
12. Joshi, Y. V. Information visibility and its effect on supply chain dynamics MS.c. Thesis, MIT, 2000.
13. Kuhn, H. W. "Extensive games and the problem of information". In *Contributions to the Theory of Games II*, Kuhn, H.W. and Tucker, A. W. ed. Princeton: Princeton University Press, pp. 193-216, 1953.
14. Lee, H., Padmanabhan, P., Whang, S. The bullwhip effect in supply chains. Sloan Management Review 1997; 38: 93-102
15. Nash, J. F. *Equilibrium points in N-person games*, Proceedings of NAS, 1950.
16. Negroponte, N. Less Is More: Interface Agents as Digital Butlers, Wired Columns, Issue 2.06 June 1994, Retrieved October 2003 from:

http://web.media.mit.edu/~nicholas/Wired/WIRED2-06.html, WIRED Ventures Ltd, 1994.
17. Neumann, J. V., and Morgenstern, O., *Theory of Games and Economic Behavior*, Princeton: Princeton University Press, 1947.
18. Norman Poire, Retrieved October 2003 from : http://www.smalltimes.com/document_display.cfm?document_id=2141
19. Parunak, H. "Go to the Ant": Engineering principles from natural multi-agent systems, Annals of Operations Research, 1997;75: 69-101.
20. Parunak, H., Savit, R., Riolo, R. Agent-based modeling versus equation-based modeling in the Proceedings of Multi-agent systems and Agent-based Simulation, 1998
21. Sarma, S., Brock. D. The Internet of Things, White Paper, Auto-ID center. MIT, 1998.
22. Scanlon, L. Behind Bars, Technology Review, April 2003, Retrieved April 2003 from: http://ww.technologyreview.com
23. Shehory, O., Sycara, K., Sukthankar, G. Agent aided aircraft maintenance, Proceedings of Autonomous Agents '99, pp. 306-312, Seattle, Washington, May 1999
24. Shubik, M. Some experimental non zero sum games with lack of information about the rules. Management Science 1962; 8: January: 215-234.
25. Shubik, M. Incentives, decentralized control, the assignment of joint costs and internal pricing. Management Science 1962; 8:April: 325-343.
26. Simchi-Levi, D., Kaminsky, P. and Simchi-Levi, E., *Designing and Managing the Supply Chain*, 2nd edition, Irwin, Inc., 2002
27. Sterman, J.D. Modeling managerial behavior: Misperceptions of feedback in a dynamic decision making experiment. Management Science 1989;35: 321-339
28. Tellis GJ, Golder PN. First to market first to Fail? Real causes of enduring market leadership. Sloan Management Review 1996; 37(2): 65-75.
29. Vickrey, W. Counterspeculation and competitive sealed tenders. Journal of Finance 1961;16 (1): 8-37.

Other Web Sites Referenced

http://ww.wrmea.com/archives/sept-oct02/0209044-2.html

http://ww.bea.doc.gov

http://www.cia.gov

http://william-king.www.drexel.edu/top/eco/game/ game.html

http://ww.legal-database.com

http://ww.aetherwire.com/CDROM/Welcome.html

http://grouper.ieee.org/groups/802

http://www.multispectral.com/pdf/APPsVGs.pdf

http://ww.technologyreview.com

http://www.csail.mit.edu/research/abstracts/abstracts03/web/02berners-lee.pdf

Chapter 2

FROM PRODUCTION TO A PRODUCT PERSPECTIVE

*A New Industrial Scenario?**

Florent Frederix
European Commission, DG Information Society

Abstract: The author tries to link the viewpoint "that future products and services will be designed to offer customers increasing value and enable manufacturers to respond faster and more flexibly to the ever increasing market demands" to the changes currently taking place, or announced in the industrial environments. The chapter further explores how current ICT trends and developments could have a decisive impact on future products and production environments.

Key words: Production, Wireless, Product

1. ICT IN CURRENT PRODUCTION ENVIRONMENTS

In the past the use of information and telecommunication technology (ICT) in production environments has taken on several forms. Probably, the most famous was the use of computers at the beginning to solve problems related to processes constraints. Most of you will for sure remember the traveling salesman problem and the issue of designing the optimal route but also the problem of efficient loading of production machines, or sequencing and batch building of products. This category of topics has received some priority from both industry and scientists since around about the start of World War II. It was then that it became apparent how large the gains were that could be made and also at a time when technology had advanced enough to be able to solve some of the optimization questions. As a result operations research became a science in its own right (Hillier and Lieberman 1974).

With the birth of low cost computers, predominantly in the late 1970s, they started to be used in production machines making them more intelligent. Some of the brainpower traditionally requested from the workers could be off loaded to machines. This use of computer systems has been increasing throughout the last 30 years in production environments and currently it is almost impossible to build a competitive production line without heavy automation.

Managers realized that not only individual production steps could be automated but that the whole sequence of production steps can be improved. Workflow systems, statistical process control, bill of materials and inventory management are some of the applications that have found a wide usage. This happened in parallel with the automation of the surrounding administrative tasks. The so called Enterprise Resource Planning (ERP) Systems played a key role to the progress realized in the late 1980s and the 1990s.

More recently external and internal Internet applications have been used to improve production processes and increase the flexibility of the manufacturing processes. To underline the importance, on a macro scale, of these changes it is sufficient to refer to Alan Greenspan, Chairman of the US Federal Reserve Board, speech (The Bay Area Council Conference, San Francisco, California, January 11, 2002) on the difference in how companies are managing the current changes in the economic landscape. His message was

> *"Thirty years ago, the timeliness of available information varied across companies and industries, often resulting in differences in the speed and magnitude of their responses to changing business conditions,.. Our economic structure changed in the mid-1990s. The crucial agent in this remarkable change was the quantum leap in information availability. ... Today, businesses have large quantities of data available virtually in real time. As a consequence, they address and resolve economic imbalances more rapidly than in the past."*

1.1 External Internet Based Business Applications

The appearance of Internet and broadband, two ICT developments of the 1990s, opened up the possibility to link independent facilities, producers and customers worldwide. This triggered another, and likely the most important, change in industrial environment of the second half of the 20th century. It became possible for the first time to build associations dedicated to the production of one product or production family by linking competent entities. The virtual enterprise was born and as an example the development in the semiconductor industry (Richards et al. 1997) did not pass unnoticed. The outsourcing, even of activities that often have been considered essential to the existence of a company (Frederix 2001), has become very quickly common practice in many industrial environments. This has created another wave of dynamics in the industrial landscape of the beginning of the 21st century, leading to a shorter time to market and faster evolving products.

Supply Chain Management (SCM), Customer Relationship Management (CRM) and Knowledge Management (KM) have been some of the enabling application environments.

"Customer Relationship Management is a term used to encompass the methodologies, software and capabilities that help organizations to structure and manage their customer relationships and interactions, with the objective of increasing customer satisfaction with the organization's products or services (Jansson 2002)." In theory, online interaction with customers (CRM) can collect and furnish a wealth of data on customers' behavior and needs. But how exactly to use this data in order to provide value added service to customers, and without getting into conflict with data protection, is to a large extent still unknown terrain, or part of experiments. CRM can also be used to collect information about how to customize products already in the production process.

Supply chain management (SCM) solutions should help businesses to reduce costs, increase revenue, and improve service to their customers by matching supply and demand through integrated and collaborative planning tools. Cost reductions are achieved for instance by integration with public and private market places and thus facilitating retrieval and comparison of suppliers. SCM also promises to reduce inventory and increase productivity by maximizing the efficiency of order processes and other administrative functions. Solutions may also support collaboration with business partners.

Yet sophisticated CRM and SCM solutions are not widely diffused. At the end of 2002 less than one fifth of small and medium sized enterprises in Europe use CRM tools and less then one tenth of these SMEs' have a supply chain solution in place. However, the implementation has gained momentum among large enterprises that already report considerable higher usage of these software solutions.

1.2 Internal Internet Based Applications

Internet technology has also been successfully used for a number of internal business processes. Nearly half of the employees work in companies that use online technologies for collaborative work purposes, such as the sharing of documents. An even higher percentage is found for larger companies; online technologies to track working hours and/or production time is another flourishing internal application used by many.

The employment of the Internet for internal applications has in general been considered the most efficient way of using this new communication technology.

2. THE EMERGING PRODUCT PERSPECTIVE

Today's manufacturing environment is increasingly competitive, with customers demanding ever more advanced and sophisticated products, greater choice and shorter delivery times. To satisfy this demand for differentiated and customized products, companies with different expertise must collaborate in order to ensure that the value chain remains flexible, so as to realize the full benefits of rapid product innovation and open competition.

In addition, manufacturers are making their products smarter by designing in added-value services as part of the customer offering. This "extended product" approach combines a product with services and enhancements that improve marketability. The customer proposition may subsist more through the benefits of the value-added elements than the physical product itself. Enhancements can incorporate tangible features that make the product more intelligent, customized or user-friendly, including embedded features like remote maintenance, or autodiagnostics. Other aspects, such as services, engineering or software, are intangible and make the offering more information, or knowledge, intensive.

Taking the product perspective is not new. It is easy to image that most of the production of vessels, clothing, houses and other goods of the past 1000 years or more have been custom made; adjusted to the wishes of the user. This customer orientation had to make place for a production focused approach to enable mass manufacturing that resulted, in the past two centuries, in the fastest increase in living standard ever seen. However, the need and usefulness of customization has never been put into question and chances are high, that in the future a much larger set of products and production environments will be affected by the development.

Looking at more recent publications (Stanford-Smith et al. 2002), one finds that the creation of value, after all the goals of most production processes, becomes more and more dependent on intangibles. This new understanding of "Value" was emphasized in the contribution of Mrs. Zobel to the proceedings of the eBusiness and eWork 2002 conference (Zobel and Filos 2002). The emergence of the Internet and network technologies has lead to a new perception of value for individuals and organizations. In contrast to traditional thinking, where value is related to "scarcity", value in the network economy is seen as the opportunity of having relationships.

Hence, the larger a community becomes, the more value a product acquires as a relationship-enabling means (Zobel and Filos 2002). Figure 2-1 illustrates the "value" advantage of service enhanced or customized goods.

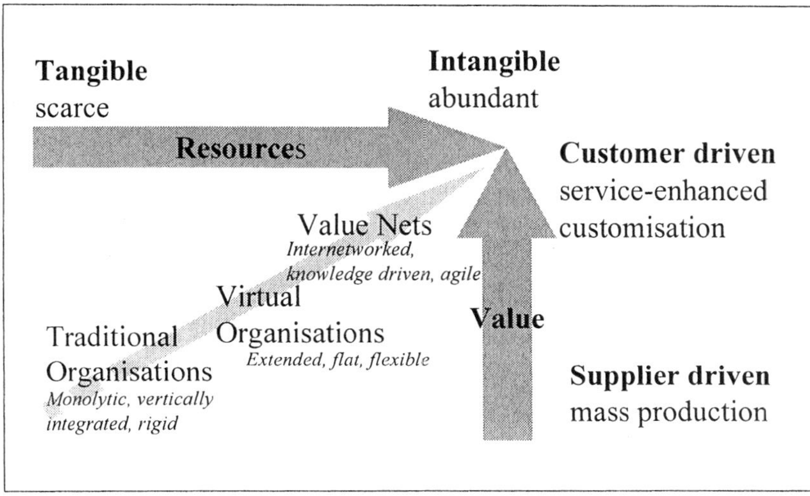

Figure 2-1. Creations of "Value" according to Zobel and Filos (2002)

2.1 Economic Importance of the Product View

The manufacturing industry has roughly 3 value creation (pre-production, production and post-production) stages. The most explicit part, the production phase, is resource driven. However, due to the economic differences, such as wages and cost of living, between Europe and Asia as well as the aging population in Europe most economic experts predict an acceleration of outsourcing of pure production activities to Asia.

This means that Europe should very carefully consider the pre-production and post-production value creation phases very and try to maximize the economic returns from these. Statistics today show that only 15% of the cost of a new car is linked to the production phase, therefore the focus on pre- and post production phases offers opportunities for profitable ICT investments.

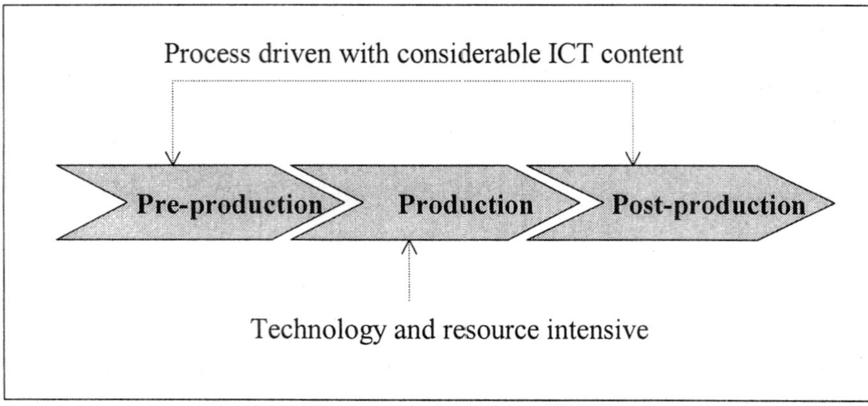

Figure 2-2. Value Creation Phases in Production

2.1.1 The Pre-production Phase

The first stage is product creation. In this stage value is dependent upon a production chains' ability (1) to be fast to the market (for advanced products); and (2) to be as much as possible adapted to the customer preferences.

'Fast to market' (1) implies the use of processes and techniques, such as Simultaneous Engineering and Concurrent Manufacturing, which deliver in a timely manner a correct design.

Customization and deferred configuration are the means to adjust the product to customer preference (2) that more and more are offered to customers as extensions of CRM packages.

2.1.2 The Production Phase

The second stage is the production of a product, or parts of it, in the most efficient way. This stage is the most resource intensive. Resources include the availability of a human workforce as well as capital to invest in the most appropriate production equipment. It also depends, to a large extent, on the use of the right technology, as it is mainly technology driven.

2.1.3 The Post Production Phase

This last stage is gaining importance. It includes logistics, distribution, maintenance, training, services and recycling activities. Figure 2-2 illustrates value creation in this phase and how the proper use of ICT, such as Internet and wireless technology, can contribute. The expected influences that ICT has on the post-production phase are set out in this chapter. More specifically Product Life Cycle Planning, Customer Relationship Management and Wireless Tagging and Control are dealt with.

2.2 Enablers for the product centered perspective

This section describes some visible ICT related changes in the manufacturing environment that may identify a more product centered perspective, it also tries to predict the adoption of the development.

Because take-up in an industry is a gradual process the reader will probably recognize some of the developments, or will be able to i dentify applications of the enablers in real life, even if their appearance is not yet fully in common practice.

2.2.1 Customer Relationship Management

Customer Relationship Management applications, described previously in the chapter, are a result of the mapping of traditional marketing, sales and after sales activities onto the Internet. However, other authors (Jansson 2002), already see it as a first step to a product centered perspective with extended products playing a more and more important role. In the next years the market expects a 6.7% annual growth for these systems (Aberdeen Group 2003). Some examples of this trend can be found at software companies that provide a constant and online software upgrade facility using a CRM portal.

CRM systems with a two-way dialogue assist field support engineers in remote diagnostic operations and even in some maintenance activities.

The connection of machines to the world-wide-web to allow tele-service and remote enabling of process manipulation is a further step in remote maintenance and control. Futman (2002) expects that by 2006 the USA will develop significant activities in this domain.

2.2.2 Simultaneous Engineering and Concurrent Manufacturing

The objective is to adjust products as fast as possible to customer taste by a reduction in time to market and an increase of innovation rate. Besides ICT technologies, the adoption of exchange standards for production and manufacturing are crucial for this process to succeed.

Simultaneous Engineering for new products and Concurrent Manufacturing are already common practice in product areas that have a high innovation rate, such as microcomputers and mobile handsets. However, more traditional industries are also moving this way.

Tele-design for manufacturing that allows collaboration with experts in distributed locations through the use of engineering networks is a way to reduce or avoid travelling and bring the product design closer to the customer. Customized product design, such as shown by the E-Tailor (2003) project, has the potential to become daily practice in several business sectors.

2.2.3 Product Life Cycle Planning

In forthcoming manufacturing environments the production cycle does not end when a product is shipped to a customer. It will be required to plan the whole life of a product, from knowledge generation/production to recycling/disposal, including the processes of maintenance and services in the planning.

Besides the ICT technological challenge such as computational power and data management systems, standards for life cycle and waste stream management are important. An intensified use of recycling to minimize waste is expected to be common practice in Germany by 2005 (Futman 2002).

2.2.4 Wireless Tagging and Control

It will be possible for any product and its parts to be tracked during their full life cycle. Machines and Products will be monitored and/or controlled remotely and wireless. Besides product and machine information, additional attributes such as location (GPS), environment (temperature, humidity, etc ...) and other sensor information (lab on a chip concept) will be easily available. Large scale implementations are expected by 2010 (Futman 2002).

The statement made by Heyman (2002) "*In the near future, every single object will be connected to the Internet through a wireless address and unique identifier. The Auto-ID Center is creating the standards that will shape this new age*" indicates that some industrial sectors, such as logistics and distribution, have the intention to follow a much faster track of implementation than predicted by the Futman (2002) study.

3. CONCLUSIONS

It is too early to conclude that the changes that are taking place in manufacturing environments will crystallize into a product perspective but signs are pointing in this direction. But when this happens then one of the most basic needs of mankind to adjust products and goods as much as possible will be possible once again: A trail that has been left a few hundred years ago at the start of the industrial revolution.
By pursuing this product perspective a wealth of opportunities for ICT companies can be created. It will also contribute to the solution of some environmental related issues (increase recycling, reduce pollution and waste, etc...). But even more important is that by increasing the value of the less resource and more knowledge intensive pre- and post- production phases Europe, with its high living standards and salaries compared to other regions in the world, has some means to remain competitive in the production domain.

ACKNOWLEDGEMENTS

This chapter is to some extent based on observations made from reading public results of projects sponsored by the European Commission. The projects consulted to draft the chapter include Futman (2002), Expide (Jansson 2002) and E-tailor (2003). The author wants to acknowledge the projects and project partners for their indirect contribution to this chapter.

DISCLAIMER

*The opinions expressed in this manuscript are the author's own and do not reflect necessarily the position of the European Commission.

REFERENCES

1. Aberdeen Group, CRM Market to reach $17.7 Billion by 2006, According to Aberdeen Group, Retrieved September 5, 2003 from: http://www.crmdirectory.com
2. E-Tailor Project result. Integration of 3D Body Measurement, Advanced CAD, and E-commerce Technologies on the European Fashion Industry. Retrieved September 4, 2003 from http://www.atc.gr/e-tailor/E-Tailor_final_low-res.pdf
3. Frederix F. An extended planning methodology for the discrete manufacturing industry. European Journal of Operational Research 2001; 129:317-325.

4. Future of manufacturing in Europe [FUTMAN], Industrial Approaches/Transformation Process annex to the published Final report (2002). Retrieved September 1, 2003 from http://europa.eu.int/comm/research/industrial_technologies/pdf/pro-futman-doc1-final-report-16-4-03.pdf.
5. Heyman D. (2002). "The Future of Business". Retrieved September 2, 2003 from http://www.autoidcenter.org
6. Hillier F.S, Lieberman G.J., Operations Research, San Francisco, Holden-Day Inc., 1974
7. Jansson K. (2002), Observatory and survey report of the EC sponsored EXPIDE (Extended Products in Dynamic Enterprises) Project, Retrieved September 1, 2003 from http://www.expide.org
8. Richards H.D., Dudenhausen H.M., Makatsoris C., de Ridder L. Flow of orders through a virtual enterprise, their proactive planning and scheduling, and reactive control. IEE Computing and Control Engineering Journal 1997; August: 173-180.
9. Stanford-Smith B., Chiozza E. and Edin, M., *Challenges and Achievements in E-business and E-work.* , Berlin, Germany: IOS Press / Ohmsha, 2002
10. Zobel R, Filos E. "Work and Business in the e-Economy. Technology and Policy Issues" In *Challenges and Achievements in E-business and E-work*, Stanford-Smith, B., Chiozza, E., and Edin, M. Ed. Berlin, Germany: IOS Press / Ohmsha, 2002.

Chapter 3

COLLABORATIVE SOLUTIONS IN THE SINGLE EUROPEAN ELECTRONIC MARKET (SEEM)

*Strategy Vision and Research Challenges**

Isidro Laso Ballesteros

New working environment Unit, DG Information Society, European Commission

Professor of eBusiness, CEPADE, Polytechnical University of Madrid. Spain

Abstract: Not only will Collaborative solutions increase productivity, but, provided that a Single European Electronic Market (SEEM) is in place, they will also foster innovation as a result of the positive effects on creativity coming from interconnecting culturally-diversified working environments. Multidisciplinary research efforts into the virtualization of resources, utility-like connectivity, self-established networks of wireless sensors and actuators, and next generation of semantic web and dynamic web services are required to complete the Single European Market and achieve the SEEM. These research efforts will pave the way for achieving the strategic vision of SEEM, i.e. an electronic environment allowing seamless synchronous and asynchronous interaction between European stakeholders, and with the virtualized environment, to achieve common objectives. SEEM will lead to a reduction of the cost of interactions, an increase in trust and confidence among companies and individuals, applications easy to install, take up and use, and an empowerment of people to innovate in what they do best.

Key words: Collaborative solutions, Single European Market, Single European Electronic Market, Value creation, Wireless, Sensors, Utility computing, Adaptive computing, Grid, Semantic Web, Web services, Individuals, Human factors, Ambient Intelligence

1. WHY COLLABORATIVE SOLUTIONS IN SEEM?

Collaborative solutions in the Single European Electronic Market (hereafter SEEM) will reduce cost of interactions, foster creativity and allow productivity gains due to seamless collaboration between human and artificial entities at European level (Table 3-1). Actually, "the Internet [is]

the closest Europe has come to realizing the dream of a single, borderless market" (BusinessWeek, 12 May 2003). These benefits will be based on interactions between entities surrounded by an Ambient Intelligence (AMI[1]), with sensors and actuators embedded in the environment capable of identifying individuals entering the environment, and not only delivering personalized services with pre-emptive capabilities, but also adapting chips to the particular needs of the moment (adaptive computing).

Table 3-1. Benefits of Collaborative Solutions in SEEM

Benefits of Collaborative solutions in SEEM	⇓ Cost of interactions
	⇑ Productivity gains
	⇑ Creativity - innovation

Collaborative solutions will be able to achieve productivity gains based on the decrease in the cost of interactions between members of a company and with its partners. Although these gains could be achieved through the current network of partners of a company, SEEM would provide new unique features, such as to facilitate the location of new partners all across Europe with, maybe, unique capabilities, which could foster new value creation. By contrast, huge increases in creativity, hence innovation, will only be generated if multicultural working environments are linked together in a seamless electronic working environment able to harness diversity in Europe, and around the world.

Consequently, SEEM aims at completing and extending the Single European Market (SEM) into people-oriented networked-centric environments, providing traditional entities (i.e. workers, companies, public administrations and consumers) with a collaborative working environment with no limits, no fear and no walls. In addition, there will be further features due to the new entities acting in SEEM, such as artificial agents, web/grid services, virtualized-entities representing the real things (not only human beings), and descriptions of human knowledge and competencies. The sum of traditional entities acting in the SEM and the new entities will be called hereafter "entities" (Table 3-2). These entities will be able to interact with one another in an Ambient Intelligence (AMI) environment to leverage the full potentiality of network-centric environments for creativity improvement, boosting innovation, and productivity gains.

Table 3-2. Entities in SEEM

Traditional entities	Workers
	Businesses
	Public administrations
	Consumers
New entities	Artificial agents representing human beings
	Web/grid services
	Virtualized entities representing real things
	Knowledge and competences description

1.1 Towards Adam Smith's dream

The greatest improvement in the innovative powers of labor, and the greater part of knowledge with which it is anywhere applied, seem to have been the effects of collaboration between workers. The Single European Electronic Market will multiply this effect through an open, network-centric, unconstraint and sustainable working environment. Apart from this effect, SEEM will also leverage the power of free trade and competition as stimulants to innovation and progress. On the supply-side of the market, SEEM will increase innovation through collaboration, and reduce friction. On the demand-side, it will better purchase decision through an improvement on the transparency of the market what will beef up quality certainty.

Both effects will deliver new features that will increase the competitiveness of all the entities acting on the SEEM, in particular Small and Medium Enterprises (SMEs). As can be seen in figure 3-1, currently several hundreds of isolated eMarketplaces and collaborative platforms exist. Today, SMEs take part in various e-marketplaces and have difficulties identifying possible SME partners. The participation in several e-marketplaces requires different software, including training, as well as license fees. The vision is to turn the Internet into one single electronic marketplace to do collaborative business, so that SMEs can take part at low financial and administrative costs.

Consequently, companies, in particular SMEs, are confronted with many technological barriers when doing e-business, including the cost of interactions. Costs related to the discovery of partners; negotiation of contracts and supervision of their transaction are strong constraints on the SMEs. The approach adopted by SEEM will be across industries and not split into vertical sectors, it will include interconnected value chains usually constrained inside separated industries. The synergies arising from these interconnections will leverage important benefits in terms of transparency of the market, leading the way to achieve Adam Smith's vision of the "Perfect free market".

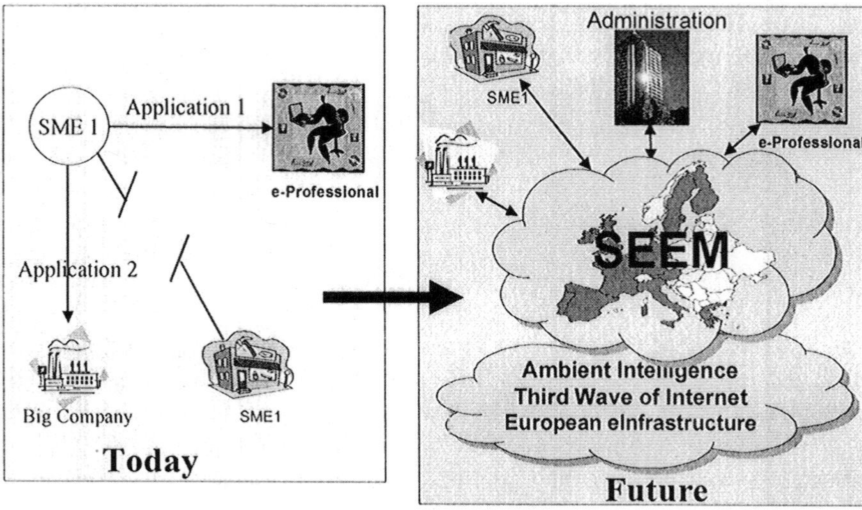

Figure 3-1. The Network is the Marketplace to Do Collaboration

1.2 RTD challenges

A number of research and development (RTD) challenges will have to be faced before the SEEM vision can be deployed in Europe and, afterwards, at global level. These RTD challenges will address the virtualization of resources, seamless connectivity and management of complexity, and will be explained in-depth later on in this chapter.

In summary, these RTD challenges can be listed in the next 7 bullet points:

1. Computing capacity: it will be necessary to develop decentralized repositories with information about individuals and enterprises. In order to have a rough idea of the amount of data, it is worth considering that there are 35 millions of SMEs worldwide, this means that, considering 5 Megabytes per SME, around 175 Terabytes will have to be handled. To this, must be added the several millions of professionals willing to participate in SEEM as individuals;
2. Virtualization: connectivity between virtualized physical environment and resources will be required, allowing mobility of the entities in SEEM. How to manage millions of seamless interactions between the European entities in an Ambient Intelligence environment?;

3. "Always-on" for seamless synchronous and asynchronous interaction, particularly important for entities on the move;
4. Self-established networks between sensors of the environment so that entities can move on those networks;
5. Web-services-like software applications capable of providing asynchronous collaboration, which is particularly important for mobile entities;
6. Web-services-like software applications able to scale to handle any kind of tasks;
7. How will collaborative solutions influence human behavior and vice versa?.

2. COLLABORATIVE SOLUTIONS

There are many definitions of collaboration, for instance Laso and Iglesias (2002) consider that collaboration includes "the relations between organizations (and their members) which define common objectives and work together to achieve these objectives, participating in previously agreed business processes". Not only does this conceptualization consider the organizational aspects as an intrinsic component of the collaboration between individuals, but it also includes the business processes, highlighting the fact that collaborating individuals cannot be considered in isolation from the virtual, or not virtual, enterprises that they work for. This approach considers that people will have to be the driving force, at the center of the creativity and innovation. This definition focuses on the need for agreement between the participating agents, pointing out the difference between collaboration and cooperation. Cooperation does not require the definition of common-objectives, i.e. two individuals can cooperate for their own interest even without having to share the goal of this joint work.

There are other definitions coming from associations of users, e.g. the National Assembly of Health and Human Service Organization (1997) defined collaboration as "the process by which several agencies or organizations make a formal commitment to work together to accomplish a common mission-related issue". It "requires a commitment to participate in shared decision-making and allocation of resources relative to activities responding to mutually identified needs".

In addition, many studies highlight the importance of collaboration. All of these studies are based on the concept of *interaction*, i.e. "*search, coordination and monitor activities done by individuals during the exchanges of goods, services and ideas*". There are several studies concluding that around 50% of all a company's activities are interactions and that they represent more than one third of the GDP in most of the developed countries (Briz and Laso, 2001). Hence, the cost of interactions for companies is huge and collaborative solutions could reduce dramatically this interaction cost.

These collaborative solutions should comprise human factors as well as Information technology (IT) components (Graf, 2003). On the one hand, humans should be put at the center of any further developments in collaborative solutions. A dynamic approach might be used so that not only are users involved in the requirement analysis and design phases to identify new RTD challenges, but also in further phases of research and development. Participation of users should not be static and the interactions with the environment need to be taken into account.

On the other hand, Information and Communication Technologies (ICT) have a role in fostering collaboration and an emerging IT market will consolidate in the coming years (Report on collaboration@work, June 2003). Recently, BusinessWeek (2003) stated that "collaborative solutions will be the next billion-dollar category". As an example of this importance, Cordis News (June 2003) reported that "technologies and application that allow for new and effective forms of collaboration will not only foster creativity and harness European diversity, but will also boost levels of innovation in Europe". Regarding the impact on interpersonal productivity, Bill Gates (Businessweek 2003) stated that "Almost everything is group-oriented. Everything has to do with the inefficiencies that exist with people working together."

There are two types of IT-based collaborative applications (Table 3-3). The first type of applications is devoted to collaborative business processes, and the second allows collaborative working environments between human beings and artificial agents. Both types of applications will be integral parts of a wider range of applications, the so-called collaborative commerce across companies' borders.

Table 3-3. Types of IT-based Collaborative Solutions

Types of IT-based collaborative solutions	Collaborative business process
	Collaborative working environments

Interest in this topic consequently arose some years ago in the research community. Actually, the majority of the IT conferences held nowadays include a considerable share of papers about IT application for collaboration, mainly for collaborative engineering.

As a result of these recent works, this is considered to be the dawn of a new R&D domain directed at the use of ICT for providing the collaborative features needed by the knowledge society. This new R&D domain will face new challenges popping up from the new ICT trends. For example, mobile work challenges the traditional concept of collaboration, generating new R&D challenges. On the other hand, the ambient intelligence vision (ISTAG, 2002) with pervasive computing environments creates new challenges in allowing collaboration between individuals and artificial agents.

3. WHAT IS SEEM?

SEEM is defined as a knowledge-based people-oriented network-centric environment allowing the collaboration between individuals and with their environment as well as amongst value chains, where individuals from any location in Europe can link into any chain without any access or interoperability problem. Hence, SEEM aims to complete and expand the Single European Market, not least because "the Internet [is] the closest Europe has come to realizing the dream of a single, borderless market" (BusinessWeek, 12 May 2003). The objective of SEEM is to achieve a network-centric environment that is:

- *open*, allowing any entity to come into and leave without barriers and catalyzing the creation and dissolution of location-independent dynamic partnerships between European entities;
- *network-centric*, bringing together available resources and competencies scattered in Europe to allow collaborative value creation relationships - consider the network as the heart of the SEEM body;
- *without* constraints imposed by IT vendors or any other technological barrier that could hinder collaboration and relationships between entities;
- *sustainable*, the virtualization of resources will beef up awareness of the limited available resources and the need to combine them in the most cost-effective way.

SEEM is based on the development of three pillars, the two first pillars are common to SEM:

- *Regulatory pillar*: To provide a certain environment for setting up collaborative relationships between European entities. Some adaptations will be required to meet new challenges arising from the new entities moving in the network environment;

- *Monetary pillar*: The advent of the euro is a milestone to increase transparency in the market and allow European-wide partnerships. Some measures will be required to assure trust and confidence to foster the use of European payment methods in the SEEM;
- *Technological pillar*: The research and development of infrastructure, description and interaction technologies capable to manage millions of interactions between the European entities in an Ambient intelligence environment will facilitate the achievement of the SEEM objectives.

To achieve the ambitious objectives pursued, SEEM will facilitate the identification and assessment of partners, and will ensure seamless collaboration between individual workers and among business processes. In addition, SEEM will be able to represent companies' characteristics and personal preferences, and to store individual worker knowledge into pan-European registries, defined as distributed common locations where metadata about specific data elements, structures, knowledge and/or service interfaces can be registered by their owners. All this information will be stored and traded in a secure way, maintaining high levels of privacy.

Above all, this platform will support persistence of data and reliable behavior of computing facilities, sensors and Web services, as a result of the use of appropriate communications protocols that allow synchronization of information and the recovery of interrupted connections.

Therefore, amongst the basic requirements of SEEM is the advent of pan-European seamless utility computing and the possibility for individuals to experience wireless interaction with human and artificial agents in their working environments. Additionally, the SEEM scenario should be based on commodity-like software so that workers and individuals do not experience any disruption in their collaborative mobile work.

3.1 Benefits of the SEEM

SEEM will have an evolutionary impact on the way that human beings interact with one another in the Knowledge Society. In addition, it will have an impact on learning process, methods of work and value creation. All these benefits are summarized below.

First, the social impact on the community will be focused on an increase of democratic values allowing the participation of all the entities in Europe in the law preparation process. In addition, SEEM will facilitate community building so that each group of persons can be in close contact with those interested in the same topics all across Europe. All in all, SEEM will deliver better services to citizens due to the setting up of European services provided by public administrations operating in close co-operation.

Secondly, regarding Learning processes, not only will SEEM facilitate the interaction student/teacher, but it will also increase the quality of the education by allowing the use of the best professors in each specific area of knowledge, location independent.

Next, the impact of SEEM on workers will be evident from the development of communities of practice gathering specialists in different areas. Moreover, SEEM will facilitate mobility and location-independent work as well as an increase in creativity and interaction between workers, decision makers, designers and engineers, etc. All these impacts will foster innovation not least because of the improved interaction between scattered competencies in Europe.

Finally, businesses will see an increase in the innovation pace due to the increase in creativity. In addition, there will be a reduction of interaction costs between the entities acting on the value chain. Furthermore, it will increase the impact and benefits of the globalization and European integration process and so productivity per capita. Moreover, there will be an increase in productivity due to the possibility of using the best competency in Europe for each specific link on the value chain. As a result, there will be better and more customized products and services available for citizens in the Knowledge Society.

4. AT THE NEXUS OF ICT TRENDS

Emerging collaboration across companies' borders will be fostered through the new ICT trends, which will allow the emergence of collaborative solutions on the SEEM. In particular, utility computing, wireless and sensors technologies, and the commodity-like software trends will facilitate the birth of the new collaborative solutions (Table 3-4). The challenges faced by this multidisciplinary R&D domain are great. Not only is it cutting across many traditionally isolated technology sectors but it also integrates some parts of these sectors to increase the effectiveness of the R&D efforts. This collaborative working environments will deliver new productivity and, mainly, innovation and creativity advantages derived form the 3rd wave of Internet, i.e. utility computing, wireless and sensors technologies (The Internet of things), and the tendency towards commodity-like software products. The following description gives details about each one of these components of the so-called "3rd wave of Internet".

Table 3-4. ICT Trends

Utility computing
Wireless and sensor technologies (including adaptive computing)
Commoditization of software

First, the utility computing trend is based on the principle of "utility". That is to say, when individuals plug into a utility network they will have access to all the virtualized resources they might need, both in terms of computing capacity and real time interaction with the surrounding working environment, as well as with other individuals. All the same, utility-like performance is expected, i.e. availability 24 hours a day, seven days a week. Not all elements of the utility-computing concept are ready for mainstream corporate use. For instance, one that still faces major R&D challenges is grid computing. Unlike a utility data center (developed by HP), which moves whole programs between machines, grid computing slices computing-intensive tasks into small pieces that are sent to dozens or hundreds of computers, taking advantage of underutilized machines. Grid computing is especially well designed for scientific or financial number-crunching jobs, but not all applications benefit from this slicing. Actually, many collaborative applications will require other kinds of utility-like services, i.e. real-time synchronous and asynchronous interactions, discovery services and context-aware location-based services. Another enabling technologies will be IPv6 (i.e. Internet protocol version 6) and next generation trust and security paradigms. In particular, IPv6 will allow automatic configuration and facilitate the existence of huge amounts of identified entities.

Secondly, wireless and sensor technologies are evolving towards "the Internet of things" (Schroenberger, 2002), allowing the virtualization of the physical world. Wireless sensors will set up networks dynamically, and the nodes of these networks will pass information from one neighbor to the next. Consequently computers will glean a connected view of the real world. With sensors, the Internet stretches to encompass global activity, whereas the Web has so far been a showcase for the human brain and the words, numbers, music and images that mankind produces. The radio frequency ID tag, called RFID, is a first step towards this aim. However, more sophisticated sensors, capable of creating their own networks and performing more difficult jobs, are required. In the future, each chip will have wireless capabilities. The advent of standard operating systems for sensor hardware and software, such as TinyOS, developed at the University of California at Berkeley, will pave the way to seamless collaborative working environments. Mobile workers will be able to receive information about their environments and make contextual searches for collaborators.

Finally, software is becoming a commodity industry due to the predominance of Semantic Web and Web services- intermediaries that allow computer systems to talk to one another using Internet technology. In addition, data will flow from one program to another in a format that will be both easy to understand and simple to integrate across many computer systems, the so-called XML. As a consequence, Web services are making software integration easier and faster in a semantic web environment where data have their meaning through the metadata and the relations and rules defined on top of them by ontologies. Therefore organizations are using Web services to keep whatever programs they are already using and just translating information into XML to talk to other systems. This could lead to less need to buy new software or to upgrade old software that works well. Hence, this commodity trend will move software to a commodity business. This commodity-like software will facilitate program independent interaction and collaboration between individuals. Regarding Semantic Web, knowledge-based collaborative applications on the SEEM require more than XML; hence, there is a requirement for the development of ontologies for collaborative mobile work. As regards web services-like software, ebXML is a first step to move web services towards asynchronous collaboration and not only synchronous enterprise application integration. Next generation web services should adapt dynamically to the changing environment allowing always on connectivity to the users and real-time interaction, and have the ability to scale in order to handle bigger tasks, being able to add virtualized resources. As a result, future web services-like software will be called grid services or maybe a new technological name.

All in all, these ICT trends will pave the way for a Knowledge Society for all, without technological barriers and enabling sustainable development, with the main goal of empowering people to innovate in what they do best, hiding the complexity of a virtualized collaborative seamless environment with a large amount of entities (tera-scale), some of them very small (nano-entities).

5. RTD CHALLENGES

After a detailed analysis of SEEM scenario and ICT trends, it is possible to identify three basic requirements: virtualization of all the resources in Europe, seamless connectivity between virtualized entities and management of complexity. The table below (Table 3-5) summarizes how the ICT trends can contribute to tackle these basic requirements.

Table 3-5. RTD Challenges

	Virtualization	Seamless connectivity	Management of complexity
Wireless and sensor technologies	Virtualizes the world	Allows mobility and interaction with the environment	Gleans a connected view of the real world (nano-scale)
Utility computing	Provides access to the "virtualized" world	Provides "always on" connectivity. "Plug & Play"	Allows access to "virtualized resources" required to handle complex problems (tera-scale)
Commoditization of software	Offers services abstracted from the "virtualized" world	Offers Software as a Service. Facilitate collaboration and application integration	Facilitate complex configurations and compositions of services

Regarding virtualization, sensor and wireless technologies will virtualize the physical world, utility computing will allow access to the virtualized world, and commodity-like software will offer services abstracted from the "virtualized" world. Virtualization of the elements of the world (physical resources, including wireless sensors and actuators, IT resources, software and human capabilities) will allow the interaction between individuals independently of their location in Europe. Therefore, workers on the move across Europe will be able to access their collaborative working environments. It will be possible to identify, assess and negotiate a collection of services and facilities that will fulfill the requirements for collaboration in each specific situation. People on the move, wireless sensors, web - or grid - services and wired physical resources will be nodes of the utility-like grid. As regards seamless connectivity, wireless and sensor technologies allow mobility and interaction with the environment, utility computing will provide "always on" connectivity, and commodity-like software will offer Software as a Service in order to facilitate collaboration and application integration. As a consequence, seamless interactive environments will be achieved, facilitating collaborative working environment, which will define membership and methods for identifying, inviting and entry of new collaborators. Besides, it will ensure that switching between synchronous and asynchronous interactions is seamless, with flexibility to consider the change in wireless network connectivity (connection / disconnection). On the whole, the entities of the SEEM will be able to interact using multiple devices and with their surrounding working environment through the use of proximity networking.

Concerning complexity, sensor and wireless technologies will glean a connected view of the real world (nano-scale), utility computing will allow access to virtualized resources to handle complex problems (tera-scale), and commodity-like software will facilitate complex configurations and compositions of services. As a result, complexity coming from heterogeneity and huge number of nano-components (quantum limit), as well as from the emergence of "disappearing organizations", and the competition between participants of the SEEM should be managed and hidden to the final user. Finally, one important challenge is the consideration of human factors and the involvement of individuals directly into the development process of collaborative applications. In addition, it is recommended to consider collaboration as a way of collaborative thinking, i.e. collaboration in the sense of thinking together. Another challenge is to capture the long term reciprocal impact of networking on human behavior and vice-versa. User acceptance of collaborative tools is another RTD challenge, the motto should be "let users experiment, do not constrain them too much"

6. COLLABORATION ON THE SEEM

While this part of the chapter explains some features of SEEM to allow collaboration, there are other features of the SEEM non-related to collaboration that will not be discussed on this chapter.

Due to the progress on the above mentioned ICT challenges, not only will SEEM enable business process collaboration between entities acting in the different links of the value chain in the life-cycle product/service, but also collaborative work between entities in creative and non-structured processes. These include business-related collaboration, such as decision taking process, as well as social and community related collaboration. In addition, it will make possible the discovery of new entities (e.g. partners) for the creation and dissolution of virtual constellation partnerships and/or communities at a very low interaction cost.

Entities acting within the SEEM would benefit, for instance, from 4 basic features provided for SEEM (Figure 3-2). The three first features can be considered as a way to achieve the fourth functionality, and are aimed at the needs of SMEs. These features are:

- *Discovery functionality*, allowing the entities a way to show their capabilities and know-how to the European community in every particular sector, and the means to locate the precise competency needed and set up an ad-hoc partnership to carry out a particular activity;

- *Contact/Relationship creation functionality*, including mechanisms to facilitate the development of trust among the parties and ways to reduce the physical distance;
- *Contract/Agreement negotiation functionality*, providing intelligent means to specify a contract electronically and negotiate it based on high trust ratings;
- *Open and network-centric Collaborative environment*, allowing the entities to collaborate in a well established security mechanism and monitoring that legal agreements are binding each and every collaborative relation between the entities. This environment will allow synchronous and asynchronous collaboration between human and artificial entities.

The features described above do not claim to be an exhaustive list of all the possible contributions of the Information Society to complete the Single European Market. On the contrary, the final contributions will be defined after analyzing the constraints imposed by the current status of the three pillars of SEEM, i.e. regulatory pillar, monetary pillar and technological pillar. Representatives of citizens, consumers, research community, ICT industry, other industries, standardization bodies, and public administrations will carry out this analysis.

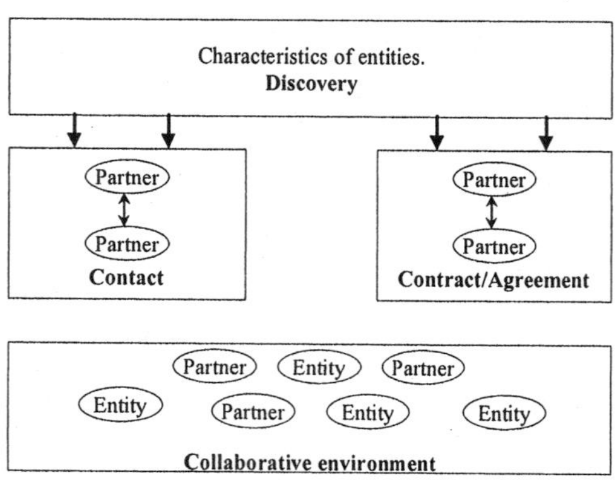

Figure 3-2. Functionalities Provided by SEEM

7. CONTRIBUTIONS OF THE ICT TRENDS TO COLLABORATION IN THE SEEM

The research to be done related to the three ICT trends will have a major impact on the future development of collaboration in the SEEM(Laso I. Report on collaboration@work, 2003).

However, as the list of features will have to be enlarged and refined after consultations with all the relevant stakeholders, the final list of contributions can not be provided now. Nevertheless, the table below provides a tentative list of the contributions of the ICT trends to achieve the features of the SEEM analyzed above in order to allow collaboration at European level (Table 3-6).

Table 3-6. Contributions of the ICT Trends to the SEEM

ICT Trend	Feature of the SEEM	ICT Contribution
Utility computing	Registry/repository	by handling Terabytes of information
	Negotiation	by allowing mobility of the entities (Connectivity)
	Run-time collaboration	Through seamless interaction and mobility
Wireless sensors	Run-time collaboration	Through seamless interaction with the environment
		By increasing mobility
Next generation Semantic Web / Web services (dynamic adaptation to the environment and asynchronous interaction)	Negotiation	By enhancing interaction between artificial agents
	Run-time collaboration	By allowing collaboration without software-related barriers. Asynchronous web services will enable mobility of individuals.
	Registry	Providing an "Intelligent layer" based on ontologies.
	Negotiation	
	Run-time collaboration	

8. SCENARIOS APPLICATIONS

While a scenario is described in the following paragraphs, it should not be considered as a design specification for SEEM. On the contrary, it should be considered as food for thought mainly because the final design specifications will come from open discussions between all the stakeholders of the Information in the Society value chain. Consequently, there will be representatives from the Society as well as from each of the steps of the value chain showed in the following figure (Figure 3-3).

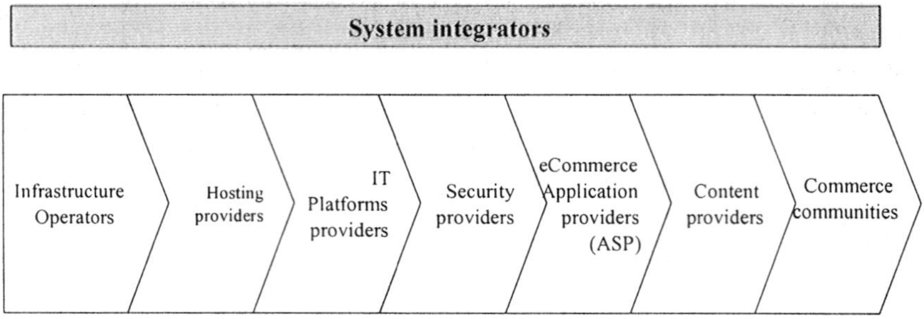

Figure 3-3. Value Chain of the Information Society (Laso and Iglesias, 2002)

8.1 Scenario 1: Textile SME

Sara SA is a textile SME, a micro-company actually, and has only 5 employees without technical knowledge about computers. Sara's owner has heard about the benefits of doing business with another partners, independent-located, in the SEEM and decides to make itself visible and available to all the other 15 millions SMEs in Europe.

He starts preparing the details of the company, which will be made available at the registry of the SEEM. Sara considers its main strength to lie in their competence in the design of new models. Sara is very well known for its modern and innovative designs and this competence is what Sara wants to boost. Hence, Sara's designers define internally which are the techniques and samples that should be shown to potential partners.

Next, Sara goes to the Internet connected device that is in the office and enters to register itself as "node" into the SEEM space. The system registers the IP address of Sara's device automatically and asks Sara's manager the main data of the company, such as legal address, public virtual identity, type of activity, and method of payment.

Now, Sara is asked to model their business and starts drawing the basic process that is done inside the company, explaining carefully the interfaces with other possible partners. Sara draws the process of designing in which they are experts, indicating that the input of the process is a requirement from a customer that can arrive as a DXF file or a scanned sketch, and the output file with the design will be in DXF format. Sara's experts indicate how long each task of the process takes and who is the person responsible for each and every task.

Then, Sara will have the possibility to enter the financial data of the company if it wants to do all the steps for the future collaborative relations by electronic means. As Sara usually sends and receives physical products, Sara's logistic responsible individual decides to include into the SEEM which is the delivery place so that Sara will be able to set up partnerships with logistics companies to fulfill the whole process in Internet.

Finally, Sara gets a certification for its device so it appears as a new node of the SEEM. Sara's authorized person can get a copy of this certification into their personalized communication device, maybe on their wrist. Now Sara competencies are available in Internet, with all the details to carry out collaborative relations with other partners.

From now on, Sara is always in the SEEM. Sara's employees will have access to all the collaborative tools available for SEEM nodes and will be able to set up peer-to-peer relations in order to carry out joint projects for new textile products.

8.2 Scenario 2: Two SMEs Setting up Collaborative Relations and Using Them

Companhia dos Texteis SA (CdT) is a textile SME which is available and visible in the SEEM, as a manufacturer located in Portugal. CdT's manager is Luisa who is planning to expand their market into Northern Europe. She decides to look up the SEEM for an automotive partner interested in their products for cars. Hence, Luisa sends her personal assistant to look for the best suitable partner according to CdT preferences.

It finds Finn plc after checking its description of business processes and all other resources and the company description that are available at the SEEM. Therefore it informs Luisa about this company which agrees in start negotiations with Finn's managers.

Luisa and Finn's managers decide to start negotiations by taking a contract template as the basis for the negotiation and the agreement is easily achieved thanks to the help of personal assistants. Afterwards they take strategic decisions to go beyond the simple interchange of data between automated business processes towards a real collaboration between their business processes.

To achieve this, they start to defining a common business process divided into several tasks, which specific roles assigned to each task and provide a detailed description of the information related to each task. Next, they decide which role will be taken by each company and also the timetable of the sequence of tasks. This common process is agreed by the CdT and Finn's top managers who authorize their employees to use the inter-company workflow automation system which is available at the SEEM as a collaboration service.

After several months working with this common business process it is clear to everyone that it is not enough and they start to think in terms of defining common unstructured and creative business processes as well. Their objective is not only to collaborate in the operative part of their businesses but in the creative and design parts as well. After agreeing on the definition of the creative relationships they decide to use collaborative tools available at the SEEM in order to set up a collaborative working environment.

The relationship between CdT and Finn is so fruitful that they decide to define common decision taking processes in order to plan and forecast their production on a common basis. Once they have taken this decision they start using collaborative decision taking tools available at the SEEM.

At this stage these companies are collaborating in virtual partnerships and they start thinking of widening this partnership towards a virtual constellation including raw material providers as well as logistics services providers. Before doing this they will have to agree on the basis of the collaboration and to define new operative business processes, creative processes and shared decision-making processes as well. In order to do so they will not start from scratch but start from the standard collaboration protocol between automotive and textile industries which includes proposed collaboration procedures with all the stakeholders in the value chain.

8.3 Scenario 3: Worker assistant

After working the whole day preparing financial analyses, Maria is enjoying a coffee together with her husband and children, relaxing and taking a breather, this being a busy period for her with many job offers from 3 different countries in Europe.

Maria is a freelance financial analyst who is part of the SEEM and is kept updated about incoming offers to work thanks to her personal assistant which is embedded in her clothes. Before joining the SEEM she had problems to get a jobs as she is living in a small village with difficult access to big cities. But now she gets so many jobs that she has to transfer some of them to her colleagues in her virtual community.

After taking the coffee she switches on her personal assistant and realizes that one new offer has arrived. It is a company, SCH, asking for her professional services to process all the financial data of the company and to prepare tax returns. She asks the system some details about SCH, it is a SME in the furniture-manufacturing sector and has a high score in the employer rating, meaning that it fulfils its legal obligations with the people working for it. Maria ask her assistant for some information about the products manufactured by this company, she gets to know that it has a medium score in the quality of products rating.

Maria is happy with this company, hence she takes a look at the company's processes and discovers that all the financial data are always transferred using "DataSheetXML" format which happens to be the same format that Maria usually works with. Therefore Maria instructs her personal assistant to negotiate the terms of the contract, Maria wants contract duration of between 1 and 3 years, with a compensation of a minimum of 200 euros per hour worked to be paid by bank transfer within a maximum delay of 30 days.

After sending the assistant to negotiate, Maria focuses on the financial analysis that she is preparing for a Northern European company. At the end of the day she receives a message from her assistant saying that both parties have signed an agreement with SCH in accordance to the instructions received from her. Maria agrees with the terms of the contract and asks her assistant to arrange all the practical details.

Her assistant finds that Maria has 2 slots of 2 hours free every Monday and Wednesday, therefore those 2 slots are booked for doing SCH financial analysis. The assistant of SCH believes that it might be necessary to allocate more time, at least at the beginning. Consequently Maria's assistant rearranges some other slots of time in order to have 6 hours available for SCH, which satisfies SCH's Assistant.

While Maria is sleeping, Maria's assistant realizes that Maria will have 8 hours free per week from next Monday on, therefore it starts moving through Internet offering the services of Maria, which are clearly described in her profile available at the SEEM.

9. CURRENT STATUS OF WORK

The foundations for SEEM have been achieved through the work done during EU's Fifth Framework Programme for Research and Development, FP5 (First report on SEEM October 2002 and Second report on SEEM March 2003), but at the same time new challenges have arisen. The SEEM cluster brought together 10 projects funded by FP5, the members of this cluster are: eBip (tile industry), OpenXchange (construction industry), RegNet (cultural heritage organizations), Bidmed (bio-medical sector), Ecos (furniture industry), SCOOP (textile industry), Opus (medical sector), Adapt (SMEs). Although the projects address primarily small parts of various business sectors their conclusions are more widely applicable. In a number of cases ebXML schemas have been used to good effect on product ids and company profiles. To exploit the findings on a wider scale and to introduce more scope, some severe obstacles are underlined. It is found in common that applications and platforms need in future to be made interoperable so providing a seamless geographic and intergenerational roaming. It is also concluded that the framework architecture and service enablers must be independent of Operating Systems and the application layers must be agnostic.

The main challenges identified by these projects are: to establish a registry of companies in the market, defined in a standard manner; to define a repository of the decision-making structure and business processes to build trust and confidence in the SEEM; to provide the possibility to achieve inter-company agreements through eContract; and last but not least to set up an environment for good collaboration between companies.

The new Research and Development Framework Programme, FP6, of the EU provides a framework that can strongly influence the necessary processes that must be shaped to meet SEEM. The advent of AMI vision in the FP6 adds a new orientation to the SEEM so that the focus is now put on people. People, and their needs, will be the drivers of any Research and Development efforts aiming at overcoming the RTD challenges faced by the SEEM.

The new Programme is ambitious, more risky than before, but at the same time has a much sharper focus. The guiding principles that will be applied are very different in the detail than for previous Programmes. It is anticipated that FP6 will be the vehicle for large integration projects and promotion of networks of excellence, providing a framework in where SEEM could be established. It is anticipated that project proposers will be high powered, skillful and resourceful teams, not fearful of the challenges ahead to lead the desired large integration projects that will have the necessary content and high potential impact and delivery for achieving SEEM. For instance, SEEMSEED is a new project funded under the FP6 due to start in January 2004, which will research RTD and policy-related challenges to achieve the SEEM.

10. CONCLUSIONS

Although collaborative solutions will bring important benefits by themselves, huge boosts in innovation will only be achieved if diverse cultures are linked together within SEEM, harnessing European diversity.

SEEM will complete the Single European Market with an electronic environment allowing seamless synchronous and asynchronous interaction between European stakeholders, and with the virtualized environment. This chapter does not claim to set up the design specifications and details of SEEM, primarily because they will only come after discussions between stakeholders of the value chain of the Information Society.

In order to achieve these ambitious objectives and benefits of SEEM, there are several RTD challenges to be tackled. These are in terms of "always on" connectivity, seamless asynchronous and asynchronous interaction between human beings and also with the virtualized network-enabled environment, as well as the capability to scale in order to handle any kind of task.

ACKNOWLEDGEMENTS

I would like to acknowledge the support of Bror Salmelin, the work and enthusiasm that my colleague Tina Mede has always put in the SEEM initiative, and the extensive support of David Cornwell, as well as the understanding of my wife Marta Iglesias.

DISCLAIMER

*The views expressed in this document are those of the author and do not necessarily reflect the opinions of the European Commission

NOTES

1. AMI vision. ISTAG, advisory group on Information Society Technologies. http://www.cordis.lu/ist/istag.htm

REFERENCES

1. AMI vision. ISTAG, advisory group on Information Society Technologies. Retrieved October 2003 from: http://www.cordis.lu/ist/istag.htm
2. Bill Gates. BusinessWeek. 27 October 2003.
3. Briz, J. and Laso, I. "Internet y Comercio electrónico". Mundi-Prensa and ESIC. 2001
4. BusinessWeek. 12 May 2003.

5. Cordis News (2003), Retrieved October 2003 from: http://dbs.cordis.lu/fep-cgi/srchidadb?ACTION=D&SESSION=159572003-10-23&DOC=4&TBL=EN_NEWS&RCN=EN_RCN_ID:20487&CALLER=EN_NEWS.
6. First report of the SEEM Workshop held at the European Commission in Brussels on October,1st_2002. Retrieved October 2003 from: http://europa.eu.int/information_society/topics/ework/seem_at_work/workshop1oct2002/index_en.htm
7. Graf, M. "Grid collaboration in the mobile environment". IBM. Speech at the collaboration@work_workshop. Retrieved October 2003 from: http://europa.eu.int/information_society/topics/ework/information/eventspages/collaborationatworkjune2003/index_en.htm
8. ISTAG (Information Society Advisory Group). Strategic orientations and priorities for IST in FP6. European Commission. 2002.
9. Laso, I and Iglesias, M.. Internet, comercio colaborativo y mComercio. Mundi-Prensa. 2002.
10. Laso, I. Et al. Report on collaboration@work workshop. European Commission. Retrieved October 2003 from: http://europa.eu.int/information_society/topics/ework/information/eventspages/collaborationatworkjune2003/index_en.htm
11. Report on collaboration at work. Retrieved October 2003 from: http://europa.eu.int/information_society/topics/ework/information/eventspages/collaborationatworkjune2003/index_en.htm
12. Second report of the SEEM Workshop held at the European Commission in Brussels on March 2003. Retrieved October 2003 from: http://europa.eu.int/information_society/topics/ework/information/eventspages/seemmarch/index_en.htm
13. Schroenberger C.R. The internet of things. Forbes. 2002
14. The National Assembly of health and Human Service Organization (1997). *The new community collaboration manual*

PART 2. CASE STUDIES

"The human-centered systems movement looks sensitivity at those forms of science and technology which meet our cultural, historical and societal requirements, and seeks to develop more appropriate forms of technology to meet our long term aspirations." Mike Cooley, Human-Machine Symbiosis, Springer Verlag 1996

Part two adds the necessary element of realism. Many solution/technology vendors and researchers claim that their tools or ideas will significantly improve value chain efficiencies and present example results tested within ideal settings or optimized settings. However, the success in real environments is a different story. Real world implementation faces the practicalities, constraints, and more complex steps, including sometimes tedious or trivial steps. Implementers together with eventual users have at first to review the infrastructure already in place, the processes operated, employee behavior, and the specific environment and so on and then by synthesis discover where the maximum potential benefits lie with new implementations.

Understanding the environment, processes and requirements is critical for the success. Different conditions and situations create a different requirement, companies have different experiences and areas of concern and the way they do business is unique to them. Even if an ICT provider has outstanding software and implementation methodology, which has potential to provide considerable benefit for a company operating in a network with others, if those benefits do not target the concerns or requirements of the company then they are worthless. So collecting and understanding specific user requirements together with a user company is vital to both parties. Success for any implementation will be measured against those specific requirements.

There are two challenges to achieve fulfillment of objectives: the groundwork that has to be done before any implementation and the implementation process itself. Without appreciating to the fullest extent the company's or user desires and aspirations and the nature of the working environment failure is inevitable on both counts.

Part two concerns a variety of different cases to appreciate the many different processes that can take place in the real change process that companies and business networks are undergoing. It takes a look at: a successful implementation story for high tech industry; research agenda created after user requirement analysis; how new concepts of collaboration can be used in a traditional industry; and how to create new forms of value by understanding user demand and behavior. Inferences are made about individual cases, where succeed and where they fail and how they will have to change in the future. Interpretation and comprehending the internals, user requirements within a company or business network, are dealt with, as is the new challenge of appreciating and understanding the externals that is what is required from a customer perspective. Examples of Small and Medium sized Enterprises (SMEs) to large businesses as they prepare next generation solutions for AVNs and eMarkets are discussed.

Chapter 4, written by Juno Chang and Min-Hyung Kang, explains a supply chain management project implemented by i2 technologies at Samsung's memory division. They review how the project team has managed changes and adapted i2's solutions to Samsung's business by explaining i2's business release approach. The chapter also describes some of i2's key solutions, the implementation challenges and benefits that have arisen.

Chapter 5, written by Flavio Bonfatti and Paola Daniela Monari, presents conditions of European SMEs and micro businesses from the perspective of their access to the forthcoming single electronic market (SEEM). Because of the extreme variety of contexts, structures and behaviors it is not easy to figure out and classify the real needs of SMEs and micro businesses with respect to the participation in the electronic market. The needs of SMEs and micro businesses can be classified into five main categories of requirement: access to the e-Market; regulations, trust and security; support to networking; problem solving; and Web mediation services. This chapter tries to answer the question according to a naive bottom-up approach consisting in the derivation of hints from the analysis of some real cases. It concludes with a scenario for a small company for the electronic market.

Chapter 6, written by Thomas Callarman, John Fowler, Esma Gel, Michele Pfund and Dan Shunk, analyzes the SCM needs for a semiconductor supply network. From their survey analysis, they determine a research agenda for semiconductor supply networks. This agenda comprises 7 clusters: supply network integrated demand; supply network design methodology and coordination; supply network integrated, enterprise-wide planning; supply network integrated scheduling and execution; coordinated business and financial models; supply network technical infrastructure; and organizational structure.

Chapter 7, written by Guido Grau and Marcus Winkler, presents a case study of one of the most crucial collaborative activities in the textile industry, product creation and development. Their findings suggest that sharing ideas and information in a business network was the key to reduce the number prototypes, the cost of logistics and the time to market.

Chapter 8, written by Christopher Lawer and Simon Knox, proposes a reverse market concept that defines markets and marketing activities from a consumer value perspective. Based on the case studies undertaken, interviews and experiences they describe how reverse market brands are able to create new forms of mutual value by organizing consumer demand and by supporting their lifestyles, need sets and complete consumption experiences. They propose a simple framework of emerging customer value drivers to identify the source of reverse market brand and customer value. They also discuss the benefits and the barriers that await brand management wishing to develop reverse marketing practices.

Chapter 9, Rainer Breite and Hannu Vanharanta, examines how hypertext based Internet applications support decision making in different purchasing situations. Three different purchasing case tests are discussed, compared and the outcomes synthesized to draw conclusions about hypertext based Internet applications in support of a purchaser's decision making process in a complex procurement situation. The results of research revealed: that the Hyperknowledge framework used is well suited for analyzing Internet applications; and that in purchasing situations different amounts and types of support are needed during the decision making process.

Chapter 4

SUPPLY CHAIN SOLUTION IMPLEMENTATION

Optimizing Supply Chain in Samsung Memory Division

Juno Chang[1], Min-Hyung Kang[2]

[1]SangMyung University,Korea, [2]KAIST Graduate School of Management, Korea

Abstract: Samsung Memory Division has gone through almost 5 years of optimizing their supply chain enabled with supply chain management solutions. We will explore how supply chain management solutions help the company improve supply chain performance and drive the process innovation. We also examine their phased approach for implementing supply chain solutions and how they have managed changes and adapted these solutions to their business. Especially, we focused on how the characteristics of the memory business impacted on their supply chain design and solution implementation

Key words: Supply chain management, Samsung, i2, Optimization, APS

1. INTRODUCTION

There is a turning point in software license revenue for Enterprise Applications Software (EAS). The new license revenue for Enterprise Resource Planning (ERP) began to lose market share, while Supply Chain Management (SCM) and Customer Relationship Management (CRM) began gained ground. Companies started to gradually increase their investment and interest in SCM and CRM from 1996. According to the report published by Gartner Dataquest Research in August 2001(Gartner 2001), the new license revenue for EAS in 1997 totaled $7.89 billion USD. ERP accounted for roughly 83% of the market. In 1998, the share of ERP fell to 78%, while the EAS market expanded by 20%.

Many companies that had implemented ERP systems have realized tremendous benefits from EAS, and they began to focus on SCM and CRM solutions. Leading high-tech companies were especially interested in SCM, and many of them turned to i2 Technologies, which had led the SCM market at that period. With 1998 revenue, i2 Technologies had 14% market share, almost 100% higher than the number 2 vendor Manugistics (AMR Research 1999).

The Semiconductor Division of SEC (Samsung Electronics Co, Ltd.) was one of those companies that had experienced overall improvement in business processes by introducing an SAP system. The Division then reviewed various plans for further improvement. In 1993, the Chairman Lee Kun Hee of Samsung Group promulgated the "Frankfurt Declaration (1995)", urging the entire group to renew itself for the 21st century. With such strong backing of the top management, the Semiconductor Division of SEC was able to carry out a series of business assessment projects with outside consulting companies, and it identified numerous low-level improvements that could be made. It designed "to-be" processes covering sales /production /development /purchasing and finally in early 1997 completed a master plan for software, hardware, and data infrastructure.

A consensus had been reached within the division for software. The major transaction-based processes would be taken care of by the ERP system from SAP, but there was a need for a new SCM solution to support supply chain management processes. Innovation was especially required in the planning area of the production organization because production planners were under increasing pressure to reduce the planning cycle time and improve responsiveness to demand changes. Furthermore, they needed to address to their heavy workloads, which were increasing due to expansion and increasing complexities of the business. In short, production plans were becoming less and less reliable. SEC had reviewed various options including in-house development of an SCM solution and an introduction of a commercial solution for its planning area. In doing so, the company had evaluated various solution packages and finally selected the i2 solution. With the introduction of a production planning solution, SEC built up effort to streamline its supply chain processes.

This chapter focuses on the description of the solution implementation effort over a five year period for SCM that was carried out by the Semiconductor Memory Division of SEC. The objective is to share information on the innovation efforts made to improve its SCM activities. It is true that there are no 'plug & play' SCM solutions; no SCM solution can support unique corporate business processes with zero customization or configuration. However, the key to a successful SCM solution implementation is to clearly identify what tasks the SCM solution must fulfill and what needs to be done by change management.

2. CASE OVERVIEW

2.1 Samsung Electronics Memory Division Overview

SEC (Samsung Electronics Co, Ltd.) is the world's largest manufacturer of memory products like DRAM, SRAM, NOR flash and NAND flash memory. In 2002 SEC's total sales of semiconductor products amounted to $7 billion USD. While significantly lower than $13 billion it achieved in the previous year, it represents 15.1 % of the world's memory chip sales –large enough to keep SEC in its number one position in the DRAM market 10 years in a row and become the world's second largest semiconductor company (Cassell 2003).

Given the extreme volatility of the memory business, SEC pursued a balanced portfolio of related electronics businesses. SEC now has four divisions - Device Solution (memory, system chips and TFT LCD), Telecommunication Network (handsets, network and fiber optics), Digital Media (monitor, PC peripheral and digital TV), and Digital Appliance (microwave oven, refrigerators and air conditioners). Digital Appliance is the smallest representing 13% of SEC's $25.1 billion sales in 2001, while the other three divisions have about the same share of sales.

Several products become global winners in respective markets - e.g., DRAM (40% of global market share), TFT-LCD (35%), CDMA mobile phone (77%) and CRT (35%). As these products compete and alternate in contributing to SEC's sales and profits, they have a stabilizing effect on SEC's financial performance (Table 4-1). Recently, the *Financial Times* (May 10, 2002) listed SEC as the world's 85th most valuable company in the world, ahead of Matsushita, Hewlett-Packard and Sony, and SEC is the world's third largest IT company after IBM and HP (Business week online 2003) . This chapter will concentrate on the Memory Division of SEC, the largest division and also the first division of the company to implement SCM. Hereinafter, the Memory Division of SEC will be referred to as 'Samsung Memory'.

Table 4-1. SEC Financial Summary (reproduced with permission from SEC)

(million US$)	1998	1999	2000	2001
Sales	$14,387	$22,810	$27,231	$25,160
Operating Profit	$ 2,221	$ 3,914	$ 5,906	$ 1,784
(Operating Margin)	15.40%	17%	22%	7%
CAPEX	$ 1,179	$ 2,973	$ 4,368	$ 3,652
R&D	$ 1,192	$ 1,391	$ 1,603	$ 1,757

2.2 Samsung Memory's Supply Chain

Memory production has 250 distinct process steps, but since some steps are visited multiple times, a chip typically goes through 600 or more steps. These steps require extremely high precision in a non contaminating environment within the highest standard of clean room design. Most steps are handled by precision machinery, but some steps still involve people dressed suitably, to protect the work from their source contamination, in what is termed in the industry as a 'space suit'. Memory production is roughly composed of four stages – fabrication, testing, assembly (including packaging) and final testing. First, raw silicon wafers are made and then processed through the reentrant fabrication system. Next, they are tested electronically to see whether they are good or bad as a semi-product. After testing, they are cut into chips and the good ones, then packaged. Finally, the packaged components are tested again as a finished product. Samsung Memory sells product in the forms of wafers, components and module types. Figure 4-1 describes the memory production process.

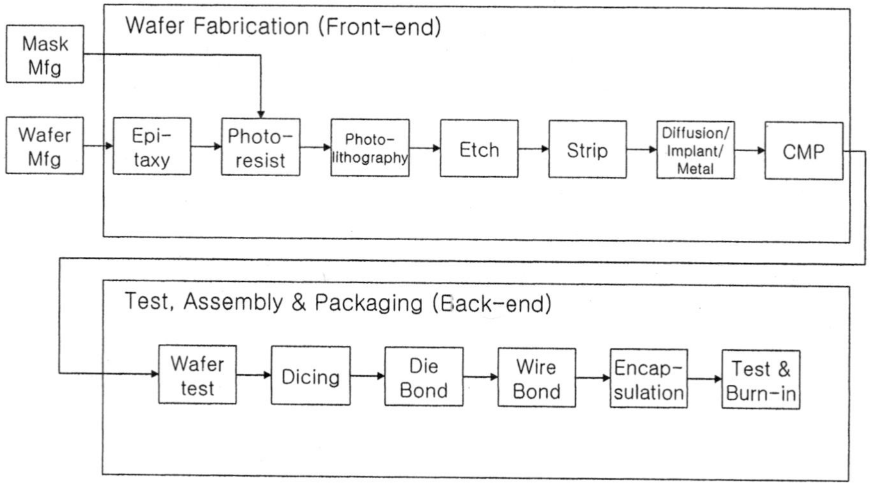

Figure 4-1. Memory Production Process (adapted from http://infras.com/Tutorial/)

The memory chip business is characterized by:

a) Cyclical demands: market demand is highly cyclical (each cycle used to be four years long in Olympic cycles, but recently it has reduced to two years);
b) Rapid price erosion: for a 64Mb chip, the price started at $60 in 1997, but reduced to $20, $10, $6 and $2.5 over the next four years;

c) High price fluctuation: the variable production cost is small and the depreciation cost is high at 36-40%, so cut-throat competition breaks out when demands are low;
d) Chunky investment: it would cost $1.75B to build a new fabrication unit that would produces 30,000 wafers per month; and
e) Fast technology advance: the price per Mb decreases at 33% per year due to technological innovations like larger wafer (8" to 12") and finer circuitry (0.15 to 0.12 micron).

The above characteristics of the memory business are closely related to supply chain design. Because of short life cycle and significant prices fluctuations, the risk of inventory becoming obsolete is fairly high. Inventory management is, therefore, very critical in the memory industry. The capital intensity and the very high rate of depreciation of manufacturing equipment drives their supply chain to maximize utilization by taking a push-type production approach. Even in cases where they take a demand-driven approach, they usually respond to changes in demand by changing their product mix instead of decreasing or increasing intrinsic rates of production. Therefore, one of the most important objectives for forecasting is to create the most suitable product mix for production. Lastly, the distribution cost, in the perspective of supply chain management for the memory business, is relatively low. The products are very small in size but are of very high value, allowing for different outlets to easily swap inventory. In other words, the space and holding cost constraints of distribution are relatively insignificant.

2.3 i2 Company Overview

A leading provider of next generation closed-loop supply chain management solutions, i2 designs and delivers software that helps customers optimize and synchronize activities involved in successfully managing supply and demand. i2 has more than 1,000 customers worldwide - many of which are market leaders - including seven of the Fortune global top 10. Founded in 1988 with a commitment to customer success, i2 remains focused on delivering value by implementing solutions designed to provide a rapid return on investment.

2.4 i2's Solution Overview

Let's take a close look at i2's solution.

2.4.1 Solution Overview

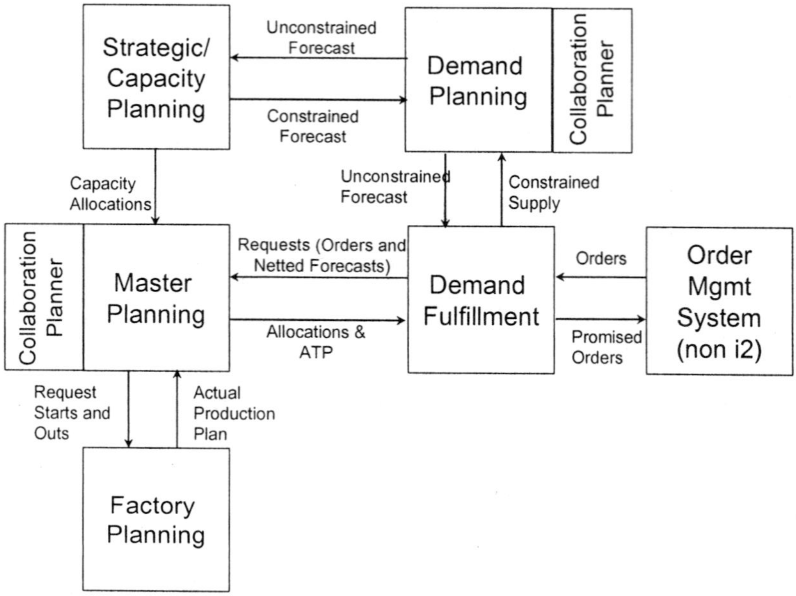

Figure 4-2. i2 Solution Workflow

i2 has solutions to support all areas of supply chain management such as planning, execution, and monitoring. In this section, the center of attention is primarily on the planning solution that is often referred to as APS (Advanced Planning & Scheduling), which has been adopted by Samsung Memory (Figure 4-2).

The Demand Planning engine supports the forecasting of demand, and demand is starting point for all planning activities. However, unlike other general forecasting functions, sales forecasting can be influenced by a variety of factors in addition to the sales strategy, size of budget, investment plan, manufacturing capacity of the company, and consumer demand. For this reason, sales forecasting often requires feedback from the strategic-level planning and supply planning functions. The Purchasing plans of the major corporate customers can also be fairly important with regard to sales forecasting, and forecast data from these customers are often accessed through a collaboration engine.

The Demand Planning outputs generated by the aforementioned process are then separated into forecasts and orders using a forecast netting engine(a packaged module from i2). The forecasts and orders are then passed to the Master Planning engine with different levels of priority. Based on the demand and material/capacity constraint information collected in this process, the Master Planning engine establishes an optimized enterprise-level production plan.

The output from the Master Planning engine is later used by the Factory Planning engine for planning at the factory level. It also serves as important criteria for the Demand Fulfillment engine, as it seeks to satisfy orders coming from the order management system. The system also engenders the relationships with the outside suppliers. The master planning engine utilizes the collaboration engine to access suppliers' supply plan/capabilities and shares the newly generated master planning output with the suppliers. So far, the flow of overall SCM solution has been reviewed. The next section will focus on the details of demand planning, supply planning, and the demand fulfillment process.

2.4.2 Demand Planning

Demand planning is a process for forecasting future demand and providing a target for supply planning. Most companies that do not operate in a pure build-to-order environment proceed with their production based on demand planning. It is, clearly, a critical process that drives overall supply planning. If the output from demand planning is inaccurate, the results will always be inaccurate: the wrong products will be delivered through the wrong sales channels at the wrong time no matter how well supply planning satisfies the demand.

The i2 Demand Management solution framework provides the functionality required for demand planning and demand collaboration. Steps that businesses should consider prior to implementing such a system include:

a) Clean up of historical data: As is typically the case for a solution implementation, the data frequently used for forecasting such as inventory data, shipping data, and historical sales data must have adequate detail and be sufficiently accurate to satisfy user requirements;
b) Determining the proper forecasting technique: When statistical methods are used for demand planning, a decision has to be made on which techniques to use;
c) Capturing all the factors influencing future demand: Various factors can impact future demand such as the industry or the region the company is operating in. If information on factors such as promotions or sales events that may affect future demand is available, forecasts will be more accurate;
d) Comparison of inputs from different players using different languages: Demand planning is not merely about forecasting demand; it is about reaching consensus on a profitable and feasible demand plan considering

how much should be sold (sales target) and how much can be produced (capacity constraints). Therefore, collaboration with various departments is required. For the sales plan and promotion budget, financial departments must collaborate, and the manufacturing departments must collaborate to determine actual production capacity. Even within the sales & marketing department, HQ and branch organizations often have different perspectives, and the discrepancy should be reconciled through a process and system infrastructure that promotes data sharing and consensus;

e) Alignment through collaboration: The internal demand planning process should be aligned with demand collaboration with other c ompanies so that demand information from corporate customers can be effectively reflected effectively in the internal demand planning process.

Corporations that take t hese factors into consideration during a s upply chain implementation can overcome some of the challenges. Implementing a collaborative demand management solution can shrink forecast cycle times and improve forecasting accuracy. This can result in reduced inventory costs and improved customer service levels which will ultimately increases the profitability of the enterprise.

2.4.3 Supply Planning

Another important component of APS (Advanced Planning & Scheduling) is the supply planning function. Supply planning creates an optimal plan for all of the processes that take place throughout the supply chain such as purchasing, production, shipping, storage, and sales. In other words, the role of supply planning is to provide an optimal supply plan using inputs such as forecast data (netted forecast excluding orders from forecast) from the demand planning engine and order data from the order management system. The supply planning process also considers the physical configuration of the modeled supply chain (including purchasing, production, shipping, storage, and sales) and the constraints and business rules for each phase.

APS supply planning provides s olutions t o a number of problems that cannot be resolved by DRP(distribution requirement planning)/MRP II (manufacturing resource planning) solutions. One of its biggest merits is that material/capacity/demand constraints are considered simultaneously in the planning process. By contrast in a conventional MRP II solution, capacity planning and material planning are conducted separately, and capacity and material cannot be considered at the same time. As a result, the system more often produced infeasible plans. In addition, there is no prioritization of demand, and customer segmentation is not possible. However, in APS, the DRP and MRP II solutions are modeled as one system, and this feature has enabled global planning to cover procurement, manufacturing, and distribution. Lastly, since global planning can now be done in much less time than in the past, users have time to simulate supply planning in many different versions as well as to carry out immediate re-planning whenever critical events occur. In summary, APS is capable of defining a larger problem set (simultaneously considering material/capacity and procurement/ manufacturing/ distribution), and yet is capable of producing results in much less time than the previous systems. This improvement is made possible mainly by innovative optimization technology from vendor like i2 and the ever improving performance of new generation computers.

In order to realize this APS, i2 uses a specific modeling approach described in the following FLO (Flow, Load, and Operation) diagram.

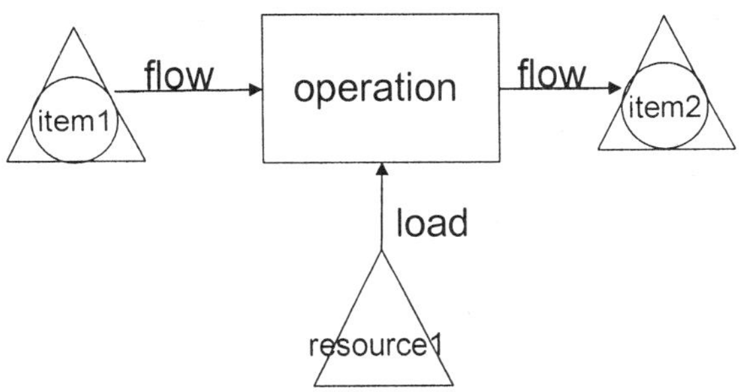

Figure 4-3. Sample FLO Diagram

As shown in the figure, modeling constructs such as Flow/Load/Operation/Item/ Resource are used to model the physical configuration of the supply chain; constraints and business rules are configured based on the attribute values of each construct. For example, operation is a construct that expresses the produce/move activity on the supply chain; its attributes include setup time, production time, waiting time, production volume, etc. that are required for unit manufacturing, and those attributes are used to configure related constraints and business rules. Such modeling constructs serve as simple building blocks that when put together are able to represent the largest and most complex supply chain networks known today.

After configuring various constraints/business rules related to supply planning in the supply planning engine, the next task is to collect demand information and create the best supply plan. Demand information is gathered by the demand planning engine and order management systems before it is passed to the supply planning engine as input along with previously-mentioned data on the supply chain model. Then, specialized algorithms create the most appropriate supply plan for the given circumstances. These algorithms use a combination of mathematical optimization as well as heuristics embedded in the supply chain planning engine. The supply plan is generated to optimize multiple business objectives while simultaneously considering the material constraints and capacity constraints, and as a result, the plan is far more reliable than any that would have been created by MRP. Also, since it is an optimized plan, it promotes a more efficient utilization of the manufacturing resources.

2.4.4 Demand Fulfillment

It is a common perception that fulfillment entails all the activities that must be done to fulfill a customer's request – procurement, manufacturing, assembly, testing, shipping, and installation (Shapiro et al. 1992). However, effective demand fulfillment begins by making reliable order promises. The i2 Demand Fulfillment module enables fast, accurate, and reliable delivery commitments to customer orders based on planned supply (i2, 2001). Other i2 modules such as i2 Supply Chain Planner can help plan the fulfillment of promised deliveries across the supply chain. Using standard integration tools, companies can integrate their order management transaction systems with i2 Demand Fulfillment, so as to embed the sophisticated order promising capability behind their current transaction systems.

Demand fulfillment covers both the planning and transaction system. Order fulfillment (matching demand with supply) belongs to the planning system and pick/pack/shipping/delivery belongs to the transaction system. This means that implementing a demand fulfillment solution requires not only the implementation of a planning system, but also its interface with transaction systems like ERP.

Demand fulfillment is the process where end-users interact with the output from the planning system. The main objective of the demand fulfillment process is to ensure reliable and optimal delivery to customers. Companies usually consider introducing demand fulfillment solutions because they cannot respond with promises they can keep quickly enough in their supply planning response to changes in demand. While demand is constantly changing, the supply chain is at best re-planned on a weekly or daily basis. Companies need a more flexible solution that can be agile in response to demand. This is where the demand fulfillment solution comes into play. The demand fulfillment solution is also capable of allocating production to satisfy each customer, such that products can be reserved for key customers even if they have not yet placed orders. The customer segmentation enabled by the demand fulfillment solution helps allocate planned supply and increases the satisfaction of key customers.

The demand fulfillment solution plays a critical role in aligning APS- the supply planning solution with ERP, the order management system. Great attention must be paid to the interface of the demand fulfillment solution with those systems. One of the many important decisions that need to be made is whether the interface between ERP and demand fulfillment engine should be in a batch mode or a real-time mode. A batch mode interface supports various data conversions through the database; in the case of the real-time mode, various workflows for each stage of the order management process must be analyzed, and a detail design of for minimizing and managing the risk of potential interruption of the integrated solution must also be developed.

There are also significant change management issues to be considered. Plan-based and rule-based operation must be firmly established. All players within the organization must fully understand the importance of respecting plans and rules at every stage of production and delivery. The reliability of the supply plan is critical to the demand fulfillment solution since responses to orders are made not only on the basis of inventory, but also the supply plan. Even if the demand fulfillment solution is fully implemented, it will not be very reliable if the sales force engages in activities that circumvent the supply plan such as ad-hoc swapping of inventories in the interests of some of their customers. If the supply plan becomes unreliable, the demand fulfillment solution that was developed on the basis of the supply plan will also be unreliable.

However, once companies overcome these challenges and implement demand fulfillment solutions, they will be able to make efficient and reliable promises and satisfy more orders with their existing supply chains. This will increase supply chain efficiency and raise the satisfaction of key customers.

2.5 i2's Implementation Approach – Business Release

i2 has its own methodology for implementing an SCM solution; its basic flow is described in figure 4-4. The first phase is the SOA (Strategic Opportunity Assessment) phase, which is when the company identifies opportunities for improving supply chains and specifying the attainable values through target KPI(Key Performance Indicator)'s. The second phase is called Business Release (hereinafter referred to as BR). This phase is for carrying out activities such as project scoping/scheduling, resource planning, solution design, solution development, user acceptance, deployment, and transition. In the BR approach, a process that can be implemented in a short period of time (6 months to 1 year) is selected as a target process based on the capabilities and the environment of the implementing company. The BR focuses on that target process, and the target process is implemented to achieve a return on investment in a short time frame. Most i2 implementation projects use this BR approach, and after going through several BR's, the SCM maturity of the company can be improved.

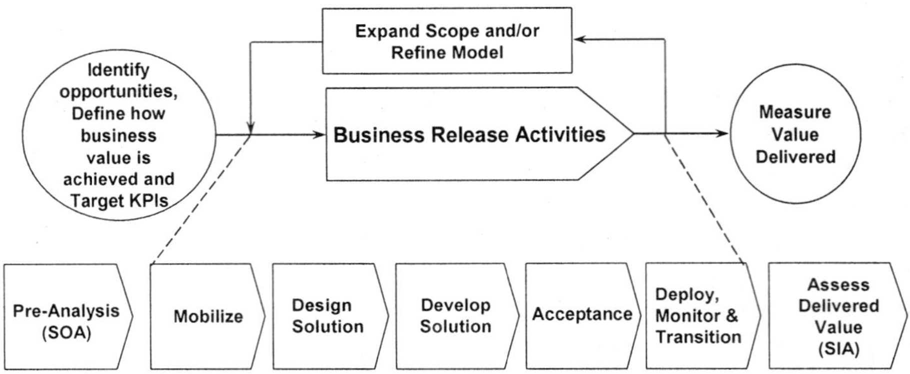

Figure 4-4. i2's Business Release Methodology

Taking this approach, additional business targets will be achieved every time a BR is completed, and this process will be repeated with each phase of the solution implementation. This concentrates the capabilities of companies in a more efficient manner. The last phase of the BR methodology, Strategic Impact Assessment (SIA), comes at the end of every BR, and it is during this phase that the values attained though a BR are measured and checked to see if the target has been attained. The results are used to establish the target for the next BR. However, depending on the implementation roadmap the BR approach can also have some drawbacks: until all BR's are completed, portions of the new and old environment will coexist within the company, and this might require temporary integration to support both environments and could also cause organizational confusion.

Because of these potential drawbacks, some companies opt for the 'big-bang approach' instead of the BR approach. In the big-bang approach, all process and system changes for SCM are implemented not in phases but simultaneously throughout the entire company. Because of its huge scope, the big-bang approach generally entails greater project complexities and delays in project implementation. It also creates pressure on the company because many different kinds of capabilities must be built up in a short period of time. The decision on which implementation approach to take should be made prudently after considering the benefits and drawbacks of each and also the business environment in which the company is operating.

Samsung Memory opted for the BR approach in its SCM implementation. In this section, Samsung Memory's SCM implementation is described in accordance with the flow of the BR approach. The last phase of the BR methodology is SIA. This phase comes at the end of every BR, and it is during this phase that the values attained through a BR are measured and checked to see if the target has been attained. The results are used to establish the target for the next BR.

3. BUSINESS RELEASE 1- MASTER PLANNING

After Samsung Memory completed its internal review of the planning solution, the project kicked off in May 1997. The scope of BR1 included modeling of only the manufacturing portion of the overall supply chain; namely, FAB, sorting, assembly, and testing. In other words, the decision was made to focus on master planning for the manufacturing function.

3.1 Challenges

The project scope of BR1 included the following:

a) Allocation of corporate-level capacity to Business Units;
b) Generation of daily starts and outs production plan by manufacturing line for all semiconductor business units;
c) Daily order response to Sales;
d) Generation of feasible plans in terms of material and capacity.

The objective of BR1 was to improve the quality of the supply plan, reduce the planning cycle time, and enhance the efficiency and responsiveness of the supply chain. There is something unusual about the second item listed above, and that is the scope of the master planning engine (hereinafter referred to as the MP engine), it implies a very detailed level of planning. At the time when the MP engine was implemented, an attempt was made to implement master planning and line-level planning as one system. However, as one can see in the i2 solution map, master planning is handled by the SCP (Supply chain Planner), and line-level planning is handled by the FP (Factory Planner). It is recommended that companies establish clarity in defining and securing agreement of the role between the master planning function and the factory domain up-front when implementing an end-to-end SCM solution.

As the implementation unfolded, Samsung Memory realized that data availability was an issue. This might be a problem with any project entailing the introduction of a new planning system. In the initial phase, planning-related data often does not exist or even if the relevant data does exist in ERP or MES (manufacturing execution system), the data is often not adequately refined. As a result, a great deal of time can be lost in gathering and cleaning data. For example, a plant's capacity information might be stored in ERP, but it needs to be continuously recalculated if it is going to be used for planning. That is because the product mix and working days, which affect capacity, change depending upon the specific time the planning is done. Furthermore, information specific to planning such as demand priority usually does not exist in ERP. Therefore, there is a need for a process for determining and calculating priority rules and applying them to the system. In the initial phase of the implementation, data cleaning and gathering was the most time-consuming task.

Corporations generally implement SCM solutions after they have implemented ERP systems. Many spend enormous amounts of time and energy cleaning data in their ERP implementations, and this easily leads them to believe that the data for the SCM is already available. However, as has been mentioned above, once they begin their SCM implementations, many companies realize that their ERP data is often unusable for SCM, and that there is a need for another round of data cleaning/transformation. It is true that having ERP makes data gathering much easier, but refining the data for SCM is a task that should never be taken lightly.

During implementation it was necessary to reach a desirable level of plan quality. Even though the team validated every unit function based on sample data, many unexpected problems emerged as a result of testing a supply chain model with a large amount of production data. Since supply chains are complex networks with significant interdependencies between elements no amount of unit testing can substitute for integrating and volume testing. Resolving these required Samsung Memory and i2 to work closely together. This engaged the Samsung Memory staff in fine-tuning the system, so that their understanding of the i2 MP engine improved enormously. This resulted in increasing their confidence in being able to handle future maintenance tasks.

Another component of the implementation required that Samsung clearly define its business objectives so that optimal plans could be developed. Initially, i2 planned to use Strategy Driven Planning, which is a rule based heuristic optimization technique used to meet Samsung Memory's requirements, a solution approach was decided upon that uses mathematical optimization with linear programs in addition to heuristics that was considered more effective when addressing simultaneous requirements of multiple business objectives. i2 has a rich heritage of optimization algorithms that are based on their work in the metals and consumer products industries. i2 has extended and adapted these algorithms to meet the unique requirements of the semiconductor industry. This has led to a development of a new solution designed to meet the needs of asset intensive supply chains including both semiconductor and metals industries.

However, this requires changes in the overall solution design and internal system interface. While the existing solution utilized the strategy-driven planning approach of the i2 MP engine, Samsung Memory's new solution is designed to create a final optimal solution through interaction between strategy-driven planning and linear programming. Although this required some retesting, from a business perspective this was the proper and reasonable course of action.

According to Dr. Adeel Najmi, who was i2's project manager of BR1, MP implementation, "Samsung Memory needed to optimize the supply chain plan to account for complex tradeoffs and multiple business priorities. These included on-time demand satisfaction, inventory minimization, resource utilization, alternate sourcing preferences, build-to-stock as well as build-to-order preferences. i2 and Samsung Memory worked together to configure i2's optimization algorithms to meet Samsungs challenging business needs." He further stated, "It has been an exciting journey partnering with Samsung Memory and bringing their vision to reality."

In order to achieve the desired level of engine performance between master planning and line-level planning functions, it was necessary to split the planning engines along the lines of shared supply chain network. All the products that shared the same material and capacity were grouped to be included in the same engine so that the initially-proposed business value of a single engine would be maintained while the performance improved.

3.2 Benefits

Implementation of the MP engine brought about significant changes that effectively shifted existing planning concepts at Samsung Memory. First of all, the decentralized and site-based planning approach used previously changed to a centralized and product line-based planning approach through single modeling of master planning. Also, the previous concept of capacity-driven planning (push type) designed to maximize facility utilization was replaced by demand-driven planning (pull type) that is geared to satisfy customer demand. More specifically, the planning cycle time was greatly reduced. In addition, as planning was conducted on the basis of customers' orders, the system was able to react more quickly to changes in demand. This enabled each branch to reduce inventory in its warehouse and resulting in increased customer satisfaction. By generating a plan in consideration of both capacity and material constraints, the accuracy and reliability of production plans are greatly enhanced and importantly reliability of delivery commitments made to customers improved.

3.3 SIA (Strategic Impact Assessment) for BR1

After implementing the MP engine (1997-1998), Samsung Memory realized several benefits such as a shorter planning cycle time, improved capacity utilization, reductions in inventories, etc. However, these benefits were enjoyed almost exclusively by the manufacturing division, which was mainly interested in increasing efficiency (or reducing costs). The sales division, whose objective is to maximize customer satisfaction (or revenue), did not get much benefit out of this implementation. Originally, the MP engine's planning frequency was proposed as "daily" to the sales division. However, following the implementation and stabilization phases, the PI(Process Innovation) team finally concluded that running MP engine twice each week was the best compromise to be made considering the manual revision after planning, data availability, etc. With this unfulfilled promise of the Master Planning project, the sales division's expectations fell short and looked to enhance the performance of the system with:

a) Daily demand changes promptly communicated to manufacturing;
b) Delivery commitments connected to the supply plan;
c) Preferential treatment for key customers;
d) The matching of forecasted demand and output of supply.

To reaffirm the lessons from the Master Planning implementation experience and to identify areas of improvement, Samsung Memory launched an SIA (strategic impact assessment) project. This process can help companies ensure that they are gaining the full benefits out of i2's solutions and identify opportunities for further improvements. Figure 4-5 shows i2's value delivery cycle with SIA. The findings of Samsung Memory's SIA showed that the MP engine was near to stabilization and the remaining opportunities for improvements were mostly in the demand fulfillment and demand planning areas. Production planning was also cited as an area where improvements could be made to complement the supply planning functions, along the delegation of roles and responsibilities recommended by i2. However, these improvements were not seen as urgent issues because Samsung Memory already had a system in place that they didn't want to change.

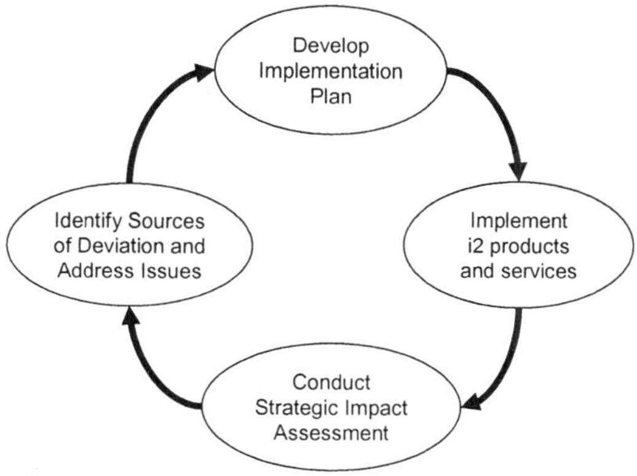

Figure 4-5. i2's Value Delivery Cycle with SIA

After reviewing the SIA findings and the sales division's needs concerning the MP engine, Samsung Memory decided to focus next on the demand fulfillment area. For the following reasons, Samsung Memory concluded that demand fulfillment was the one area where the sales division's concerns could be most directly resolved:

a) The demand fulfillment engine can communicate and apply the change in demand more frequently (daily);
b) The demand fulfillment engine's allocation logic is capable of retaining some supply plan for key customers even when they have not yet placed orders so that key customers can have the benefit of "temporarily" dedicated supply;
c) Generic product modeling in the demand fulfillment engine is capable of providing some level of flexibility in supply/demand match to resolve (or mitigate) supply/demand mismatch problems.

4. BUSINESS RELEASE 2 – DEMAND FULFILLMENT

The direction for BR2 was determined on the basis of the BR1 SIA results. As mentioned previously, the scope of BR1 was limited to supply chain optimization for the manufacturing function of the company. For BR2, the company had to decide whether to include all of the supply chain or just focus on problems with the Sales Division. The company ultimately decided to pursue more comprehensive objectives in BR3 and BR4; for BR2, the focus was going to address the most urgent goals that had not been resolved in BR1.

4.1 Challenges

Back in 1999, Samsung Memory's Sales & Marketing Division was using a legacy OMS (Order Management System). It was a kind of CBO (Custom Bolt On) built on top of the previously implemented SAP R/3 SD module. The basic mechanism of the legacy OMS was to determine allocations with forecast-type orders, and when actual sales orders were placed by customers, the ordered quantities were deducted from the allocations by the relevant forecast-type orders. There was no process for consensus forecasting and no clear conceptual division between forecasting and orders. Additionally, the rules for determining demand priorities were uncertain.

The Demand Fulfillment engine (hereinafter referred to as the DF engine) had the capability to match demand with planned supply. The project team believed that by using this capability, changes in demand could be communicated in an adequate timeframe. This tool could provide demand change information daily even if the supply-side remained unchanged. Also, the project team expected that the service ratio for committed delivery dates could be improved by applying detailed order priorities based on order type, account type, due dates, etc. Lastly, by introducing a new concept of allocation, the team believed it could pave the way for the introduction of another new concept called 'allocated ATP,' which could be used to ensure availability of products for key customers on a first-in first-out basis. Looking back, BR2 proved to be effective in establishing a foundation for BR3 and BR4, which were implemented later to achieve an overall improvement in the Sales & Marketing processes.

One of the unique characteristics of this solution implementation was the introduction of postponement in supply-demand match using the generic product concept. Previously, Samsung Memory had many product codes for the same product depending on the sales channel. Once the sales channel for the product is determined and the product code is marked depending on the sales channel, the product could no longer be taken by another channel if it was needed or it could be taken only after a cumbersome label change process. This product code-marking step was the last step in the manufacturing process, and it took very little time to change the marking. However, according to the existing process, the MP engine was operating on a weekly cycle. This meant that marking decision had to be made up to one week in advance and locked thereafter.

To resolve this issue, in the DF engine, the production plan beyond a certain period of time (called frozen period -a period that could not be modified by changes in demand to promote steady manufacturing) was made not at the specific product level, but at the generic product level, that is a higher-level grouping of specific products sold to customers based on their characteristics.

Product	12/20	12/23	12/25	12/28	12/30	...
Spec. Prod A1	40	50	30			
Spec. Prod A2	30	40	20			
Generic Prod A				90	100	...

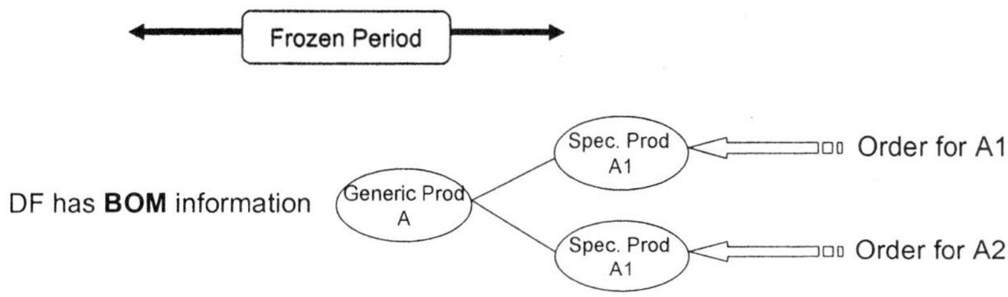

Figure 4-6. Frozen Period and Generic Product

As shown in figure 4-6, as of December 28th, or after the frozen period, supply planning was done at the level of generic product A, which is a grouping of specific products A1 and A2. When orders for specific products A1 or A2 were placed, they were promised based on this generic product-level supply plan. In this way, the process of order fulfillment is more flexible.

As mentioned earlier, the concept of 'allocated ATP' was also introduced on the process side. This concept enabled Samsung Memory end-users to retain control over some portion of the supply plan for their key customers. It was an unfamiliar concept, but in essence, it could be used to function like the forecast-type orders that existed in legacy OMS. To minimize the impact on the process, the end-users continued to create forecast-type orders; the DF engine generated allocated ATP based on the values extracted from those forecasts orders. The users gradually became familiar with the concept of allocation and then began to think about the necessary business rules.

4.2 Benefits

The DF project had to be completed in a shortened period of time with minimum impact on the existing processes and systems. The benefits of the project turned out to be fairly significant considering the amount of investment:

a) The supply plan generated by the MP engine is more flexibly matched with demand;

b) Delivery fulfillment has improved;
c) The committed delivery dates change less often;
d) A concept of allocation was introduced.

Thanks to such positive results, the second DF project was implemented to extend the generic product concept to a wider range of products. By using the i2 CTP (Capable-to-Promise) solution, it became possible to swap supply plans between module products (PCB boards with memory chips inserted in units) and chip products. With this CTP solution, a sub-component-level supply plan and capacity information can be used for order promising, as well as an end-item-level supply plan. This has increased flexibility in matching the supply plan with demand. As the next step, Samsung Memory began a review of the demand planning area.

In summary, the DF project began with various restrictions for introducing APS in the sales area. However, by achieving tangible results in a short period of time, it made a significant contribution to improvements in the demand fulfillment process and also served as a starting point for improvements to the overall sales & marketing process.

4.3 SIA (Strategic Impact Assessment) for BR2

As mentioned above, the introduction of the DF solution (1999) had increased flexibility in matching the given supply plan with the demand. Based on this achievement, the Sales Division began looking for ways to further leverage the concept of SCM. For a more objective assessment, Samsung Memory carried out a pilot project with Accenture Consulting (then Andersen Consulting) aside from the SIA with i2. Five key capabilities were recommended to improve supply chain performance in the sales area. They were:

a) Customer-Driven Planning;
b) Streamlined Order Management;
c) Global Inventory Management;
d) Global Account Planning;
e) Integrated Performance Management.

Process and systems were reviewed from the perspective of those five capabilities and it was concluded that there was room for improvement. The evaluation recognized a lack of organization in the sales & marketing division forecasting process. The forecast was usually done manually using Excel and by fax or mail. The problem was made more complicated by the forecasts that were handled like sales orders in the OMS, and gave rise to the following issues:

a) (In this case), forecasts are not true consensus forecasts. Sales personnel enter forecast orders mainly to secure extra quantities of product that need to be sold on top of the actual ordered quantity. Because of this, forecast quantities constantly change as actual sales orders arrive and formulation of a tracking forecast becomes virtually impossible;
b) Since forecasts are handled in the same way as sales orders, the process for entering forecast information becomes unnecessarily cumbersome: detailed information required for sales orders needs to be entered for forecasts as well (for example, forecasts only require weekly figures, but in this case, sales people need to designate daily due dates);
c) Forecast orders are very much more likely to change than sales orders, and they are often cited as the cause of erratic planning by the MP engine;
d) Widespread data entry by manual means leads to a high incidence of data entry errors.

Another issue was that the Headquarter(HQ) system was not integrated with the branch systems. All of Samsung Memory's overseas branch systems were divided into seven SAP instances, and data sharing between HQ and the branches was impossible:
a) The branches and HQ did not share a single system. The branches' orders coming into the HQ system were actually additional quantities that those branches needed in addition to their local inventory;
b) Branch sales people often created dummy orders just to get more supply (from HQ);
c) Because of the two problems mentioned above, HQ could not accurately monitor the demand from branch customers.

5. BUSINESS RELEASE 3, 4 - SALES & MARKETING INNOVATION

To overcome the problems identified through SIA in 4.3 and to realize the recommended key capabilities, Samsung Memory kicked off its Sales & Marketing Innovation Project. From this point in time, i2 and Accenture jointly participated in this project.

5.1 Challenges

The following tasks were selected as specific execution strategies for Sales & Marketing Innovation:

a) Introduce a demand planning solution dedicated to consensus forecasting;
b) Replace the legacy OMS that mixed up forecasts with orders with a standard OMS (utilizing SD module of SAP R/3 system);
c) Implement a distribution planning solution connecting HQ warehouses with branch warehouses for integrated global planning;
d) Reorganize the sales/marketing organization structure leveraging the Global Account Management concept;
e) Introduce a dedicated reporting solution to deploy an SCM performance monitoring solution.

5.1.1 Introduction of Demand Planning Solution Dedicated to Consensus Forecasting

Figure 4-7 describes a general forecasting process recommendation.

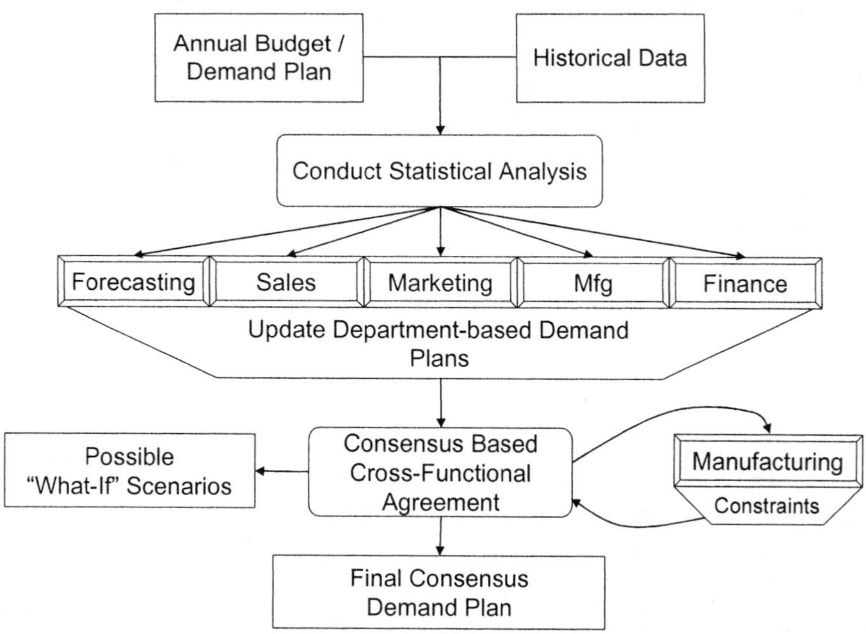

Figure 4-7. General Forecasting Flow by i2

Statistical analysis is carried out based on the annual budget/demand plan made by the management and the historical sales data. Based on these baseline forecasts, the departments involved in demand planning, forecasting/market research, sales, marketing, manufacturing, and finance, create their own forecasts. Finally after reaching a cross-functional agreement when considering various what-if scenarios and manufacturing constraints, a consensus demand plan is created. This is an example of the top-down approach. In the case of bottom-up approach, forecasts created by individual departments are validated through a statistical analysis and the final consensus demand plan is created in consideration of the annual budget/demand plan.

Between the top-down and bottom-up forecasting approaches, Samsung Memory's forecasting process was closer to the bottom-up process. In Samsung Memory's case, customer demand entered by branch sales people goes up to the upper organizations through the bottom-up approach and the consensus forecasts are created along the way in consideration of upper-level constraints such as a demand guideline. The demand guideline was developed by a high-level product group based on the weekly cross-functional meetings between the production and sales organizations. It is designed to share capacity constraint information of production and market trend information of sales.

The biggest challenge in implementing the demand planning solution was the real-time conversion of module-level forecast quantities into chip-level forecast quantities. It is important to note that the demand guideline explained above, had chip-level quantities not module-level quantities. Whenever HQ sales people make changes to module-level forecast values arriving from Branches, they have to check the converted total chip quantities to consider the demand guideline.

In order to satisfy this requirement, i2 decided to model the relationship between modules and chips in the OLAP engine of the demand planning engine, and developed functionality for real-time conversion of any entered values. This functionality can be used by all silicon chip memory manufacturing companies that sell both modules and chips as finished products.

5.1.2 Replacing the Legacy OMS with a Standard OMS (SD of SAP R/3)

The legacy OMS in Samsung Memory was designed to mix forecasts and orders all together. If the forecasting process was to move to the demand planning engine, the legacy OMS had to be changed significantly. For example, the concept of "order split" that allowed sales orders to take away the allocation secured by forecast orders would not be supported. A decision was then made to replace the legacy OMS with the standard SD functionalities of the SAP system. This meant that the as-is system flow had to undergo significant change.

In the as-is process flow from MP to DF and then to the legacy OMS, the DF engine received the supply plan from the MP engine and demand information from OMS and matched the demand with the supply only once a day. The changes that were made after matching were handled in real-time by the legacy OMS system. With the legacy OMS taken off-line, the DF engine had to respond to the demand in real-time. i2's standard modules namely, High Availability (HA) and Rapid Optimized Integration (ROI) were implemented to provide a continuous, real-time order promising solution.

HA is a technical solution that enables the DF engine to stay live 24 hours a day; ROI is a solution that supports short time period development of an interface (online mode or batch mode) between the i2 solution and the external systems by utilizing a template concept. Replacing the legacy OMS required a great deal of change in the process, but the changes that had to be made on the system side were no less challenging. Integrating the seven SAP systems around the world into one DF engine was a daunting task how it was achieved is explained in detail in 5.1.3.

5.1.3 Implementing Global Planning System Integrating HQ and the Branches

To connect the Branch system and HQ system, the scope of supply planning was expanded to include branch warehouse. As mentioned in 3.1, the initial Samsung Memory Solution (internally called Global Planning System or GPS) scope was limited to the manufacturing area. This meant that only the HQ warehouse was included in the supply planning scope. In order to achieve global planning that also reflects inventory information of the Branches, a Distribution-Planning engine that was capable of considering the branch warehouses had to be implemented [Fig. 2-8]. Using the Distribution-Planning engine, the response buffer on the supply chain turned into the Branch warehouse, and the supply chain of HQ and Branches became integrated both in process and system. The Distribution-Planning engine was capable of daily re-planning, reflecting the changes in inventory and demand, because the memory business did not have any capacity constraints. Demand fulfillment previously responded only to the demand coming from HQ's legacy OMS based on HQ's supply plan. However, as the scope of the supply plan was extended to all Branches, the DF engine had to respond to the demand arriving from Branches as well. Previously, when Branch sales personnel received orders from their customers, they first checked the local inventory in their Branch warehouse, and they only entered an order in the HQ system when there was not enough local inventory to meet the demand. However, after the change, the HQ DF engine was able to respond directly to orders from the Branches. This required significant change to both of the process and the system of the Branches. Towards the end of the project, the project team had to visit each of the Branches to carry out system integration and process-change management activities.

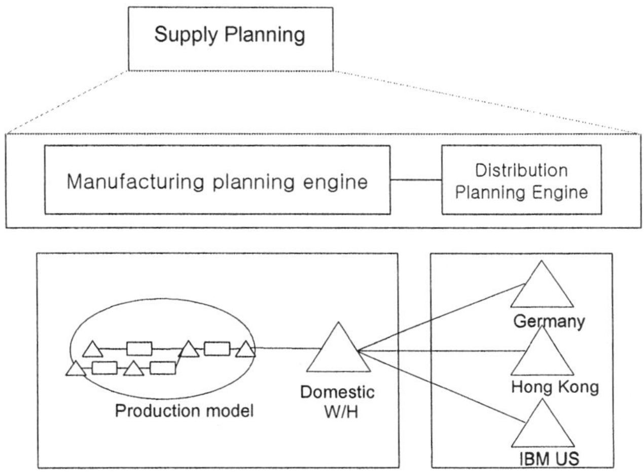

Figure 4-8. Planning Scope Extension using the Distribution Planning Engine

5.1.4 Sales/Marketing Reorganization Leveraging the Global Account Management Concept

The Global Account Management (hereinafter referred to as GAM) concept was pursued because there was previously no integrated account management for global customers who purchased globally. Based on the as-is sales & marketing process, those customers were handled by regional Branches. Under the GAM concept, new customers and small & medium-sized customers are managed by the regional Branches, but global accounts are managed by HQ with due responsibilities and rights. This was an attempt to introduce a more customer-oriented organization structure, but this plan faced strong resistance from the Branches. Sales managers in the Branches who were affected by the plan strongly resisted the loss of rights. However, thanks to the on-going change management efforts carried out under the strategy to become a demand-driven/market-driven company, the HQ global account team now handles global accounts globally instead of the branch marketing organizations. Such an organization change is strong proof that the global SCM is not simply about developing or integrating systems.

5.1.5 Implementing an SCM Monitoring Solution by Introducing a Dedicated Reporting Solution

The SCM monitoring solution was introduced to provide a consistent global view of various SCM data to HQ and the Branches. By centralizing scattered SCM data from ordering to shipping into one system, managers and working-level users were all given ready access to all SCM-related information. In addition to sales/production/inventory/ forecast information, they also had access to the KPI's (key performance indicators) for different areas so that they could continuously monitor and measure SCM performance. The SCM monitoring solution was developed in-house by purchasing a commercial OLAP engine and client and then customizing it with Samsung-specific requirements. As the SCM-status and performance data were easily shared across the company, the level of interest in SCM increased, and the foundation for continuous improvement in SCM performance was established.

Finally in June of 2001, the system went into production, and stabilization efforts were made throughout the second half of the year. To summarize the changes brought about by this project, introduction of the GAM (Global Account Management) concept and full-scale forecasting process were major changes on the process side. The concept of GAM meant the transfer of responsibility for global account management to HQ. This change faced strong resistance from Branch sales organizations that previously had that responsibility. The forecasting process was an extra workload for the sales people as they did not have responsibility for this process and system beforehand. However, thanks to a strong effort to persuade people to accept the changes for the sake of the company and for global optimization, the changes gradually took root in the affected organizations. In this process, the support team, named the "SCM helpdesk," played a critical role. The power users who participated in the project did not go back to their original job immediately following the completion of the project, but instead were asked to operate a 24-hour helpdesk to resolve whatever problems that occur in the initial phase of system operation. The helpdesk team resolved numerous questions and concerns from the end-users, paving the way for widespread acceptance of the new system.

On the system side, the biggest challenge was to carry out global planning by simultaneously collecting order and inventory information from seven separate SAP systems. Furthermore, the central DF engine was required to provide real-time response to customer orders arriving from the

Branches. The biggest issue was the integration between those distributed ERP systems and the central planning system. The integration between SAP R/3 SD module, a transaction system and the DF engine, a planning system, turned out to be more complex than expected. The questions of how to integrate SAP functionality with the i2 engine and how to handle customization all depended on the specific circumstances of the project. Therefore, it was important to have an accurate analysis of the status of the project.

In the case of Samsung Memory, a significant-level of customization had already been made in SAP, so the existing i2 ROI solution (a solution for integrating the i2 engine with a standard external system) needed customization. During an implementation of this kind, the trade-off between the needs for standardization to promote quality assurance of the solution and the needs for satisfying diverse customer requirements should be carefully considered.

Figure 4-9 is the system flow before and after Sales & Marketing Innovation project

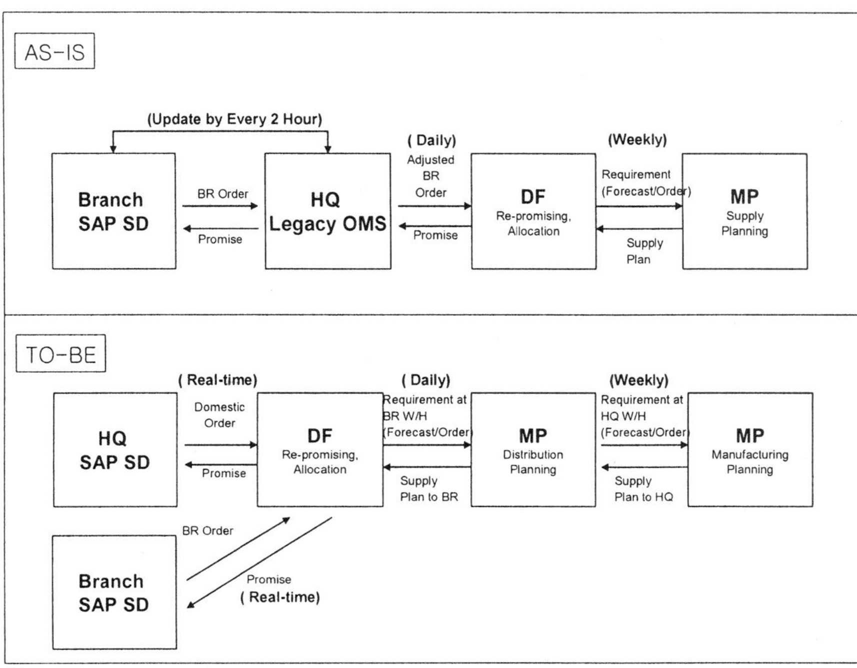

Figure 4-9. System Flow before/after Sales and Marketing Innovation Project

5.2 Benefits

The benefits achieved by the Sales & Marketing Innovation project can be summarized as follows.

The first benefit was that the sales organization ultimately accepted the forecast management process. Previously, forecasts were managed in the order management system as forecast orders. By separating forecasts from actual orders, the forecasting process was allowed to play its intended role of forecasting the future (or sellable) demand, and not used to simply secure additional inventory. In addition, access to forecasting was provided across the company to enable forecasting performance management.

The second benefit was the improvement in the order fulfillment process. Previously, order fulfillment was done at the level of decentralized SAP systems. Even if HQ had a demand fulfillment solution implemented, the Branches were still promising their orders on the basis of limited local inventory and planned receipt information available in their SAP systems not yet integrated with the planning. As a result of this project, the Branch OMS's were integrated with the DF engine in real time, and the Branch sales personnel could promise orders based on the HQ supply chain as well as their inventory information. HQ sales personnel also had the advantage of having more accurate customer demand information, which was not manipulated by Branch sales personnel, so they could use it for global planning and monitoring.

The third benefit of the project was that the whole of Samsung Memory organization now had a global visibility of demand information. Introduction of the demand planning solution had centralized all the forecasting information (forecast history/order/inventory/allocation) into one system; no longer was the same information stored in multiple systems with different values (data inconsistency). Having visibility also facilitated global planning and matching of global demand with global supply. It also became easier to manage and measure the forecasting performance of individual departments and team members.

Additionally, a systematic performance measurement approach was introduced by the project. Thanks to the improved data visibility made possible by the Demand Planning solution and SCM monitoring solution, a large amount of clean source data was collected. Based on this data, key performance indicators (KPI's) e.g., forecast accuracy, forecast fulfillment ratio, and on-time delivery ratio, were defined and continuously managed for supply chain innovation.

Such benefits had a direct impact in realizing the key capabilities recommended as a result of the earlier pilot project led by Accenture Consulting. This was possible because from the initial design phase of the project the project had never lost sight of the strategy to achieve the given goals. This strategic direction proved to be very useful in the change management phase that followed the system implementation phase. The strategic direction was also one of the criteria used in selecting the KPI's for specific performance measurement. The results and benefits achieved by the project led to the overall performance improvement of Samsung Memory, further improvement is being made as a result of on-going supply chain innovation efforts

Previously, orders created by the Branches were interfaced to the HQ legacy OMS every two hours. This meant that Branch sales personnel sometimes had to wait as long as two hours for a delivery response. However, after the Sales & Marketing Innovation project, orders entered by the Branches received real-time responses directly from the DF engine. The scope of the MP engine was also extended to include the Branch warehouses. Distribution-Planning for the horizon integrating the HQ warehouse with the Branch warehouses was updated daily.

6. CONCLUSION

In this chapter, the SCM implementation of Samsung Memory has been described in a chronological manner. This seemed appropriate given the fact that Samsung Memory's approach was a typical phased approach. At the end of each phase, a thorough analysis was carried out based on what was achieved and what remained to be done. The challenges and benefits were studied and explained after each phase. One of the most basic lessons learned from Samsung Memory's journey was that the SCM innovation could not be achieved just by implementing a good solution, it needed to be bolstered by appropriate change management activities. The SCM solution is a type of decision-support solution. In other words, the excellent production planning provided by the MP engine would merely serve as a reference plan if it were not applied in actual production. In fact, SCM solutions are often used to do just that – to provide reference. One could say that the limited usage of SCM is the result of unsatisfactory quality of plans created during the initial phase of solution implementation. The real cause, however, was primarily unsuccessful change management. One of the typical examples of the inadequacy of change management would be to have DF create allocated ATP, and allow allocations to be assigned to other sales channels without any discipline.

Another important lesson from Samsung Memory's SCM implementation experience is that the nature of the industry should be thoroughly considered and reflected in the implementation process. The concept of SCM may be more or less the same across industries, but the detailed characteristics of the industry are often critical in the implementation. As mentioned previously, it is important to choose a solution that offers industry-level templates. It is also important to analyze the pros and cons of the big-bang and phased approaches then choose an approach best suited in light of the circumstances of the implementing company.

It is regrettable that the success of Samsung Memory as a result of the five-year effort could not be analyzed with KPI's and shared publicly because of the proprietary and confidential nature of the information pertaining to the implementation. If the opportunity arises in the future, it would be desirable to propose appropriate KPI's from the perspective of SCM and use those KPI's to analyze Samsung Memory's performance.

ACKNOWLEDGEMENTS

The authors would like to acknowledge the support of Dr. Adeel Najime of i2, and Mr. D.G. Lee of Samsung Electronics Corporation for reviewing this document.

REFERENCES

1. AMR Research, Supply Chain Management Software Report 1998-2003, 1999.
2. Article, Strategic and Future, Agenda fur das 21 Jahrhundert, 1995
3. Business Week online, Info Tech Scoreboard, Retrieved March 2003 from: http://bwnt.businessweek.com/it100
4. Cassell J., (2003), iSuppli's Final 2002 Semiconductor Market Shares, Retrieved March 2003 from: http://www.isuppli.com/marketwatch/index.asp?p=details&id=156
5. Gartner Dataquest Report, Dynamic Shifts in Spending for Enterprise Applications Software-Market Share and Forecast, 2000, SOFT-WW-EX-0013, Aug. 9. 2001.
6. i2, "Module 11 Demand Fulfillment" In Intelligent Supply Chain Management, 2001
7. Shapiro, B.P., Rangan, V.K., Sviokla, J.J. Staple yourself to an order. Harvard Business Review 1992; July-August: 113 – 121

Chapter 5

SPECIAL NEEDS OF SMES AND MICRO BUSINESSES

Flavio Bonfatti[1], Paola Daniela Monari[2]
[1]University of Modena and Reggio Emilia, Italy, [2]SATA Applicazione Tecnologie Avanzate SRL, Italy

Abstract: The chapter is aimed at analyzing the present condition of SMEs and micro businesses in the perspective of their access to the forthcoming single electronic market. The study of some real cases coming from recent RTD projects allows to identify the main needs of small companies and to derive requirements for the development of appropriate technological supports.

Key words: SME, Single electronic market

1. INTRODUCTION

The main challenge for SMEs in the next few years is the need to become credible actors in the global electronic market. This implies developing visibility to customers and partners, ability to identify and involve suppliers and partners, aptitude to exchange amounts of knowledge, and qualification to collaborate with companies belonging to different sectors and established in different countries. While in the traditional market most SMEs base their survival on steady relations with (often) local actors, in the forthcoming electronic market they risk being left aside unless they extend their knowledge base and adopt advanced collaboration means.

As a matter of fact, from the electronic market perspective it is hard to classify industrial sectors as less or more RTD intensive. Even in traditional sectors, such as building or footwear, it is possible to identify companies that are very well positioned and others that, although exporting most of their production, are not equipped with adequate communication and collaboration supports. Then, the critical factor is not the sector but, in most cases, the company size. To this purpose, and taking into account the openness to new technologies, we can roughly classify companies into three main categories:

- Medium to large (rich) companies: They have no problem to position themselves in the traditional market, and are already investing to remain key actors in the electronic market. They play a proactive role, by influencing market rules and technological development, or behave as fast followers, meaning that they tend to adopt new technologies as soon as these appear in the market;
- Small to medium sized companies: They are aware that the electronic market might potentially increase their business, and of the risks to remain excluded from it, but they cannot individually afford the challenge. They are used to adopt new technologies only when forced by big customers or by market maturity, and in this decision they often fall prey of opportunistic advisors;
- Micro businesses: They have very limited investment capabilities, all focused on the core business, and their main expectation from the electronic market is to become visible to larger companies that might engage them as partners or subcontractors. They adopt only elementary and cheap technologies, and use them as simple substitutes of traditional means.

The companies belonging to the last two categories are particularly interesting for a number of reasons:

- Critical role in the European economy: SMEs and micro companies represent the 95% of the European companies and employ more than 70% of the manpower. Then they are the solid platform of the European economy and, obviously, there is no electronic market without their participation;
- Reaching the critical mass: In the perspective of the global (single) electronic market, the involvement of SMEs and micro companies is the condition to reach the necessary critical mass of actors that makes eBusiness an irreversible phenomenon;
- Neglected by the technological development: For business reasons innovation is mainly addressed to large companies or to individual consumers, while the category of SMEs and micro companies is normally disregarded. The only solutions come from local developers that, however, cannot introduce real breakthroughs. Then, there is an actual need for technological development of solutions that are simple, direct, easy to adapt and use, and cheap.

A strong point of the intended companies is their attitude to associate. Associations know very well the member companies, and they can correctly interpret their needs and sometimes promote the smooth adoption of innovative solutions. This is therefore a potential powerful lever to overcome the limitations of SMEs and micro companies, but specific actions must be undertaken in the right direction

Because of the extreme variety of contexts, structures and behaviors it is not easy to figure out and classify the real needs of SMEs and micro businesses with respect to the participation in the electronic market. Of course, it is well known that they can only afford simple and cheap solutions, but which are the problems those solutions are asked to solve? This chapter tries to answer the question according to a naive bottom-up approach consisting in the derivation of hints from the analysis of some real cases.

The chapter is subdivided into three main sections. Section 2 examines some cases of SMEs and micro businesses acting in different industrial sectors, while section 3 carries out a similar analysis by considering real cases of the service sector. For the sake of readability, the analyzed cases have been slightly generalized and made independent of the respective regional and economical contexts, so that it is easier for the reader to recognize analogous cases in his or her home country. On the basis of this analysis, section 4 translates the identified needs into requirements for technology and application development.

2. REAL CASES IN INDUSTRY

Most SMEs and a significant percentage of micro businesses act in the industrial sector. They produce high quality niche goods that are fundamental in an increasingly customized market, and have the necessary flexibility to adapt to the changing environmental conditions. In this section we examine some significant cases that have been studied in recent RTD projects.

2.1 Components and Compounds

A large number of SMEs are specialized in the manufacturing of components belonging to a variety of sectors such as mechanics, fluid-power, machine tool, electronics and control. In some cases, they join forces to increase the overall production capacity and extend the offer to include compounds. As a rule, components and compounds are sold in the global market to other industries focused on the assembly of complex machines.

AXIAL, ENGINES, VALVES and GEARS are four autonomous SMEs producing fluid-power and mechanical components, namely engines, valves and gearboxes. Their size ranges from 40 employees and 4 million Euro of yearly turnover (AXIAL) to 150 employees and 30 million Euro of yearly turnover (VALVES). They decide to constitute an equal-rights network to design, engineer and manufacture a compact motor-wheel for crawlers, small diggers (2 to 3.5 tons) and snowmobiles, the motor-wheel being constituted by an engine, a valve and a gearbox. Big Japanese companies presently lead this market. Two basic conditions are agreed upon: each SME remains autonomous and maintains its own customers, and the subdivision of work between AXIAL and ENGINES, the two engine producers, is based on a precise balance rule.

The case of EUROMACHINES (25 employees and 18 million Euro of yearly turnover) is slightly different, as it is a producer of finished machine tools (metal-sheet bending, punching and cutting machines) for the worldwide market. In addition to suppliers and subcontractors, it has two small companies as main partners: NUMERIC to provide the CNC component and MECH to ensure the final assembly and test of products. Even though independent of each other, the three companies collaborate on the basis of a strong agreement assigning to each of them a precise role, including the commercial function and brand of EUROMCHINE.

In both cases every partner is free to decide time by time how much of its capacity it makes available to the network. The network takes into account this parameter and others like lead-time or past performances, to ensure a balanced workload distribution to partners. This is particularly important in case of SME networks that including competing partners, that is, partners are able to supply the same products or services. Neutral workload distribution according to a clear agreement is the condition to maintain a good level of trust and confidence within the network.

SMEs acting in the component industry live on subordinated or peer-to-peer relations with customers and partners. Their daily activities are affected by a number of problems:

- Visibility to the market: The first problem of the component industry companies is how to reach new customers and survive in the global competition. It means searching for new channels to represent competence and offer in addition to the traditional ones (fairs, publications, and so on);
- Adaptation of regulations: Clear relations with customers, partners and suppliers are the necessary condition for a successful collaboration. Because of the changing market and the dynamics of such relationships, it is important to periodically adapt agreements and collaboration protocols;

- Co-design and know-how protection: The intended companies are characterized by a strong competence in their specific application field. When collaborating in the design of a complex product they must take part in co-design activities without renouncing to protect their respective know-how;
- Giving room to negotiation: Many problems are still managed through direct discussions and negotiations between representatives of the involved companies. It means that, although provided with structured communication channels, they must leave sufficient room for informal negotiations;
- Involvement of occasional partners: Depending on a single customer order, it is possible that the network needs to include additional partners to supply uncovered work capacity. The availability of a registry recording occasional partners and their past performances can help in this phase;
- Damping down network perturbations: Problems (such as delay or material waste) coming from a network node affect the work of the other partners. Good network organization and re-planning ability is very important to damp down these perturbations and solve them before they reach the customer.

2.2 Building and Plant Construction

AUTOMAT is a very dynamic medium-sized company (90 employees, 30 million Euro of yearly turnover) realizing large automated warehouses and selling them all over the world. The construction of an automated warehouse is split into two distinct phases: first, it is designed and its components are manufactured by AUTOMAT and related suppliers and subcontractors, then it is assembled and tested on the field at the customer site. The former phase is quite similar to the manufacturing process of other complex products, while the latter is very special and critical.

AUTOMAT realized its last automated plant near Magdeburg, in Germany. The plant is 90 m long, 24 m wide and 16 m high. It is constituted by 12 arrays of cells, each array hosting a maximum of 1050 pallets, and 6 lifts moving along ways between cell arrays to store and pick pallets. The entire plant is completely covered to realize a closed building only accessible by load/unload conveyors. Plant components are steel structural work, special coverage tiles, electromechanic devices, electric circuitry and electronic controllers.

The installation and test phase at the customer site requires the opening of a construction yard lasting months, and the need to solve very critical logistic problems for the materials to move and the personnel to work in an alien environment. It is easy to see that similar problems occur to the many building construction or restoration companies operating far from their home site. Once again, many of them belong to the SME category and in some cases, such as in the restoration sector, they are micro companies.

The problems that AUTOMAT and other similar companies must manage are several and concern nearly all the phases of the product life cycle:

- Issuing of winning offers: The first stage of a project is the compilation of an overall plant design and the preparation of the relative bid. Part of the know-how and production capacity is split among suppliers and subcontractors that in some way must be involved in this task;
- Co-design and co-manufacturing: These phases of the product life cycle call for careful work planning, strict negotiations, information and document exchange, component acquisition or production, material testing and storing, work in progress monitoring, and exception handling;
- Transportation and logistics: The constitution of suitable work conditions at the installation site requires the solution for difficult logistical problems. Among them it is very important to ensure that on time delivery of materials and components are made available when actually needed;
- Involvement of local partners: The identification of reliable local partners for provision of personnel, materials and services is a critical condition for the success of an initiative. The construction is carried out in new locations and the experience from previous project is only partially reusable;
- Monitoring the work progress: It is necessary to set up a suitable organization to keep the construction progress under control even when it takes place very far from the company home site. This is particularly important in the test phase when the plant is complete but not yet available to the customers.

2.3 Food Production and Distribution

Consider the case of RAVIOLI, a medium-sized company specialized in the production of prepared pasta dishes for the large distribution in Europe (supermarkets, low-level restaurants and refectories). There are a lot of such SMEs in almost every European country, proposing industrial versions of local or universal dishes: they differ from the large food industry in the ability to offer selected niche specialities. In the analogy with RAVIOLI, they act at the regional level but in many cases they also have customers in other countries, and their export potential can be highly improved by the electronic market.

In turn, RAVIOLI buys rough materials (flour, sugar, spices, dried vegetables and so on) in big lots from Asia, America, Africa and European countries, and purchasing is adequately planned to take advantage of falls in international prices. Also, because of the perishable nature of the intended products, they are manufactured after receiving the customer or agent order and, to keep the lead-time short, RAVIOLI prepares semi-finished components in advance.

Specific problems of RAVIOLI and other SMEs acting in the prepared dish industry are:

- Managing a scattered trade network: The products are generally sold through a non-exclusive network of trade agents behaving as intermediaries with respect to customers. Communication and co-ordination difficulties arise from this scattered organization in a number of culturally different contexts;
- Worldwide purchasing: It is not easy, for an SME, to optimize the acquisition of materials from the global market. In spite of its limited resources, the company must operate a very sophisticated and informed purchasing function;
- Role of warehousing: Another very important company function is that assuring an efficient storing and picking of rough and semi-finished materials. This is done at special warehouses that are often provided as a service by transport and handling companies;
- Relations with final consumer: This is a further problem for RAVIOLI. At present both the trade networks and the large distribution selling the products filter the opinion of consumers. Then, it is difficult for the manufacturer to undertake actions to improve customer satisfaction and enforce the company brand.

2.4 Bidding in the Biomedical Sector

Hospitals and health-care organizations are used to launch calls for tender to purchase the huge quantity of materials and products they need to satisfy the demand on a six-month or yearly basis. The lots correspond to wide spectra of goods and services (often more than 200 item types) including simple disposable and sterile instruments, complex machines and their maintenance service. Then, only large manufacturing companies or trade organizations can answer the calls with complete offers, but often including goods acquired in turn from specialized SMEs.

In a certain Italian industrial district there are several of such SMEs, whose size ranges from 15 to over 100 employees. They supply medical devices, accessories and services including sterilization, design, molding, assembly, testing and packaging. About twenty of them have decided to constitute the BIOMED Consortium so as to directly manage the bidding phase and to exclude brokers. The Consortium estimates that this organization dramatically increases the probability to issue winning bids since customers apply selection criteria that are, in descending order of importance, product technical quality, company profile and price. The quality is already high, the Consortium is credible and the exclusion of brokers ensures a 30% of extra margin.

The participation in the Consortium is based on the 'complementary principle', meaning that its members cannot compete with each other at least in bid preparation. To this purpose, clear rules are adopted that establish what are the products that every member can offer and what are the supply conditions (price, production capacity, notification time, quality and the like) it ensures to the Consortium.

As such, the BIOMED Consortium acts on two different time horizons. On the long term, it simply intends to create an organization able to behave as a large company in the regional and continental markets. On the short term, it manages the single call for tender and undertakes the necessary actions, including the temporary involvement of external partners, to realize a winning bid.

For its long-term life the Consortium must face a number of problems:

- Publication of a unitary catalogue: The Consortium intends to give a unitary view of the goods offered by its members through the publication of a catalogue where offers are classified and homogenized. For the sake of readability, individual offers must be mapped onto a general and possibly standardized product and service coding system;
- Update of supply conditions: The Consortium members remain autonomous and can autonomously decide to change time to time the supply conditions of their offers. Specific rules and constraints must be agreed upon to separate individual initiatives from general behavior in the framework of the Consortium;
- Qualification at potential customers: The Consortium's visibility in the market is typically reached by qualification procedures at hospitals and health-care organizations. The result is that the Consortium becomes an acknowledged supplier in place of, or in addition to, single companies;
- Opening to the international market: The return of the effort spent in organizing and operating the Consortium is obtained faster by extending the market to other regions. This strategy has two main implications: adapt to new and different legal frames, and act in a multi-language environment.

Then, whenever the Consortium is invited to tender, further problems arise:

- Management of legal issues: In addition to general applicable laws every call for tender introduces specific regulations that are particularly strict in the medical sector. These calls for proper legal advice, but neither the Consortium nor its members, because of their limited resources, have an internal legal office;
- Plan the bidding effort: The bidding procedure requires strict planning of activities. Auction constraints must be met, and the appointment of a bid responsibilities. Relevant benefits come from taking advantage of previous experiences and behavior templates;
- Search for external partners: While preparing the bid, it would be useful to temporarily involve partners not belonging to the Consortium, for instance to acquire missing products or to interpret local regulations. Keeping a repository of previous relations makes this activity easier and faster;
- Monitor the bidding process: An important responsibility task is monitoring the progress of the bidding procedure so as to meet the strict offer deadline. In case of complex bids with the participation of several Consortium members, the use of a workflow management support is highly recommended;
- Apply for financial support: The supply products and services can be a high cash flow, but payment time is often quite long. The Consortium however is used to asking for financial support from banks, according to collaboration protocols that tend to become steady over time;
- Monitor the delivery process: In case of a winning bid, the Consortium must subsequently plan and operate an accurate delivery process. Goods are manufactured (or acquired) in due time, products and services are delivered according to rough plans and explicit periodical customer requests.

2.5 Managing Flows of Materials

BOLT&NUT is a medium-sized company (22 million Euro of yearly turnover) supplying the hardware stores of a wide European region with a quite complete spectrum of bolts, screws and the like. Most of the BOLT&NUT products are acquired in turn from traditional manufacturers, and only very special models are produced internally. A huge automated warehouse, with a modern test room to check the quality of incoming products mainly occupies the BOLT&NUT plant.

The trade organization of the BOLT&NUT Company is based on a net of autonomous sales agents each covering a certain geographic area according to an exclusive sales agreement. The agents manage their own local warehouses they use as buffers with respect to the final distribution phase to hardware stores. The orders coming from agents are partially based on demand forecast and partially customized so as to cope with special and urgent requests of the final customers. The percentage of customized orders is presently in the order of 40%, but it is rapidly increasing and is expected to reach the 70% in few years.

BOLT&NUT is managing most of the transport activities from its suppliers and to its agents. To this purpose it has established long-term agreements with a number of transport companies so as to obtain special prices and conditions. The number of transport orders BOLT&NUT issues every day ranges between 150 and 300, each involving one or more parcels. The incidence of handling and transport costs is about 7% of the yearly turnover, and BOLT&NUT is not very happy with this situation.

The industrial district where BOLT&NUT is located hosts some tens of micro business (less than 2 million Euro of yearly turnover) producing a variety of mechanical components. Some of them work (partially) for BOLT&NUT but in general they sell to the market. They are specialized in niche production and very appreciated for the high quality of their products. For this reason their customers are spread across Europe, but also in America and in Far East. They issue few transport orders per day (typically, some units), then they have no power to negotiate favorable transport prices and conditions.

The problems that BOLT&NUT and the micro companies have to face on the transport side are indeed many:

- Direct management vs. outsourcing: Directly managing the transport service can ensure additional margins, provided that the logistic solution is efficient. Outsourcing (mandatory for small businesses) removes a problem but increases costs considerably;
- Complexity of price lists: Every carrier has its own price list structure and prices are very hard to compare by the company logistic function. So, both medium and small businesses must rely on general rules (e.g. that destination is always assigned to a carrier, TRSP) that are seldom optimal;
- Aggregation of transport orders: Optimizing the transport services also calls for aggregating transport orders according to origin and destination, transported materials, picking and delivery time, and the like. Unfortunately, the aggregation criteria differ by carrier, price lists and the company (especially if micro business) cannot manage them;
- Pricing of transport orders: As a consequence of price list complexity, a large percentage of transport orders are issued without indication of the

service price. The price will be known only when invoiced by the assignee carrier, with no real possibility to control its correctness;
- Even worse on the long distance: The cited problems become harder to overcome when the transport services have to move goods to (from) far regions. In fact, there is little chance for an SME to negotiate with carriers established in other countries that are difficult to identify and select;
- Incidence on full product cost: The difficulty, for the company logistic function (if any), to deal with heterogeneous data from different carriers and to split the transport cost among the parcels, makes it hard to determine the incidence of the transport services part of product costs.

3. REAL CASES IN THE SERVICE SECTOR

There is a significant percentage of SMEs, in particular micro businesses, supplying services of different types to end consumers, or to other businesses. The intended services range from professional advice (administrative, legal, fiscal, technical, aesthetic, etc.) to transport of persons and goods, from modeling and design (dress, furniture, building, electricity, safety, hydraulics, etc.) to repair and maintenance (same sectors as design), up to training and other cultural supports.

Service supply companies must face up to a fragmentary market that is different from small company's industrial activities, which establish steady relations with larger customer companies used to involving suppliers in their own organizational growth. In this section we examine a couple of cases that have been studied in recent RTD projects.

3.1 Associated Hauliers

The transport sector is populated by a huge number of micro companies owning one or two trucks and carrying out qualified services at a low price. They survive provided that they fully operate their trucks with no idle periods. This need is in contrast with the limited resources they can spend in marketing and organizational activities, and also in searching for customers. A typical case is a family company where mum keeps the company office (phone calls and faxes, invoices and few other functions) while dad and son drive the trucks. Risks are very much reduced by associating with other hauliers.

MOVITRANS is a co-operative of 180 small hauliers performing generic transport services on the short and medium distance. Each MOVITRANS member owns a minimum of one and a maximum of five trucks, and makes available its entire transport capacity to the co-operative. MOVITRANS collects regular and occasional orders from customers and assigns missions to the different members according to a strict balance criteria. More precisely, when assigning a mission it buys the transportation service from one of its members and pays it on the basis of an agreed price list. In turn, it buys fuel, tires, as well as paying for maintenance and repair services at low price and resells them to its members.

The main condition for the co-operative to keep working is assuring a homogeneous workload for all its members, taking into account the number and type of trucks owned by each of them. In hard times the members share the difficulties but remain in the market, although with reduced income, while in good times it is a task of the co-operative to search for additional transport capacity that external companies can provide.

As a haulier association, MOVITRANS must manage several problems:

- Collection of customer orders: When the customer calls for a transport job, a MOVITRANS operator must confirm in the shortest possible time the service, possible conditions and the expected price. Currently this is done with no support by ICT tools, and a few errors sometimes occur;
- Mission assignment: This critical activity is carried out based on personal experience of the operators and access to a file of previously assigned tasks. In spite of this, drawbacks are infrequent and soon recovered under haulier claim;
- Involvement of external companies: Whenever MOVITRANS members cannot satisfy the customer order, external transport companies are contacted for the service. The historical record of previous collaborations can be very helpful in this activity;
- Exception handling: It may happen that a planned service cannot be done because of problems associated with the assigned haulier or their truck. Then, the mission assignment function must re-plan the service and engage another haulier in the time remaining;
- Mission report acquisition and invoicing: The co-operative collects reports from hauliers on the missions they have been engaged on, including possible changes with respect to the plan. On this basis the co-operative generates the invoices that in turn lead to payment of its members.

In turn, each haulier is asked to manage other problems:

- Mission acceptance: For a number of reasons the assignee of a certain mission can reject it. This must be immediately communicated to the co-operative so as to let the mission assignment function find an alternative solution;

- Information on mission execution: In order to minimize the communication flow, a haulier informs MOVITRANS only about exceptions and anomalous situations during its mission execution. The communication is split into two stages: an event is notified immediately, while evidence and details are supplied later;
- Generation of mission reports: A haulier concludes a mission by issuing a report on which basis MOVITRANS determines the amount to pay. Whenever the report differs from the planned mission (payload, distance, time, etc.) the haulier must justify the changes.

3.2 A Business-to-consumer Perspective

Imagine a rich city of 180.000 inhabitants like many others in Europe. The demand of small services addressed to individuals and the home has dramatically increased, especially in urban areas. This is because new family structures made up of a few busy persons who have no time or skill to manage the many, small daily problems that can occur. These individuals split their time between home (sometimes a condominium) and work (sometimes a small business) and have similar needs in both environments. What they ask for is the supply of services that can be booked without spending a considerable number of phone calls, are executed on time in the desired day, and their price and supply mode are known in advance.

With respect to this demand, the present condition of service supply is often critical. Suppliers are either an expensive micro-company, to be booked weeks in advance, or informal workers operating in the black economy. In either case their number is insufficient to satisfy the demand and the backlog is increasing. This could present an opportunity for increased employment to meet this demand that may benefit disadvantaged persons such as youths, immigrants, women and unemployed over-fifties.

Craftsmen and other service suppliers alone are not able to overcome the problem. However, one of their associations has recently constituted the SOSS Company, meaning Smart Organization for Small Services, to organize and stimulate responses. At present, with only four employees SOSS represents about 60 service suppliers offering 200 different types of services to households and 150 to condominiums and small businesses. Services offered include baby sitting, transport of persons and goods, house cleaning, repair and maintenance, gardening and many other activities. SOSS ensures the qualification of suppliers and a booking service that allows the customer to chose day and time and select the preferred supplier on the basis of price and code of practice.

In the operation of the booking service, the SOSS company deals with these problems:

- Search for suppliers and services: The choice for citizens and small businesses must be continuously enriched with new services, taking into account seasonal needs and demand drifts. The SOSS Company is charged with finding new suppliers through periodical search and qualification;
- Support to service suppliers: Because of their very small size, service suppliers have no resources to spend in organizational efforts. Their involvement is subordinated to the SOSS Company. SOSS's support to the service supplier ensures identification of new services, proposals for special offers, and the like;
- Communication with customers: The success of the initiative is strongly conditioned by the level of trust and confidence that the SOSS company is able to establish with its customers. This requires focused information campaigns and the set up of information channels to collect feedback after a service execution;
- Claim management: Either service suppliers or customers can raise claims whenever the behavior of other party has created obstacles to a full and correct service provision. Claim management, including clear and timely response, is a further condition to establish the desired trust;
- Performance evaluation: An important indication the SOSS company can give to service suppliers is the analysis of their individual behaviors, and of the system behavior as a whole. This extended knowledge base is useful to plan further investments and service improvements.

In turn, the single service supplier must manage these problems:

- Offer update: The offer includes service description, code of practice, possible options and constraints, price and other minor parameters. Some of them can change over time, and it is the supplier's responsibility to ensure the publication of the updated situation;
- Calendar management: The SOSS system forces the service supplier to represent its availability through the declaration of time slots with an estimated duration of the respective services. Moving from an informal behavior to a clear organization of work is a benefit but also at a cost;
- Reaction to service booking: In principle, the selected supplier is bound to execute the service notified by the SOSS system. Anyway, in case of problem the supplier is asked to raise any exception in the shortest possible time, so as to let the customer change the booking.

4. DEVELOPMENT REQUIREMENTS

The analysis of the cases presented in the previous sections suggests that the needs of SMEs and micro businesses can be classified into five main categories of requirement: access to the e-Market; regulations, trust and security; support to networking; problem solving; and Web mediation services. In this section we examine each of them in some detail.

4.1 Access to the Electronic Market

The first set of requirements concerns the possibility, for the intended companies, to reach a clear visibility in the electronic market with respect to potential customers, and to use the same channel for identifying partners and subcontractors. After years when many companies have developed their own (static) Web sites, they then begin to understand that nothing has actually changed. What is really needed is a unitary public infrastructure through which every company can present itself and its offer, and also search for other companies.

The electronic market infrastructure must include, at least, a unified catalogue of products and services and a standard registry of companies. The unified catalogue is needed to homogenize the offers of the different companies acting in different sectors, and make them comparable with each other. To this purpose:

- The unified catalogue should be based on one or more widely accepted standards, either generic or specific of single sectors, properly mapped onto each other. This knowledge is quite steady and can be used as a sound reference for extending the unified catalogue to cope with special needs mentioned below;
- The unified catalogue should be multilingual, meaning that associations and companies in different countries will use home language terms in specifying and classifying products and service types, while the same information must also be read and understood by the searching company in its respective home language;
- The unified catalogue should be extended to include product and service coding methods that are valid in particular sub-sectors or regions and whose scope corresponds, at least, to the single association. These additional coding methods must be mapped in turn onto the items of the basic standard structure;
- The offer of products and services will remain mostly directed to potential customers established in the same region. Then, even though access is open to other regions a unified catalogue results at least for local usage;

- The offer of services is often addressed to potential customers interested in a specific sector. It is, however, frequently the case that businesses put together contributions from partners and subcontractors acting in different sectors.

The establishment of customer-to-supplier and partner-to-partner (dynamic) relations calls for the constitution of a registry where all the companies are represented together with their offers, typical behaviors, document formats and so on. To this purpose:

- The registry should be conceived as an open and interoperable infrastructure with respect to other generic and specialized registries, and with respect to the legacy systems (if any) of the involved companies. In order to make company description understood to any party, it must be based on a widely-accepted (syntactic, semantic) standard framework;
- Company registration should include the company profile, offered products and services, supply modes and conditions (capacity, code of practice, price list, calendar and so on), used or preferred contractual forms, used or preferred negotiation procedures, used or preferred format of exchanged documents (bid, invoice, technical material, reports and so on);
- The registry should be multilingual, meaning that the companies in the different countries will keep using home language terms in specifying profiles and offers, while the same information must be read by the searching company in its respective home language;
- Associations to help the registration of new companies according to the specific qualification procedure should access the registry. A new company's profile and representation is either the job of the Association or the Company itself depending on the Association's role and power;
- The registry should be accessed by any company to complete and update its profile, offer and collaboration preferences. Moreover, a company could access the registry to search for possible subcontractors or partners that might help it in meeting at best a customer request;
- Potential customers that need to search for suitable suppliers and establish with them transparent and disciplined collaborations should access the registry;
- Customer-to-supplier and partner-to-partner relations are often established within the region. Then, even though open to access from other regions the registry must be particularly efficient in case of local usage. Also, local habits might suggest using additional parameters to characterize companies.

4.2 Regulations, Trust and Security

SMEs and micro companies require simple and practical solutions to the problems they face when preparing to join an electronic market. Therefore, the focus is on specification and development of pragmatic legal and security mechanisms and procedures, and also the adoption of appropriate trust models and techniques to engender the development of trusted electronic business relationships.

In order to enable SMEs to effectively participate in an electronic market, they need re-assurance from a legal perspective, which can be provided by tightly focused contractual support for the formation of trusted, ICT-enabled business relationships. Having established such a relationship, an SME will want to undertake an eBusiness relationship in a secure manner, where the level of security is directly related to the transaction being undertaken. To this purpose:

- The first condition is given by the possibility, for the intended companies, to behave according to few easy collaboration protocols while interacting with customers, partners or subcontractors. It means formalizing the distributed business processes in terms of actions and reactions, alternatives and decision points, and templates of the documents associated to the different actions;
- Among the exchanged documents, bids and contracts are of paramount importance as they fix the collaboration contents and conditions. In particular, both long-term and short-term regulation must be defined, the former establishing the intention to become partners and the latter referring to the execution of a single collaboration activity. SMEs can obtain real benefits from accessing a library of contractual templates coding common ways to regulate the electronic market relationships;
- Trust and confidence in the envisaged collaboration environment call for facilitated secure operation of eBusiness relationships. This requires, in particular, the adoption of suited (and possibly sector-sensitive) security models and policies and their implementation through simple and cheap techniques.

4.3 Support to Networking

In order to access an electronic market it is necessary to bridge the distance between any public infrastructure and the back-end system (if any) available at the single SME. Taking into account the limits of the intended companies, a fundamental condition for overcoming the present obstacles is hiding the complexity of the collaboration infrastructure under a set of simple functions supporting the networking activities. These networking applications will behave as an intermediate platform managing distributed relations and activities and coping with any peculiarities and constraints of SMEs. The definition and realization of such an easy networking application takes into account a number of critical requirements:

- The networking application should be designed as a tool suite for a single company. Its functions will cover the different steps of the collaboration life cycle, covering company self-representation to distributed work scheduling and partner search, up to bid negotiation, signature of contract, service progress monitoring and final invoicing;
- Depending on the resources available at the single company (simple browser, server, skilled IT personnel and so on) the networking application can be run on a local computer or provided as a service, for instance by the Association. The former case is not very frequent, and normally occurs when the company already runs its own legacy system while the latter is probably the preferred solution, in the short term, for most of the intended companies;
- The basic condition to make the networking application easy is confirming the linguistic adaptation capability of the work environment. The networking application should be natively and coherently designed following a multilingual approach with choices concerning a unified catalogue and standard registry;
- When joining the collaboration system, the self-representation stage should be done as a guided configuration process. This means that the user is just asked to define the offer by selecting items within the Association catalogue, - to code its practice by selecting business process schemas from those associated to the identified items, - to declare the preferred operation modes by selecting collaboration protocols from those associated to the business processes, - and so on;
- When searching for suppliers or partners, the user should be conditioned to specify the profile of the company under search by selecting items and behaviors from the public infrastructure. Should the searched company belong to a different Association (and, possibly, to a different sector and region) the operation must be completely transparent to the user;
- When negotiating for the supply of a product or service, relative collaboration protocols should belong to those declared by the company and also be well known to the user. They generate messages and document exchanges whose timing, contents and options are already coded and automatically submitted to the user by the networking application;
- While taking part in the execution of an order, the networking application should guide the user to perform the actions that are assigned to it by the

relative collaboration protocols. These actions will include sending messages, answering messages, taking decisions on the submitted matter, choosing one of the available options, and the like;

- Generally speaking, the networking application should create an easy environment where the user can find the current status of all its open processes, the history of previous activities and a simple support to decide which (and when) new activities will be carried out. By recording favorite suppliers and partners, preferred profiles or commonly used document formats, the networking application should introduce further facilitation to SMEs and micro companies.

According to the above requirements the networking application should include, at least three main functions, namely a distributed scheduling function, a workflow management function and a messaging function:

- The use of a distributed scheduler is foreseen, first of all, whenever the company receives a customer request that implies the involvement of partners. Distributed scheduler inputs are the business processes to plan (activities, actors, roles, precedence and alternative rules) and the capacities made available by the company itself and the candidate partners;
- The output of the distributed scheduler feeds the workflow manager. Time by time the workflow manager scans the planned and still active collaboration protocols and proposes to the interested user the execution of the next step. Also, the workflow manager intercepts the events occurring in the collaboration environment (e.g. messages, decisions, answers, etc.) and automatically updates the state of the running interactions;
- In turn, the execution of the workflow manager triggers the composition and sending of proper messages to the involved partners. The automation degree of the composition and sending functions strongly depends on the choice of the single company and, obviously, on the level of trust it has reached with respect to the collaboration environment;
- The messenger is a critical component of the easy networking application. It is based on the idea that messaging is the basic communication channel to other parties and, in turn a possible obstacle to interoperability. The intended companies are strongly oriented to the use of mobile communication devices as a way to overcome the lack of well-organized back offices. Mobile devices can be used in most stages of the collaboration life cycle, from contract negotiation to service execution.

4.4 Problem Solving

In addition to formalized collaboration protocols, SMEs and micro businesses inevitably rely on person-to-person relations and communications. In this context, distributed and ad-hoc problem solving as well as collaborative decision making form an essential part, and takes into account on average more than one third of daily working time. Problem solving techniques range from coded negotiation procedures to free and possibly synchronous dialogs, and demand an essential new quality of information, communication and knowledge transfer with colleagues, partners, suppliers and customers.

But instead of this obvious potential, the deployment of ICT to optimize and accelerate this process is rare, due to a lack of new developments and appropriate SME tools. The need is for the development of well-founded and easy-to-deploy ICT tools, in order to increase problem solving competence and solution (innovation) development ability of SMEs acting in an electronic market. To this purpose:

- It is required to increase firstly distributed and ad-hoc problem solving competence of SMEs with an easy-to-apply and case oriented methods toolbox, in the sense of on-line guides (e.g. comparable with Microsoft's Wizards) of best appropriate problem solving or assisted decision making (Problem Solving Moderator);
- Then, it is required to improve and accelerate distributed problem solving, decision making and knowledge transfer processes with an easy-to-deploy and completely Web-based ICT platform (Problem Solving Workbench);
- Finally, the focus is on enhancing problem solving and innovation capacity by smart agents, pushing personalized and automatically profiled information, respectively knowledge (information with context and linked expertise) to individuals or distributed teams (Problem Solving Assistant).

4.5 Web Mediation Services

As we have seen so far, there are important potential benefits, for SMEs, to join an electronic market and become active players. However, the cost/benefit ratio of this dramatic innovation, and the habit to associate and collaborate, could be strongly improved by the access to further eMarket services that integrate and enrich the basic infrastructure functionality. Some of such Web services are already offered by commercial/proprietary packages that make demand and offer of certain services meet with reciprocal advantages.

Examples of mediation services are those recalled in Section 2.5 (managing flows of materials) and in Section 3.2 (a business-to-consumer perspective). In both cases the offers of products or services come from small companies, while customers are individual consumers or small businesses in turn.

5. CONCLUSIONS

Let us conclude with a vision of what the condition of SMEs might become thanks to their planned access to a forthcoming electronic market. It corresponds to the behavior of Small Ltd, a small company associated to Small Association, a certain day of year 2006:

- Small Ltd is invited by Small Association to join the public infrastructure and become a credible actor in the electronic market. Small Ltd accepts and receives a URL, a user code and a password;
- Through the URL Small Ltd accesses the Small Association site, and it is recognized as a newcomer and guided in the representation of its features. Among minor requirements, it means coding its offer in the Small Association registry. To this purpose it navigates the unified catalogue and identifies the items that represent at best its production. It also means selecting, among the collaboration protocols and contractual templates coded in the Small Association site, those that better correspond to the usual way of negotiating and interacting with customers, partners and suppliers;
- A few hours later Small Ltd receives an unexpected bid request for the supply of a specific service from Foreign Co, a potential new customer established in a different country. The request is issued according to one of the collaboration protocols selected by Small Ltd, and written in its home language;
- On the basis of the same collaboration protocol, the system suggests to Small Ltd the list of possible reactions it can take (accept, ask for more info, reject, negotiate, etc.). Small Ltd suspends the answer because the requested service calls for the collaboration of partners able to perform special activities that Small Ltd cannot cover;
- Small Ltd uses its easy scheduler to plan the activities of the requested service, indicates as additional resources its usual partners and fixes scheduling conditions and constraints. On the basis of the scheduler outcome, Small Ltd uses the system to automatically send the relative bid request to the scheduled partners, including Partner Ltd for a certain activity;

- Small Ltd decides that it is time to search for a new partnership in alternative to Partner Ltd. Then it sends to the Small Association site a request for candidate partners meeting a specific profile (region, company size, desired activity, supply time, etc.). The Small Association site broadcast the request to homologous sites of other associations;
- Within the fixed deadline Small Ltd receives indications of three candidate partners discovered by the system in the respective registries. One of them, New Ltd, is particularly interesting and Small Ltd continues to negotiate with it until a bid has been obtained;
- Small Ltd takes its time to compare the two bids coming from Partner Ltd and New Ltd. The day after it decides and concludes the negotiation with the selected company (while the system automatically closes the negotiation with the other);
- Now all the data needed to answer the Foreign Co request are available. Small Ltd reacts accordingly and the negotiation continues up to the agreement on the contractual framework and the electronic signature of the contract. On this basis, also the contract with the scheduled partners is finally signed;
- Following these contracts, a number of data and document exchanges take place during the following weeks, according to the specific collaboration protocols. Small Ltd is involved, on the one side, in the envisaged collaborative work with the selected partners and, on the other side, in communications with the customer Foreign Co. Both these relations are regulated and supported by the workflow manager.

REFERENCES

This bibliography is aimed at giving an overview of the RTD projects from which the real cases discussed in the chapter are derived. Each project is described by some publications.

EP 10542 EASIER

Introduction of a new development methodology for PLC software, based on an object-oriented representation model. Usually, the PLC software development cycle is quite rough, pays little attention to the requirement specification and design phases, and spends about 40% of the time in the installation and test phases. The proposed methodology modifies in depth this situation, and is specifically conceived for the development of control software for industrial plants:

- F. Bonfatti, G. Gadda, P. D. Monari: "Re-usable software design for programmable logic controllers", ACM SIGPLAN Workshop on Languages Compilers and Tools for Real-Time Systems, La Jolla, June 1995, and ACM - SIGPLAN Notices, 11, vol.30, November 1995;
- F. Bonfatti, G. Gadda, P. D. Monari: "Bridging structural and software design of PLC-based system families", First IEEE International Conference on Engineering of Complex Computer Systems (ICECCS '95), Ft. Lauderdale, Florida, November 1995;
- F. Bonfatti, G. Gadda, P. D. Monari: "An improved process for the development of PLC software", International Conference on Software Engineering (ICSE 97), Boston, May 1997.

EP 20723 PLENT

Development of a set of advanced software tools to support distributed planning in "equal-rights" SME networks, through a centralized and transparent mechanism of task assignment and operational control. Central element of the system is the Co-coordinating Unit (CU), which maintains knowledge of network products, node competencies, distributed manufacturing processes, and assigns workloads to the different nodes by applying network-agreed criteria. The CU keeps also track of the distributed plan execution, to support the evaluation of the real performances of the single nodes and of the network as a whole:

- F. Bonfatti, P. D. Monari, P. Paganelli: "Co-ordination functions in SME networks", 2nd International Conference on Architectures and Design Methods for Balanced Automation Systems (BASYS 96), Lisbon, June 1996;
- F. Bonfatti, P. D. Monari, R. Montanari: "Information flows and processes in an SME network", in Re-engineering for Sustainable Industrial Production, L. M. Camarinha-Matos ed., Chapman & Hall, 1997 (Proceedings of the OE/IEEE/IFIP International Conference on Integrated and Sustainable Industrial Production - ISIP 97, Lisbon, May 1997).

EP 25360 COWORK

Development of a tool suite to support distributed design in small-medium enterprise networks. The tool suite is quite cheap and easy to use, and preserves the know-how of each enterprise of the network. The project aims at favoring the systematic exchange of messages and documents in the different co-design phases, from partner search to product detailed design:

- D. Biondi, F. Bonfatti, P. D. Monari, F. Giannini, M. Monti: "A Product Data Manager supporting a new co-design methodology for SMEs"

International Journal of Computer Applications in Technology (IJCAT), 2001;

- D. Biondi, F. Bonfatti, P. D. Monari, F. Giannini, M. Monti: "A product modeling tool for the definition and the exchange of product data in a manufacturing Co-design development", JCAD, special issue on Product Data Representation and Management, 2002.

IST 25125 SOSS

Design and development of a Web-based booking platform to support a new entity, the Local SOSS organization (LSO), which co-ordinates the provision of services to person and home provided by a wide variety of small suppliers. Services of interests are those for everyday life, such as plumbing, wall painting, and elderly transportation.

- M. Barbi, F. Bonfatti, P. Monari "An Internet-based Platform to Support a Smart Organisation for Small Services", e2002 International Conference, Prague, October 2002.

IST 28413 INDIA

Study, development and prototyping of a software platform putting a company in condition to manage at best the planning of all the external activities, by promoting a fair competition among its partners. Particular attention is paid to non-productive activities, especially transportation and warehousing. The project extends the results of previous projects (in particular GNOSIS-VF and RUMS), focused on process modeling and distributed planning, respectively:

- B. Bucci, F. Bonfatti, P. D. Monari: "Web-based tools for constitution and operation of enterprise constellations", e2002 International Conference, Prague, October 2002;
- F. Bonfatti, P. D. Monari, E. Marri, A. Virgulin: "Enhanced modelling and planning for the Extended Enterprise", e2003 International Conference, Bologna, October 2003.

IST 28618 BIDMED

Selection and adaptation of a set of innovative software tools to support groups of companies joining their forces to participate to big calls for tender. The bidding activity is co-coordinated by a virtual structure, the Smart Bidding Organization (SBO) which, for each call, entrusts a Bid Manager representing it. The bio-medical sector has been chosen as application field:

- G. Bonini, G. Pignedoli, F. Bonfatti, P. D. Monari: "Smart Bidding Organisations", ECEC 2002 International Conference, Modena, April 2002.

IST 32521 CITRO

Selection and adaptation of a set of innovative software tools to support a cluster of individual truck owners, with the help of a proper methodology. The project intends to extend and experiment, through pilot cases in Italy, Hungary and Spain, the results of RTD projects (in particular TROP and TRASS) that studied in detail issues related to matching demand and offer of transportation services:

- F. Bonfatti, P. D. Monari, E. Montanari: "Experiences in ebXML-based interoperability", e2003 International Conference, Bologna, October 2003.

Chapter 6

CREATING A RESEARCH AGENDA FRAMEWORK FOR SEMICONDUCTOR SUPPLY NETWORK INTEGRATION

Thomas Callarman, John Fowler, Esma Gel, Michele Pfund, Dan Shunk
Institute of Manufacturing Enterprise Systems, Arizona State University, Tempe, Arizona

Abstract: Semiconductor manufacturing is at the heart of the electronics industry, which is the second largest industry in the world, trailing only agriculture. Supply Chain Management (SCM) has been a common buzzword in industry for the last decade and has been discussed in the literature since about 1985. In this chapter, the SCM needs for the semiconductor supply chain (or more accurately supply network) are analyzed. First, an overview of SCM is provided. Then, the challenges of SCM for the semiconductor industry are discussed. Next, a survey that seeks to determine a research agenda for semiconductor supply networks is described and results analyzed. Finally, the survey results are synthesized into a research agenda comprised of seven clusters. The first cluster is concerned with the management of demand and the second is focused on designing the supply network. The next two research clusters describe issues associated with supply network planning and execution, respectively. The last three research clusters are concerned with the infrastructure needed to successfully manage the supply network. Cluster five focuses on the financial infrastructure, cluster six on the IT infrastructure, and cluster seven with the organizational structure.

Key words: Supply Chains, Semiconductor manufacturing, Planning

1. INTRODUCTION

The purpose of this chapter is to summarize current research literature in the area of Supply Chain Management (SCM) in the semiconductor industry. Literature specific to the semiconductor industry on this topic is sparse, but where possible it has been included and SCM applications for semiconductor manufacturing have been cited. This chapter provides a current view of supply chain management research as a basis for formulation of target areas recommended for further research in semiconductor manufacturing and its supply chain.

SCM has been a popular topic of research in the fields of logistics and purchasing since the mid-1980s. It has since blossomed into a topic that spans a wide spectrum of investigative activity in both business and engineering. SCM has only recently begun to be explored as a broader business strategy capable of integrating all elements of the supply chain, ranging from raw material suppliers to the end-consumer, crossing multi-disciplinary and multi-company boundaries. SCM approaches use global optimization methods for decision making, which results in cost-effective solutions for the operation of the whole supply chain (e.g., the use of shared resources among supply chain partners, reduced duplication and waste making use of core competencies wherever in the world they may be located), rapid information flow, and ultimately better customer responsiveness and satisfaction. Companies that are able to successfully implement the best practices of SCM can achieve significant and sustainable competitive advantage in the marketplace.

2. OVERVIEW OF SUPPLY CHAIN MANAGEMENT

Supply chain management has been a buzzword in industry for the last decade. As a result, there are many differing views, definitions, and even schools of thought on this very complex topic.

2.1 Definitions of Supply Chain Management

Given the nature of the rapid interest in supply chain management, it is not surprising that over the last 30 years many definitions have been given for supply chain management. Most have evolved over time and differ depending on whether they are coming from a logistics or a purchasing or a manufacturing perspective. In this chapter, the authors cite a few of the more current definitions, but do not attempt to select a best definition of SCM. Instead, the authors identify critical strategies and characteristics necessary for effective implementation of the SCM concepts in the semiconductor industry.

Most authors agree that the basis for SCM began with Jay Forrester's 1961 work on Industrial Dynamics (Forrester 1961), which is appropriate in terms of this discussion since Forrester is also the inventor of the DRAM. Berry, Naim, and Towill state, "Industrial dynamics is the application of feedback thinking and control engineering concepts to the study of business and organizational systems (Berry et al. 1995). Forrester used simulation software as a tool that made it easier for others to create and use dynamic models of organizations. The work of Berry et al. goes on to describe the efforts of Forrester in further detail.

Hines claims that it was the popularity of Michael Porter's Value Chain model that led companies to think beyond their borders to achieve competitive advantage (Hines 1993). Regardless of the actual roots, it was the mid-1980s contemporary logistics literature that actually coined the term Supply Chain Management (Jones and Riley 1985). Initially, the concept focused on optimizing outbound product movement and inventory logistics between a supplier and a customer. This logistics view was quite narrowly defined around the physical movement of goods, but did lead to improved efficiencies within the firm such as business process re-engineering, enterprise requirement planning, and just-in-time manufacturing. Thinking has since expanded to view SCM as the interconnection between key processes both within firms and between firms, linking all the internal processes from new product development to customer order performance. Thus creating a supply network that enables firms to engage in competitive advantage through co-evolvement rather than competition (Bechtel and Jayaram 1997).

2.2 Supply Chain Schools of Thought

Bechtel and Jayaram published a review of SCM literature in 1997 and provided a categorization of supply chain "Schools of Thought" (Bechtel and Jayaram 1997). This analysis included the following schools: a) Chain Awareness school; b) Linkages/Logistics school; c) Information school; and d) Integration school.

The "Chain Awareness" school was primarily focused on making others aware of the importance of the supply chain and the management of it. Jones and Riley (1985) provided one of the first definitions of SCM when they stated that "Supply Chain Management deals with the total flow of materials from suppliers through end users." Papers in this school appeared in the late 1980's and early 1990's.

The papers in the "Linkage/Logistics" school were concerned with the flow of materials through the supply chain with a special emphasis on the linkages between the elements of the supply chain. Materials generally flow from suppliers to customers in the supply chain. The papers in this school appeared mostly in the early 1990's.

The "Information" school was primarily concerned with the flow of information through the supply chain. Early papers paid quite a bit of attention to the criticality of information flow and the fact that much of the information flow is in the opposite direction of the material flow (i.e. from customers to suppliers). Later papers indicated that information is bi-directional. Papers in this school began to appear in the early to mid 1990's.

Finally, the "Integration" school began to take a more holistic view of the supply chain and considered all flows in the supply chain. This school focuses on how the flows interact and how the supply chain can be integrated to provide the "best" outcome for the entire chain.

Current literature by and large expands on the "Integration School" concepts and examines value creation throughout a synchronized and integrated supply network with additional focus on alliances and strategic partnerships between supply chain constituents.

Integrated supply chain management is generally captured as an alignment of buyers, suppliers, and customers and their respective business processes to achieve an advanced form of competitive advantage. Monczka and Morgan define an integrated supply chain and value chain as a connected series of organizations, resources, and activities involved in the creation and delivery of value, in the form of both finished products and services, to end customers (Monczka and Morgan 1998). Ellram (1991) states that supply chain management is simply a method of using information to collectively manage the flow of product from source to final customer with the goal of improving end-customer service and satisfaction at reduced overall cost.

Supply chain opportunities are identified in terms of shortening product development time, reducing total product costs, streamlining supply/demand processes, compressing time throughout the system, and quickly and flexibly responding to customer needs (Monczka and Morgan 1998, Bechtel and Jayaram 1997, Ellram 1991, Slywotzky 1998). However, recognizing and developing such opportunities can be quite complex. In particular, effective implementation requires a framework that can both analyze existing supply chains and provide guidance during future chain development.

2.3 Characteristics of Supply Chains

It is difficult to grasp the concept of supply chain management from a definition alone. This is due to the fact that supply chain management is methodology that can be applied at different extremes. Cooper and Ellram (1993) provide a useful description of characteristics that can be used to differentiate between SCM and traditional approaches. These characteristics can be found in table 6-1. It is important to keep in mind that the different characteristics can be present at many levels or extremes. The table can be a useful tool in understanding SCM efforts.

Table 6-1. A Comparison of Traditional Approaches and SCM Approaches by Cooper and Ellram (1993)

Element	Traditional	Supply Chain
Inventory Management Approach	Independent efforts	Joint reduction in channel inventories
Total Cost Approach	Minimize firm costs	Channel-wide cost efficiencies
Time Horizon	Short term	Long term
Amount of Information Sharing and Monitoring	Limited to needs of current transaction	As required for planning and monitoring processes
Amount of Coordination of Multiple Levels in the Channel	Single contact for the transaction between channel pairs	Multiple contacts between levels in firms and levels of channel
Joint Planning	Transaction-based	On-going
Compatibility of Corporate Philosophies	Not relevant	Compatible at least for key relationships
Breadth of Supplier Base	Large to increase competition and spread risk	Small to increase coordination
Channel Leadership	Not needed	Needed for coordination focus
Amount of Sharing of Risks and Rewards	Each on its own	Risks and rewards shared over the long term
Speed of operations, Information and Inventory Flows	"Warehouse" orientation (storage, safety stock) Interrupted by barriers to flows; Localized to channel pairs	"Distribution Center" orientation (inventory velocity) Interconnecting flows; JIT, Quick Response across the channel

2.4 Components of the Supply Chain

The supply chain consists of three main parts: purchasing, manufacturing, and distribution. Gupta (1997) explains that the amount of value added by each of the supply chain parts changes based on the industry. Depending on where the most value is added, the supply chain is either supplier dependent, manufacturing intensive, or distribution oriented. In a supplier dependent supply chain, most of the value comes from purchased parts. Heavy manufacturing adds most of the value in manufacturing intensive supply chains. An example of the heavily manufacturing dependent supply chain is the semiconductor industry. Lastly, in distribution oriented supply chains, value is added by delivering the product to the customer, because if product is not available, the customer will usually buy a different brand (1997).

Gupta (1997) emphasizes that in heavy manufacturing, such as semiconductors, a very important task is balancing the flow of material through a very complicated set of machines and workers. It is also important to receive the raw materials from the supplier at the right time. When this flow is balanced, the parts can be finished on time, which means they can be delivered to the customer on time. This is how all the elements of the supply chain interact with each other and how they affect one other (1997).

Cooper and Ellram (1993) present an excellent analogy for understanding how important it is for all the department/areas of a company to work together through supply chain management. The authors liken an effective supply chain to a well-balanced and well-practiced relay team. A relay team is more competitive when it practices with the same core lineup so that each runner knows how to be positioned for the handoff. While relationships are the strongest amongst those who directly pass the baton, the entire team must be coordinated to win the race. This is the major benefit of supply chain management.

In their chapter on semiconductor SCM, Ovacik and Weng (1995) state that the goal of SCM is to minimize total costs and maximize throughput of the supply chain. One of the keys to SCM is that the supply chain must be considered as a whole. Hence, when considering an individual decision in the company it is necessary to pick the solution that optimizes the entire supply chain, not just a local optimal decision.

2.5 Types of Decisions in SCM

According to Fox, Chionglo, and Barbuceanu (1993), SCM functions occur on three levels depending on the length of time that the decision affects: strategic, tactical, and operational. In particular, strategic level decisions address top-level issues like long-term planning (e.g., enterprise production planning) and usually has a time horizon of a quarter to three years. The tactical level, which usually operates in weeks or months, includes tasks such as forecasting, creating a master production schedule, and determining plant level materials requirements. The lowest (most detailed) level is the operational level that has a time horizon of shifts or days. At the operational level, inventory is deployed and shop floor level scheduling of equipment and people is performed (Fox et al. 1993, La Bouff 1997). Since SCM encompasses all parts and levels of a company, many levels of management will need to be involved in the implementation and on-going phases of SCM.

2.6 Implementing SCM

The decision to implement formal SCM in a company is a huge commitment, with the potential for immense payoffs. Cooper and Ellram (1993) put forth that the three main reasons for forming a supply chain are:

- to reduce inventory investment in the chain;
- to increase customer service; and
- to help build a competitive advantage for the channel.

A formal methodology of SCM is a complete new way of thinking for most companies. As a result, it may require the entire re-engineering of the company and its' interactions with other companies to implement SCM and this may take many years. Scheller (1994) lists three key re-engineering steps as: implementing pull systems (i.e., having customer demands determine the systems), implementing a production system that is flexible with quick turnovers, and enabling a balance between make-to-order and make-to-stock modes of production.

Most of the literature on implementing SCM only provides high-level steps for implementing SCM. One example is La Bouff (1997), who describes four steps needed in order to implement SCM: (1) define your service goals, (2) define your capabilities, (3) define your business model, and (4) define a closed loop process. Most literature agrees that it becomes critical to think of suppliers as partners as the supply chain becomes more advanced in the company. This will be necessary in order to implement new supply chain concepts that will meet the needs of the customers better (Hewitt 1994, Fox et al. 1993, Ovacik and Weng 1995, Stevens 1989).

An in-depth paper on implementing and managing a supply chain comes from Stevens (1989), who describes a four-stage model that moves from complete functional dependence to inter-organizational integration. Hewitt (1994) has expanded Stevens' model further by adding a fifth stage. The fifth stage integrates the internal and external supply chain process management.

3. CHALLENGES IN SEMICONDUCTOR SUPPLY CHAINS

The semiconductor manufacturing process is very complex. Due to the large number of processing steps combined with re-entrant product flows, each wafer may visit the same manufacturing step many times throughout its fabrication (Uzsoy et al. 1992). Also, because of the long processing times required to manufacture today's complex semiconductor products, there usually exists a long supply pipeline, or high levels of work-in-process inventory (La Bouff 1997).

Another difficulty is that the semiconductor market experiences erratic changes in demand, which can make it hard to decipher a true demand change from normal fluctuations. Semiconductor manufactures are far from the end customers who are pulling demand. As a result, it may take a while before the information gets to the semiconductor manufacturers and it may be magnified when it does (La Bouff 1997).

Furthermore, it is hard to know exactly what the work in process inventory of each product type is; the exact processor speed of a product is not even known until final tests and as stated above, the flow time of typical semiconductors is rather long (2.5 to 4 times the raw processing time). Due to these factors, wafer lots are usually not affiliated with a specific customer and semiconductor companies generally run a make-to-stock system (Uzsoy et al. 1992). Having said that, a semiconductor manufacturer is also faced with the challenge of satisfying a variety of customer types such as original equipment manufacturers (OEMs) and distributor sales. There may be strict requirements from some of customers, like OEMs, and very loose requirements from others (La Bouff 1997).

Consequently, the semiconductor supply chain is very complex, spanning multiple manufacturing sites in various locations around the globe. Uzsoy, Lee, and Martin-Vega (1994) point out that most semiconductor supply chains are vertically integrated with multiple fabs, assembly and test sites, and distribution centers that are usually in different parts of the world. For most semiconductor companies, wafer fabrication plants are located in the U. S. while assembly and test plants are located in East Asia. Because of the large geographical separation of facilities it is important to look at the semiconductor supply chain as a whole (Uzsoy et al. 1992, 1994). Another factor that complicates the supply chain in the semiconductor industry is the move towards foundries and outsourcing of manufacturing or design to other companies.

The chain includes multiple suppliers, sub-suppliers, contractors, internal manufacturing, distribution, and customers. For simplicity, most reviewers have limited the scope of their analyses to only a few broad categorical links in the chain (Frederix 1996, La Bouff 1997, Ovacik and Weng 1995). Most of the literature specific to semiconductor supply chains focuses on production planning and scheduling, with notable exceptions coming from Lee and Billington (1995), Uzsoy, Lee, and Martin-Vega (1992, 1994), and Frederix (1996). Ovacik and Weng (1995) write, "The (semiconductor) supply chain is defined to be a network consisting of nodes corresponding to facilities where products are acquired, transformed, stored, and sold." The main nodes of the semiconductor supply chain are: Suppliers, Wafer Fabrication, Wafer Sort, Assembly and Test, Finished Goods Inventory, Distribution, and Customers (La Bouff 1997).

However, Ovacik and Weng (1995) emphasize that a semiconductor company is part of a larger supply chain. For example, the larger supply chain could be thought of as a supply chain that supplies computers to retail customers. This supply chain would consist of the suppliers that provide bare silicon wafers and other raw materials to the semiconductor companies. Semiconductor companies feed the integrated circuits to the circuit board assembly companies. They feed the circuit boards to the Original Equipment Manufacturers (OEMs) that assemble the final products, and distribute the computers to the retail stores that actually sell to the end customer. This supply chain sounds complicated, and it proves even more complicated considering that each member of that supply chain has numerous suppliers and customers. Currently this supply chain is too large to work with. First each company needs to optimize their own supply chain with their direct suppliers and customers. Then eventually the entire supply chain can begin to be analyzed (Ovacik and Weng 1996).

Frederix defined a supply network (Figure 6-1) including both internal and external (foundry fabrication and/or contract assembly and test) manufacturing sites (Frederix 1996). This is a good representation of the types of choices available at each level within today's virtually integrated semiconductor manufacturing industry. In reality, the supply chain becomes even more complex as each member of the chain has additional material and equipment suppliers and customers that are not comprehended in this simplified model. Any given company in a typical supply chain may have hundreds, sometimes thousands of independent suppliers.

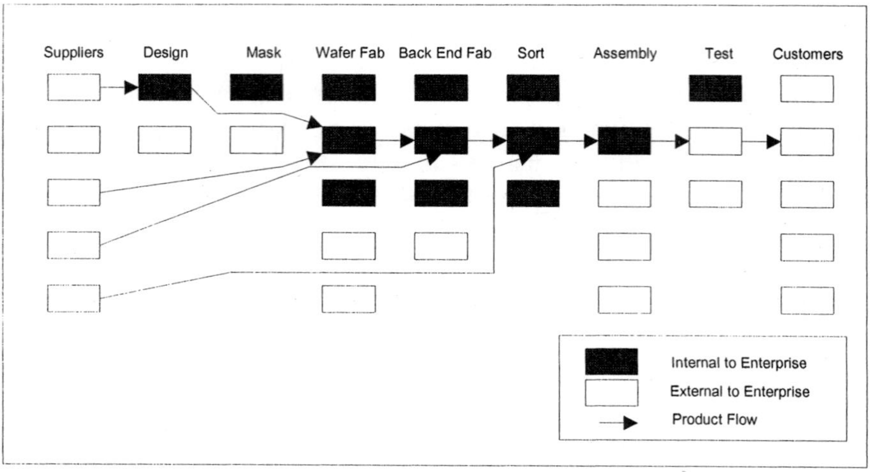

Figure 6-1. Representation of the Semiconductor Supply Chain (adapted from Frederix 1996)

To reduce the complexity of the problem, some modelers have taken an aggregate view of the supply chain making various assumptions about the internal workings of each chain member, the capacity of the facilities, etc. Supply chain analyses are generally restricted to one product or one family of products ignoring all the other products that have to flow through the same supply chain and their interactions with the products in consideration.

However, the true complexities of the supply chain are not readily apparent when analyzing the network from this macro view. The semiconductor supply chain, even the internal manufacturing subset, is very complex. As an example, Temple (1998) describes the internal manufacturing logistics of one product flow for Texas Instruments. The product actually makes 22 physical hand-offs, uses 5 planes and 6 trucks between wafer fab and final shipment to the customer. As discussed above, each macro-element of the total supply chain has similar intricacies.

Usually, most emphasis in the semiconductor industry is on the manufacturing component of the supply chain. Extensive time and effort are usually spent in planning (including scheduling) within manufacturing. Only recently have most semiconductor companies realized how important it is to manage the entire semiconductor supply chain. Decision-making, especially in planning, should consider the global implications the decision will have on the entire supply chain. Recently, semiconductor companies have begun to realize that many of the back-end components of the supply chain are particularly important because they are closer to the customers (Frederix 1996, Ovacik and Weng 1995, Uzsoy et al. 1992, 1994).

According to Uzsoy, Lee, and Martin-Vega (1994), another fault of the semiconductor supply chain is that in most companies, different departments (i.e., planning, marketing, engineering and manufacturing) are evaluated on different performance measures that sometimes are conflicting. Since they are evaluated separately, problems that arise are usually dealt with separately, using solutions that optimizes one area without regard to how it affects the rest of the supply chain. Even within manufacturing, most research in this area does not consider production planning and scheduling problems together or with any other manufacturing problems (Cooper and Ellram 1993).

4. DEVELOPMENT OF A RESEARCH AGENDA FOR SEMICONDUCTOR SUPPLY NETWORKS

The above discussion outlines the challenges in the management of semiconductor supply chains or more appropriately called semiconductor supply networks. Due to several of these characteristics, semiconductor SCM problems tend to require a more sophisticated set of techniques. This brings the need for a forum on the development of a research roadmap. For this purpose, a team of researchers from several universities during the fall of 2000, addressed the strategic question "What should the research agenda be for 21st Century Semiconductor Supply Networks (SSN)?".

In order to develop the agenda, first fundamental concepts were extracted from ten industry-accepted documents on the vision for the management and operation of semiconductor supply networks. This resulted in the initial concepts addressed in the survey described in section 4.1. These ten documents are:

- The Next Generation Manufacturing Report (NGM), US View (1997);
- The Intelligent Manufacturing Systems (IMS), World View (2000);
- Proceedings of the International Conference on Modeling and Analysis of Semiconductor Manufacturing (MASM 2000);
- Industry publications and predictions – Clockspeed by Charles Fine (1997);
- Industry publications and predictions – Bullwhip by Hau Lee (1997);
- Benchmarking Semiconductor Manufacturing from UC-Berkeley (2003);
- CORDIS: IST European Technology Forecast (1999);
- National Technology Roadmap for Semiconductors (NTRS) (1997)

- International Technology Roadmap for Semiconductors (ITRS), published 1999 (ITRS, published 1999);
- The SCOR Model (Supply Chain Council, 1999)

To reflect the views and vision of semiconductor industry in the research roadmap, we employed a survey, development of that is outlined in the next section.

4.1 Survey Development and Methodology

The objective of the survey was to capture the expert opinion on the key characteristics of semiconductor supply network performance. NGM Report (1997) was used as the document to put structure into the questionnaire. A team of hundreds of manufacturing experts in the mid-1990's to capture what Manufacturing will look like in the 21st Century, e-business environment, wrote NGM. NGM used four categories to structure various concepts (drivers, attributes, imperatives, and metrics). To further refine these categories, additional concepts presented in the above 10 research articles were added to the appropriate SCM concept. Below, we list the various concepts that the survey addressed under the four categorical headings.

Drivers of the Semiconductor Supply Network (What are the forces that shape the future competitive environment?):

- Accelerating pace of change in technology;
- Globalization of markets and business competition;
- Rapidly expanding technology base;
- Global wage and job skills shifts, tight labor markets;
- Increasing customer expectations;
- Strong emphasis on environmentally safe processes;
- Constantly changing needs and market opportunities;
- The need for concurrency in all operations;
- Ubiquitous availability and distribution of information;
- Short lead times;
- Need for "instantaneous" information transformation to knowledge to make effective decisions;
- Distorted demand and market information due to independent supply chain practices;
- Global issues dealing with trade, customs, etc.;
- Increasing capital costs / capital intensity;
- Business Model changes resulting from outsourcing with fabless operations;

- Underlying cyclical nature of the industry;
- Shift in market to communications products;
- Shift to commodity pricing.

Attributes (What must we design into the research agenda?):

- Customer responsiveness – external view;
- Physical plant, equipment and infrastructure responsiveness and reconfigurability – internal view;
- Human resource responsiveness and flexibility;
- Effective partnering and strategic alliance capabilities;
- Rapid innovation and development;
- A global perspective on knowledge sharing and social inequalities;
- A deep understanding of supply chain dynamics and its impact;
- Effective information sharing;
- Demand stream visibility;
- "Trusting" environment;
- Effective channel alignment;
- Flexible, coordinated activity with decision methodologies for allocation issues.

Imperatives (What must the capabilities of the supply network be?):

- Flexible manufacturing capability with group technology;
- Mass customization (high volume – high mix) capability;
- Available-to-Promise (ATP) visibility;
- Information and knowledge modeling overlays;
- "Real" cost modeling overlays with correct predictive values;
- Virtual Agile Information and Execution System (VAIES) on a global scale;
- Links from the process characterization to the enterprise characterization;
- Data capture and consistent interpretation of factory SECS I, II and GEMS data;
- Capability to reconfigure the supply chain with different business models;
- Extensibility, flexibility and scalability (EFS);
- Reduced ramp time;
- Knowledge supply chains;
- Rapid product/process realization;
- Effective innovation management;
- Change and learning management;
- Next-Generation manufacturing processes and equipment;
- Pervasive modeling and simulation for analysis and decision making;

- Adaptive, responsive information systems;
- Extended enterprise collaboration;
- Seamless integration of supply chain partners thru appropriate information technology channels;
- High operational efficiency and quality;
- Reduction of variability and uncertainty;
- Inter-technology generation flexibility;
- Coordinated planning on an SSN scale;
- "Win – Win" collaborative buyer – supplier environment.

Metrics (Critical performance metrics driving the Semiconductor Supply Network):

- Supply chain reliability
  - On time delivery (percentage)
  - Order fulfillment lead time (days)
  - Fill rate (percentage)
  - Perfect order fulfillment (percentage)

- Flexibility and responsiveness
  - Supply chain response time (days)
  - Upside production flexibility (days)

- Expenses
  - Supply chain management cost (percentage)
  - Warranty cost as percentage of revenue (percentage)
  - Value added per employee (dollars)

- Assets/utilization
  - Total inventory days of supply (days)
  - Cash-to-cash cycle time (days)
  - Net asset turns (Turns)

- Others
  - Yield (percentage)
  - Yield ramp time (days)
  - Process development time (days)
  - Fraction of value–added time (percentage)

Once the questionnaire was created, a series of workshops and interviews were conducted to solicit data from a group of experts from academia (ACAD). In addition, industry experts were chosen from fields in integrated design and manufacturing (IDM), Packaging (Pack), and Distribution Service Providers (DIST). Detailed participation information is provided below:

- Thirteen experts from ACAD;
- Nineteen experts from IDM;
- Three experts from PACK;
- Ten experts from DIST.

Once the survey results were collected, each element was analyzed to determine high priority components of supply chain network performance. This was done by consolidating the top four items for each of the categories (drivers, attributes, imperatives, performance metrics) for each of the 4 types of experts (ACAD, IDM, PACK, and DIST). Ties were included, where necessary. Results are shown below (Table 6-2).

Table 6-2. Prioritization of Key Drivers, Attributes, Imperatives, and Metrics

Concept	ACAD	IDM	PACK	DIST
DRIVERS				
D1. Accelerating pace of change in technology	Hi		Hi	
D2. Short lead times	Hi	Hi	Hi	
D3. The need for "instantaneous" transformation of information to knowledge for making effective decisions	Hi	Hi	Hi	Hi
D4. Business model changes resulting from outsourcing with fabless operations	Hi		Hi	Hi
D5. Increasing customer expectations		Hi	Hi	
D6. Constantly changing needs and market opportunities		Hi	Hi	
D7. Globalization of markets and business competition.			Hi	Hi
ATTRIBUTES				
A1. Customer responsiveness – external view	Hi	Hi	Hi	
A2. Rapid innovation and development	Hi		Hi	
A3. A deep understanding of supply chain dynamics and its impact	Hi	Hi	Hi	
A4. Demand stream visibility with effective information sharing on a global scale	Hi	Hi	Hi	Hi
A5. Flexible, coordinated activity with decision methodologies for allocation issues.	Hi		Hi	
A6. Effective partnering and strategic alliance capabilities.			Hi	Hi
A7. Human resource responsiveness and flexibility.			Hi	Hi

Concept	ACAD	IDM	PACK	DIST
IMPERATIVES				
I1. Flexible manufacturing capability with group technology	Hi		Hi	
I2. Mass customization (high volume – high mix) capability	Hi		Hi	
I3. Available-to-Promise (ATP) visibility	Hi	Hi	Hi	
I4. Capability to reconfigure the supply chain with different business models	Hi	Hi	Hi	
I5. Extensibility, flexibility and scalability (EFS)	Hi	Hi	Hi	Hi
I6. Seamless integration of supply chain partners thru appropriate information technology channels	Hi	Hi	Hi	
I7. High operational efficiency and quality	Hi		Hi	
I8. Pervasive modeling and simulation for analysis and decision making.		Hi	Hi	
I9. Virtual Agile Information and Execution System (VAIES) on a global scale.		Hi	Hi	Hi
I10. Change and learning management.			Hi	Hi
I11. Adaptive, responsive information systems.			Hi	Hi
I12. Extended enterprise collaboration.			Hi	Hi
METRICS				
M1. On time delivery (percentage)	Hi	Hi	Hi	
M2. Supply chain response time (days)	Hi	Hi	Hi	Hi
M3. Total inventory days of supply (days)	Hi		Hi	
M4. Cash-to-cash cycle time (days)	Hi	Hi	Hi	Hi
M5. Net asset turns (Turns)	Hi		Hi	
M6. Order fulfillment lead times (days)			Hi	Hi
M7. Perfect order fulfillment (percentage)		Hi	Hi	

4.2 Findings from the Survey

The above findings drove our vision of the semiconductor supply network of the 21st century and yielded the requirements for Research in Semiconductor Supply Networks (RSSN). Our basic visionary concepts derived from the established drivers are included in the following list.

A total enterprise coordinated view must be developed. This notion captures the need for coordination and visibility in "real time" especially in the rapidly changing technology space.

This view must include the characteristics of adaptability, reconfigurability and scalability in a matter of hours. This may be our most difficult research challenge. Here, one needs to integrate all the business models, planning and execution on a global scale with heterogeneous products families flowing through heterogeneous processes.

The planning and execution of this global supply network must first be tested in a digital test bed that will allow "optimized" performance to be approached that meet the business and financial desires. By creating a technological infrastructure that is based upon a "model-based enterprise" approach, we believe that all the concepts and "what ifs" can be tested prior to going "live!"

Managing the "web-inside" extended enterprise requires information technology commonality of syntax and semantics throughout the world. This notion of "copy consistently" across the extended enterprise will be based upon standards and advanced web technologies. We believe that the migration from an "SQL" environment to an "XML" environment with operational reference models developed by consortia must be monitored and tested for their applicability and business implications.

The distributed, collaborative, horizontal, outsourced supply network requires fundamental rethinking of all organizational implications. Many issues arise here – from the "win-win" desired performance of the complete supply network to the rewards for "thinking" at the individual contributor level. Although technological advancement of the supply network may have the glamour, organizational and cultural issues will still be major barriers to success.

5. STRUCTURING THE RESEARCH CLUSTERS

In order to develop a research agenda for RSSN, we took the prioritized business drivers from our survey and used them to develop a prioritization of the issues that form the semiconductor supply network vision. This prioritization is then used to determine a list of areas where further research is needed to achieve this vision.

The result of our clustering exercise yielded seven distinct research clusters, which are defined below with respect to the:

- The goal of the cluster (the "long term – what" we wish to accomplish!);
- Key drivers, attributes, imperatives, and metrics;
- Fundamental research agenda questions.

5.1 Cluster 1: Supply Network Integrated Demand

5.1.1 Objectives

The cluster goal is to understand and incorporate the dynamics of the "buy" side into all design, planning, execution and financial decisions. This will likely lead to strategies for demand management and visibility at all levels into the final customer demand.

5.1.2 Key Drivers, Attributes, Imperatives, and Metrics

D1, D2, D3, D4, D5, D6, D7, A1, A2, A3, A4, A5, I3, I4, I5, I6, I8, I9, I10, I11, I12, M1, M2, M6, M7

5.1.3 Fundamental Research Agenda Questions

The fundamental research agenda questions for this cluster are broken into two segments: (1) the notion of "proactive" demand creation, (2) the notion of robust responsiveness to demand fluctuations.

<u>"Proactive" Demand Creation Questions</u>:

- When should the company introduce the next generation product?;
- How can we make our customer successful in the customer's customer environment?;
- How should the Supply Network integrate / coordinate / synchronize with the marketing strategies?;
- How should the Supply Network integrate / coordinate with the design functions?.

<u>Robust Responsiveness to Demand Fluctuations</u>:

- How do we achieve demand stream visibility throughout the Supply Network?;
- When product supply is less than product demand, how should we manage the "allocation" process?;
- Can product substitution provide robustness?

5.2 Cluster 2: Supply Network Design Methodology and Coordination

5.2.1 Objectives

The cluster goal is to enable rapid design/redesign of the SSN to respond to changes in market conditions, business climate and technology. In the SCOR model, and other approaches, this is referred to as the "Design Chain".

5.2.2 Key Drivers, Attributes, Imperatives, and Metrics

D1, D2, A1, A2, A3, A4, A5, I1, I2, I3, I4, I5, I6, I8, I9, I11, I12, M2, M6

5.2.3 Fundamental Research Agenda Questions:

The fundamental research agenda questions are:

- How can semiconductor supply networks be designed to include aspects such as extensibility, flexibility, scalability (EFS), and mass customization;
- How should virtual supply networks be designed, built and run tò meet the requirement of dynamic reconfigurability given different business models?;
- How can we develop a "Semiconductor Supply Network Physics" that emphasizes the aspects of language, control, planning, architecture and economics?;
- How should we factor in the "constraining" resources into the design methodology? For example, if we cannot get new key processing equipment fast enough from the suppliers, how should this be factored into the overall capital and business strategy?;
- How can knowledge management facilitate building a smaller, smarter, faster, more economical factory?;
- What organizational forms and cultural change management techniques will facilitate organizational and supply chain modularity?;
- What methodologies should be used to determine the location of facilities?

5.3 Cluster 3: Supply Network Integrated, Enterprise-wide Planning

5.3.1 Objectives

The c luster g oal is to d evelop a m odel-based p lanning f ramework t hat converts high-level SSN objectives to actionable plans that are coordinated across and "fair" to all SSN partners. The cluster "begins" with a supply network designed and present, and now we wish to plan how to "win!" In some cases this may use existing models, such as the Economic Order Quantity as the basis for rule development.

5.3.2 Key Drivers, Attributes, Imperatives, and Metrics

D1, D2, D3, D4, D5, D6, D7, A1, A3, A4, A5, A6, I4, I5, I6, I8, I9, I10, I11, I12, M1, M2, M3, M4, M5, M6, M7

5.3.3 Fundamental Research Agenda Questions

The fundamental research agenda questions are:

- How can we synchronize the overall semiconductor supply network?;
- If there are multiple companies involved, the issues of collaboration, buyer/supplier success, information sharing, and overall value to each must be addressed;
- How can forecasting more effectively drive the enterprise-segment decisions of the supply network?;
- How do we address the linear capacity additions (versus the "chunks!")?;
- How do we respond more quickly to forecast changes by designing robustness in the design and system development processes?;
- How do we manage the multiple dimensions of enterprise-wide supply network planning effectively?;
- How many tools to buy? When?;
- How many raw materials are needed? When?;
- Where should we position inventory?;
- What are feasible performance targets?;
- When should we subcontract?;
- How do we manage the logistics?;
- What information should be shared? In what form? How often? (This is cross-listed with the technology infrastructure cluster);

- How can a pervasive modeling and simulation "model-based enterprise" structure add value to the planning process? What should this model-based enterprise structure be?;
- How do we select, create, and manage strategic alliances?;
- How do firms receive/provide appropriate value in the supply chain?;
- How can we effectively manage multi-firm strategic planning processes?;
- What incentive structures will yield the best outcomes for all parties?;
- How do we create the culture of collaboration while guarding intellectual assets?;
- How do we model and understand the life cycle aspects of planning? How can we take a "slice in time" view of planning by considering various products in various families in various states of maturity?

5.4 Cluster 4: Supply Network Integrated Scheduling and Execution

5.4.1 Objectives

The cluster goal is to develop a model-based execution framework that facilitates the realization of coordinated SSN plans in both a timely and cost-effective manner. The cluster "begins" with a supply network designed and present, the business plan defined and the production plan in place. We now wish to plan how to "make this happen!" Recent attempts have made to link ERP systems for this purpose.

5.4.2 Key Drivers, Attributes, Imperatives, and Metrics

D2, D3, D6, A1, A2, A3, A5, I1, I2, I3, I5, I6, I7, I8, I9, I11, I12, M1, M2, M3, M4, M5, M6, M7

5.4.3 Fundamental Research Agenda Questions

The fundamental research agenda questions are:

- How much visibility into the status of supply network partners is needed/desirable?;
- How can supply network goals be transformed effectively and efficiently into factory actions?;
- Vice versa, how can we get local actions to hit the global goals? e.g. from factory – to enterprise – to vertical market segment;

- How should factories make internal decisions that align or complement the supply network goals?;
- How does a factory "best" match factory output to orders?;
- How can a pervasive modeling and simulation "model-based enterprise" structure add value to the execution process? What should this model-based enterprise structure be?;
- How do we manage in a multi-project environment?

5.5 Cluster 5: Coordinated Business and Financial Models

5.5.1 Objectives

The cluster goal is to establish a structured business model that will allow the assessment of various business strategies and provide financial visibility of the decisions. This will allow the use of such financial measurement tools as the income statement and balance sheet both for the individual business and for the entire integrated supply network.

5.5.2 Key Drivers, Attributes, Imperatives, and Metrics

D1, D3, D4, D7, A1, A2, A3, I4, I6, I8, I12, M1, M2, M3, M4, M5, M6, M7

5.5.3 Fundamental Research Agenda Questions

The fundamental research agenda questions for this cluster are broken into two segments: the classic business and financial model questions dealing with topics such as cash flow and the "manipulating the game" questions dealing with means by which the game can be changed and measuring the corporate value of same.

Classic Business and Financial Model Questions:

- What should the capital acquisition strategy be?;
- If we can't get new key processing equipment fast enough, how should we factor this into our overall capital strategy and our business strategy?;
- How do we ensure that all partners in the supply network are fairly rewarded?

Manipulating the Business and Financial Model Game Questions:

- What is the value of manufacturing / supply network cycle time reduction?;
- What markets should the company be in? How do we support the development of this strategy? How do we implement?;
- How do we select, create, and manage strategic alliances?;
- How should we strategically make/buy?;
- What are the financial models that reflect how best to design / how best to measure the supply network designs and strategies? And vice versa?;
- How do we develop supply network strategies to support these decisions?

5.6 Cluster 6: Supply Network Technical Infrastructure

5.6.1 Objectives

The cluster goal is to create an enterprise-wide information infrastructure with "web-inside" technology that will house the model-based enterprise framework and will make available those data elements required to make timely SSN decisions. This cluster also addresses the role of the university in establishing the research foundation for the supply network infrastructure. This is also the link to IT in the supply network.

5.6.2 Key Drivers, Attributes, Imperatives, and Metrics

D1, D2, D3, A3, A5, I1, I2, I3, I4, I5, I6, I7, I8, I9, I11, I12, M1, M2, M3, M4, M5, M6, M7

5.6.3 Fundamental Research Agenda Questions

The fundamental research agenda questions for this cluster are also broken into two segments: (1) the classic information technology questions dealing with topics such as instantiating Rosetta Net and SCOR methodologies enterprise-wide, and (2) the "manipulating the game" questions dealing with means by which the game can be changed and measuring the corporate value of same.

Classic Information Technology Research Questions:

- What needs to be done to create a consistent enterprise-wide information infrastructure?;
- The creation of a universal inter-enterprise "vocabulary" (or set of formal definitions of the processes and product information) to be shared within

the supply network (requires the adaptation, adoption and complementary design of extensions to current semiconductor initiatives);

- The creation of maps of information flows within the supply network (addressing "what" information is traded? "Where" is it going? "When" is it available and used? "How" should it be presented?);
- What needs to be done to create a technological "web-inside" philosophy and capability? (Cross-listed in organizational structure cluster);
- What needs to be done to create a "Scalable Systems Environment?" (Cross-listed in design, planning and execution clusters);
- What needs to be done to design enterprise-wide, consistency of business processes (and the integration between the business processes)?

Manipulating the Business and Financial Model Game Questions:

- What information should be shared? In what form? How often? (Cross-listed with the planning cluster);
- What needs to be done to efficiently model current data models to understand the specifics with respect to data source, data extract and data transformation?;
- How can IT "change the game?";
- How can we measure the business benefits of this electronic business and enterprise integration "stuff?";
- How can we extract from the Business Strategy the strategic differentiation role of information technology?;
- What needs to be done to build a business-process-based architecture that is based upon the fundamental technological premise of "web-inside"?;
- How can the technical infrastructure support a pervasive modeling and simulation "model-based enterprise" structure? What should this model-based enterprise structure be?

5.7 Cluster 7: Organizational Structure

5.7.1 Objectives

The cluster goal is to create the organizational structure and culture required supporting a profitable firm performance in the semiconductor supply network. As the integrated enterprise view is implemented, it is likely that conventional management structure and strategies need to be modified.

5.7.2 Key Drivers, Attributes, Imperatives, and Metrics

D1, D3, D4, D7, A3, A5, A6, A7, I10, I12, M1, M2, M3, M4, M5, M7

5.7.3 Fundamental Research Agenda Questions

The fundamental research agenda questions are:

- What should the Horizontal/Outsourcing Integration process be?;
- How should one negotiate with supply network constituents?;
- Can "Win – Win" strategies be found?;
- Do "Win-Win" strategies depend on "Trust-based" relation-ships?;
- How do we engage in "Co-opetition" –competition and cooperation- with our partners/competitors?;
- How do we effectively manage the transitions that surely occur?;
- How do we best shift from a Process Focus to Product Focus?;
- How do we achieve the cultural integration of Mergers and Acquisitions?;
- How do we shift from "need to know" attitudes to a culture of knowledge sharing and rapid innovation?

Classic Organizational Infrastructure questions are:

- How should we manage the "overlapping of extranets?" and the "overlapping/integration of vertical markets?";
- What should be the Knowledge Management/Structure/ Directories/Reuse organizational facilitation structure? (Cross-listed in technology infrastructure cluster)? ;
- What cultural features support knowledge reuse?;
- What performance management systems facilitate knowledge reuse?;
- What knowledge directories facilitate knowledge reuse?;
- How do we establish "enterprise-impacting" reward structures?;
- How do we establish "enterprise-impacting" organizational forms with modular structures? ;
- How do we create internally consistent processes, structures, and cultures that support a strategy of rapid innovation?;
- What are the organizational and operational implications of the "web-inside" philosophy? (Cross-listed in technology infrastructure cluster);
- What are the organizational implications of complexity theory in the evolving semiconductor supply network?;
- Should SSN's be centrally or locally managed?;
- What are the limits of modularity?;
- How do we stimulate innovation? ;

- How do we create and manage software supply chains?;
- What organizational forms and cultures will simultaneously support technological innovation and a strong customer focus? ;
- How do we best involve suppliers in innovation and product design?;
- What are the cultural implications of collaborative technologies? ;
- How do we create, structure, and manage diverse communities of technical expertise throughout the semiconductor supply chain?

6. CONCLUSIONS

The conclusions from the creation of this research framework for the semiconductor supply network is that a rigorous view of a particular industry sector has led to a robust description of "what" needs to be addressed, and developed by the research community today and tomorrow. The value proposition, prioritization and the "how" for each of these issues must come next – but the overall, holistic requirements are captured. We believe this will serve as a simple, yet effective, launch pad for the orchestration of semiconductor supply network research.

REFERENCES

1. Bechtel, C., Jayaram, J. Supply chain management: A strategic perspective. International Journal of Logistics Management 1997; 8(1): 15-34.
2. Berry, D., Naim, M. M., Towill, D.R. Business process re-engineering an electronics supply chain. IEE Proceedings, Science, Measurement, and Technology 1995; 142(5): 395-403
3. Cooper, M. C., Ellram, L. M. Characteristics of supply chain management and the implications for purchasing and logistics strategy. International Journal of Logistics Management 1993; 1(2):13-24.
4. CORDIS: IST European Technology Forecast, published 1999.
5. Ellram, L. M. Supply chain management: The industrial organisation perspective. International Journal of Physical Distribution and Logistics Management 1991; 21(1):13-22.
6. Fine, C., *Clockspeed*, Sloan Management Press, 1997.
7. Forrester, J. W., *Industrial Dynamics*. Cambridge, Mass: The M.I.T. Press, 1961.
8. Fox, M. S., Chionglo, J.F., Barbuceanu, M., The Integrated Supply Chain Management System, Internal Report, Dept. of Industrial Engineering, University of Toronto., December 1993.
9. Frederix, F. Planning and scheduling multi-site semiconductor production chains: A survey of needs, current practices and integration issues. Conference on IT and Manufacturing Partnerships: Delivering the Promise, Galway, Ireland, 107-116, 1996

10. Gupta, S. Supply chain management in complex manufacturing. IIE Solutions 1997; 29: 18-23.
11. Hines, P. Integrated materials management: The value chain redefined. International Journal of Logistics Management 1993; 4(1): 13-22.
12. Hewitt, F. Supply chain redesign. International Journal of Logistics Management. 1994; 5(2): 1-9.
13. International Technology Roadmap for Semiconductors (ITRS), published 1999.
14. Jones, T. C., Riley, D. W. Using inventory for competitive advantage through supply chain management. International Journal of Physical Distribution and Materials Management 1985; 15(5): 16-26.
15. La Bouff, J. Semiconductor manufacturing, A supply chain management perspective, presented at the Conference on Global Supply Chain Management, January 23, 1997.
16. Lee, H., Billington, C. The evolution of supply-chain management models and practice at Hewlett-Packard. INTERFACES 1995; 25: 5: 42-63.
17. Lee, H., Padmanabhan, V., Whang, S. Information distrortion in the supply chain: The bullwhip effect. Management Science 1997; 43: 546-558.
18. MASM, Proceedings of the International Conference on Modeling and Analysis of Semiconductor Manufacturing (MASM 2000).
19. Monczka, R. M., Morgan, J. What will happen and what you should know, Purchasing Magazine. 1998; January 15: 78-85.
20. National Technology Roadmap for Semiconductors (NTRS), published 1997.
21. Ovacik, I. M., Weng, W. A framework for supply chain management in semiconductor manufacturing industry. IEEE/CPMT International Electronics Manufacturing Technology Symposium, pp. 47 – 50, 1995.
22. Slywotzky, A. The profit zone: Managing the value chain to create sustained profit growth. Strategy and Leadership 1998; 26(3): 12-16.
23. Scheller, B. Supply chain management extends to plant floor. INTECH 1994; 41 February: 38-41.
24. Stevens, J. Integrating the supply chain. International Journal of Physical Distribution and Materials Management 1989; 19: 8: 3-8.
25. Supply Chain Council, The Supply Chain Operations Reference (SCOR) Model, 1999.
26. Temple, S. Sand to circuit: Extending the supply chain, presented at i2 Planet'98 conference, September 1998.
27. The Next Generation Manufacturing Report (NGM), Agility Forum, 1997.
28. The Intelligent Manufacturing Systems (IMS), World View, published 2000.
29. Uzsoy, R., Lee, C.Y., Martin-Vega, L. A review of production planning and scheduling models in the semiconductor industry Part I: System characteristics, performance evaluation and production planning. IIE Transactions 1992; 24: 4: 47-60.
30. Uzsoy, R., Lee, C.Y., Martin-Vega, L. A review of production planning and scheduling models in the semiconductor industry part II: Shop floor control. IIE Transactions 1994; 26: 5 September: 44-55.
31. UC-Berckeley, Benchmarking Semiconductor Manufacturing, Retrieved September 2003 from: http://esrc.berkeley.edu/csm/Csm35long497/

Chapter 7

DESIGNTEXNET

A Web-Based Service for Product Development in Networks of Textile and Apparel Enterprises

Guido Grau, Marcus Winkler
Management Research Department, ITV Denkendorf, Germany

Abstract: In textile and garment industries the process of product development is one of the most crucial and competitive elements. The sharing of ideas and information in networks becomes more and more important, in order to reduce the number prototypes, the cost of logistics and the time to market.

Key words: Network of textile, Garment companies, Product development, Information sharing, Web-based ad hoc workflow

1. INTRODUCTION

The sectors of textile and apparel have to cope with consequences of globalization ("pricing competition"), ascending market power of suppliers and customers (concentration) and changing needs of customers. Especially in Germany, but as well in Europe and abroad, the following points have to be faced (Boudon and Robinet 2001) [1]:

- Increasing competition regarding costs and prices;
- transfer of production to low-wage countries;
- small range of manufacture and little added value as well as;
- high income elasticity of demand and short product life cycles.

Nevertheless there is a trend of increasing productivity (production per employee) in Germany, which is higher than the shrinking of production since mid 1990 (EPPA 2002) [2]. Within the European context the productivity of Germany is more or less average (7th place), according to a study by

Euratex from 1999 (Euratex 1999)[3]. At the same time there are opportunities for companies by applying new technologies, developing new products for different use (e.g. technical textiles) and adjustment and re-design of existing forms of coordination.

The prerequisites for increasing productivity are mainly capital investment in modern machinery, computerization as well as information and communication technology (ICT). This applies especially to small and medium enterprises (SMEs) of textile and apparel industry. The industry is highly fragmented and heterogeneous and continuing industrial change within the sectors has had little impact on the proportion of small companies. In addition companies present a traditional corporate image and connections within the industry are mainly based on traditional means.

The textile and apparel sectors are based on four main stages of industrial activity (fiber manufacturers are not included), spinning mills, weaving or knitting mills, finishing companies and apparel companies. These are often located in regions of so-called clusters (IKB 2002)[4], in which all activities have to be coordinated. A general strategy to strengthen their position is to produce to the highest quality standards, increase their flexibility and to find profitable market niches.

In the following case study for the companies one of the most crucial co-operative activities, product development, is presented. The case depends in particular on the creativity of designers and the speed of exchange for required information, so as to shorten the time to market. The study is an outcome of the research project CREATIV 5 (Winkler and Grau 2001)[5] and addresses product development processes, in which intermediate product(s) can be represented through digital images.

2. THE TRADITIONAL PROCESS OF PRODUCT DEVELOPMENT

The process of product development is often initiated by a customer, who desires a specific design that should either be developed in a cooperative way, or should be customized from an existing design. Product development follows a process network that includes a printing company, a designer, an engraver and an apparel company. By and large the roles are fixed between the partners that have to exchange data and information several times about the (intermediate) product(s) to be developed or customized.

In general the process of product development for a printed fabric follows a specific route (Figure 7-1), but does not show explicitly how often information has to be exchanged between the different partners. Typically new ideas or a design stem from the customer, an apparel company, but sometimes they can also be created by a new collection originating from the printing company. A design, based on a new idea, is accomplished through the aid of CAD-technology, then repeated (including definition of size) and separated. Separation here means the splitting of a multi-color pattern into different schemes, one for every print color. The printing color variations create several digital colourits of the same design that after simulation transfer the colourits into physical fabrics, e.g. by inkjet printing. The first sample is checked by the customer and if it is not according to their wishes the process will be repeated for perhaps several more times.

After design acceptance the engraver prepares the shapes needed for the printing process and a second sample (printed on traditional machinery) is checked by the customer.

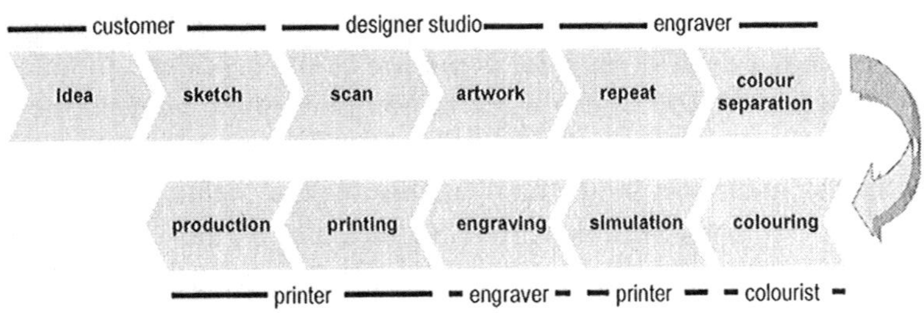

Figure 7-1. The Process of Product Development of Printing Fabrics (Winkler/Grau, p. 666.)

3. FLEXIBLE WORKFLOW FOR PRODUCT DEVELOPMENT

Usually the process of product development requires access to archives, databases of recent design models and retrieval of models according to a structured system of queries. This facilitates a search for similar or already existing samples, designs or sketches that most of the textile and apparel companies do not have in a digital form.

Further requirements to develop DesignTexNet[6] were:

- Shared workspace for network partners and information tracing;
- A simple web-based solution (no client installation is needed), comprehensive for "creative" people and easy to host and maintain;
- Highly flexible and extendible application that allows an ad hoc workflow.

ITV Denkendorf, taking into account these requirements, developed within the framework of the European project CREATIV the platform system DesignTexNet. It is an easy to host and to maintain web-based client server application that offers a centralized shared workspace, support to realize an ad hoc workflow and an image database.

In principle the cooperation between the partners in a product development network follows the scheme as set out here.

Someone, in most cases a customer, starts a new development project by creating a so-called project that describes the aim of the development and also specifies some constraints. Additionally the customer selects actors to be partners in the project. Each user of the platform has a specific role to play (customer, designer, printer or engraver). And depending on the situation new partners may be added later.

Next the initiator of the project adds a new detail; this is similar to one working step. Each detail keeps the depiction of the task done by one of the partners and a description of the work to be done by the next partner in the process sequence. In addition to the depiction of a task an object can be created on the platform and then linked to the detail that keeps the description of an intermediate product that the partner has developed. Intermediate products are:

- *SKETCH* is a workout of an idea, e.g. the scanned version of a paper-based draft drawing. Sketches are made by the customer to illustrate his wishes, or they have been bought at fashion fairs, or designer studios;
- *DESIGN* is the realized pattern, made by a designer with a CAD application that is ready to become colour separated and therefore is the basis for engraving shapes;
- *COLOURIT* is a set of colours pertaining to a design. Different colourits can be simulated on the basis of their CAD data.

Each of these objects has its own specific attributes and can keep, in various qualities, an image of an intermediate product. Therefore the initiator of the project can specify ideas in a more structured and achievable way. The remaining task is to select one of the project partners, for a specific task and to ask DesignTexNet to inform that partner. The system will then add the task to the to-do list of the following partner and notify by a covering e-mail. At this stage an iterative process starts, which will continue until the whole development process is completed.

Each partner follows more or less the same procedure:

- Firstly receive and digest information on the task to do;
- Secondly retrieve data from the platform needed to fulfil the task;
- Thirdly accomplish the task; here it is often necessary to use additional software tools (e.g. CAD application for designers);

- The fourth step is to add a new detail to the project on the platform, describe the work done so far, the work to do and to select the next partner.

In this way DesignTexNet builds up an archive of development projects that keeps the descriptions of all steps completed and detailed information of all intermediate products. This archive can be used by partners in the network to search for intermediate products from different projects. These might be used to shorten current development.

The summing-up of the process of product development, including the different partners, is shown in figure 7-2. The process ends exactly before the engraver starts to engrave the shapes for production process.

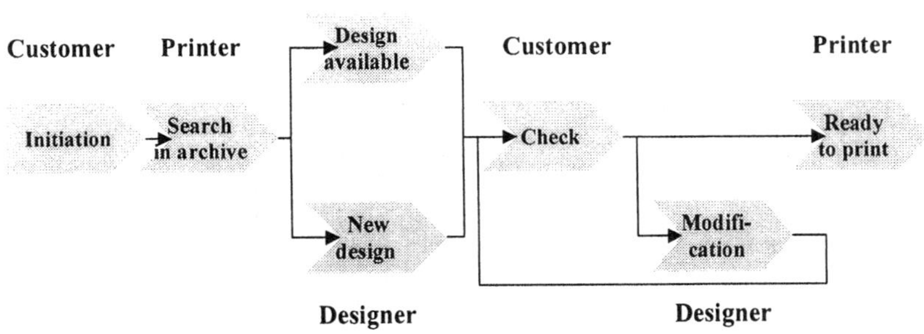

Figure 7-2. Sequence of the Flexible Ad hoc Workflow

3.1 Developing a New Design: an Example

In general there are three phases of product development that can be identified[7]. These phases correspond with the three different intermediate products mentioned before:

- Phase 1 starts with the initiation of the product development process (development of sketches, or identification of similar products that already exists in archives);
- Phase 2 begins with an existing idea of the product (one or several sketches or reference products) and ends before the production of the shapes by the engraver;
- Phase 3 deals with the creation of the (mostly country specific) colourits.

The following example includes the phases 1 and 2 and illustrates an overview a customer, an apparel company, gets when accessing the software. All the projects the company is involved in are displayed (Figure 7-3).

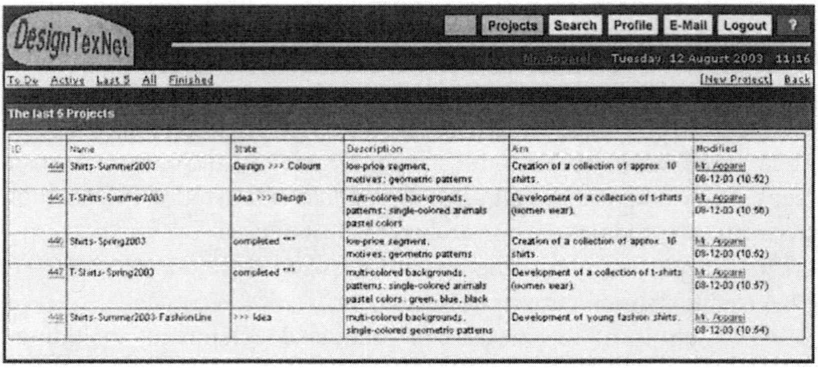

Figure 7-3. Screenshot – Overview for Apparel Company

In order to initiate a new product development a customer has to create a new project and specify some new information, such as the name of the project, its current progress status or phase (">>> idea", "idea >>> design", "design >>> colourit", "shapes made", "in production", "done"), the objectives, or the aim and the description of the product to develop (Figure 7-4).

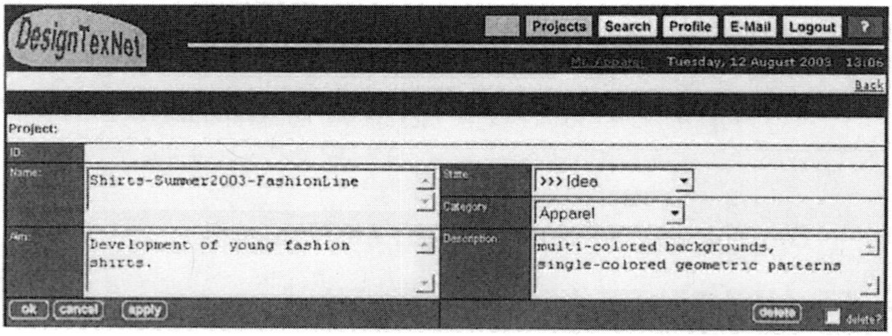

Figure 7-4. Screenshot – Creation of a New Development Project through Apparel Company

The second step in phase 1 is to identify at least one partner for cooperation in the network. To achieve this the customer selects partners from a list that contains all registered DesignTexNet users with name, company name and the role the user can fulfill in a design network.

In a further task the customer must create the first working step (called: detail) of the new project. In the input form for the detail the customer to specifies the work to do in the next step and selects and asks one of the previously chosen project partners (in this example: a printing company) to proceed.

At this stage the initiation process is complete.

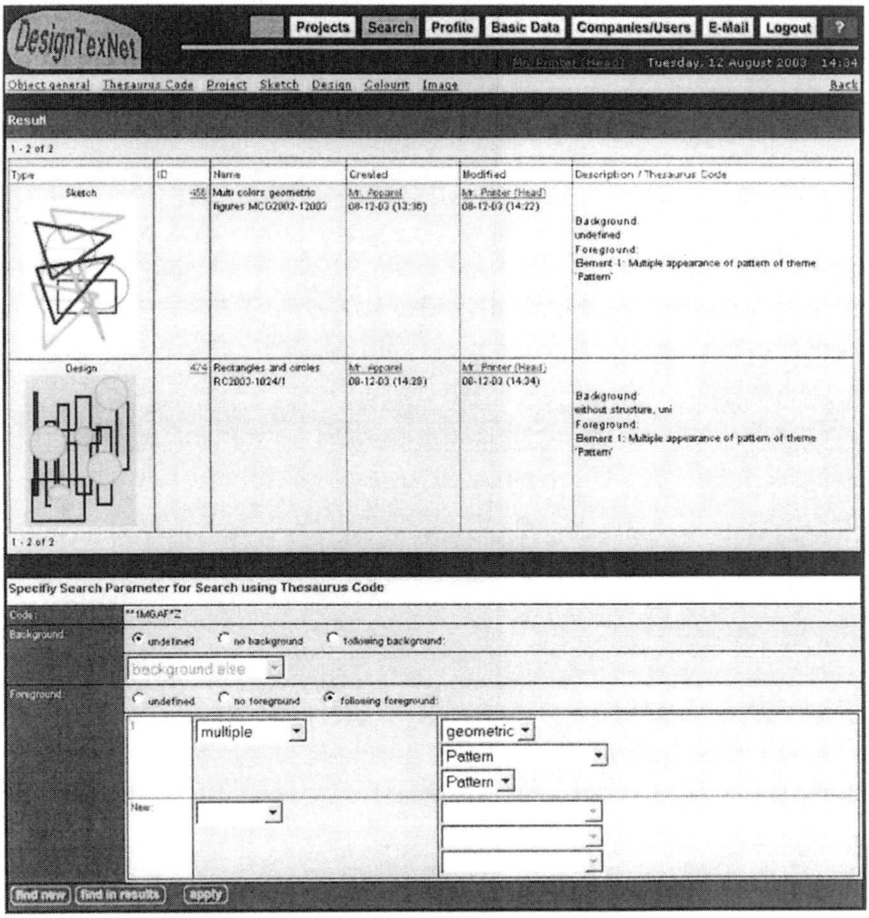

Figure 7-5. Screenshot – Result of Query from Printing Company (No Satisfactory Results)

DesignTexNet sends an e-mail automatically to the printing company that informs the responsible person that there is some work to do on the cooperation platform. The e-mail also contains a link to the platform that leads the person directly (only interrupted by the login process) to a page, where informed is located about the task to be processed. Additionally access to the project description made by the customer made during project initiation is granted.

Next there are two possibilities, either the design needed is already available in the archive, or it is not and has to be developed by a designer. Figure 7-5 illustrates a search for this design. In the upper part of the page the search results are displayed and in the lower part the form for the query definition is shown, so that there is the possibility to modify the query if the results are not satisfactory.

In the example described here the printing company gets unsatisfactory results, because all designs found in the archive do not 100 per cent match the requirements of the customer. Nevertheless there is one design that could be used as a basis for a new development.

The printing company decides to add a designer as a new partner in the project. Then it includes a new working step to the project containing the description of the work done to date (the search) and refers to the design found that could be used as an inspiration for the new development stage. Finally the new task to do (Figure 7-6) is described and the designer is informed.

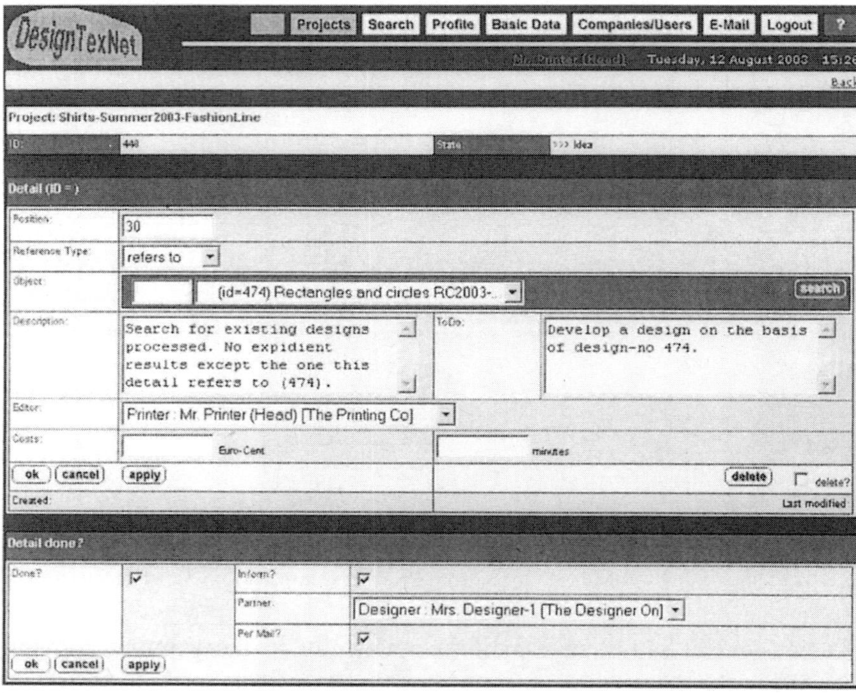

Figure 7-6. Screenshot – Printing Company Describes Work to do (For Designer)

The designer accesses the platform the same way as the apparel company and the printing company by entering name and password. On entry the instructions are found and in this example it is to develop a design based on an existing one, according to the given specifications by the printing company. Usually the designer then uses a CAD-application to develop the design.

The result of the work is exported, e.g. as JPEG-file, by uploading to the platform. It describes the work done and asks the apparel company to check the result (Figure 7-7). Furthermore there are fields that can be filled if required, like costs and the time needed to complete the task.

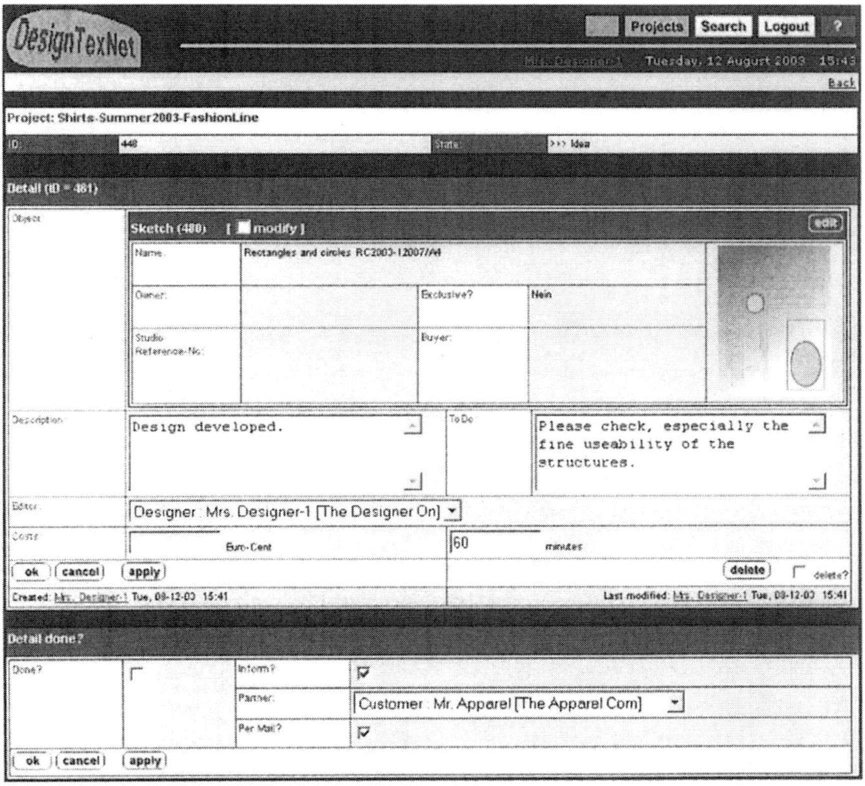

Figure 7-7. Screenshot – Designer Fills in the Workload Done Addressed for Apparel Company

The customer then checks the design and information provided whether the work has been accomplished satisfactorily or not. In this example the apparel company wants some different colourits, purple on blue and yellow on green (Figure 7-8). Then the designer is then informed about the new job requested by sending an e-mail.

The designer then develops the colourits and uploads the result(s) as before. Next the apparel company checks the results and may require further modifications, or is content with the result. If the apparel company is pleased with the result, it informs the printing company via e-mail that the development process has been successfully completed.

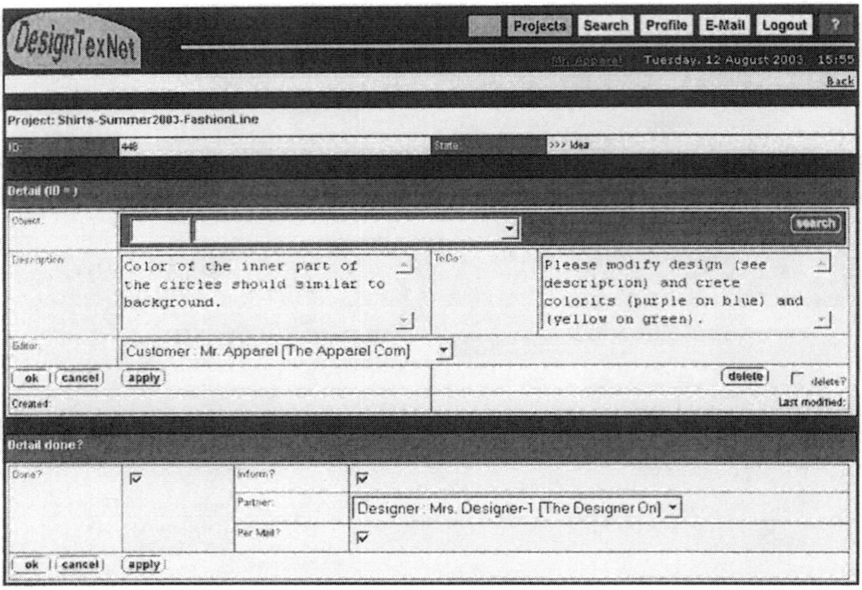

Figure 7-8. Screenshot – Apparel Company Makes Remarks for Designer

At this stage the printing company then selects an engraver for the next task and requests the shapes to be made for production of prototypes that the customer needs for the final check.

After completing all the tasks described above, the process supported by DesignTexNet is finished. All tasks done are filed in the archive, so that a total project overview can be produced either on screen (Figure 7-9) or in a printable format which contains in addition to the one on the screen information:

- the project description;
- the list of involved partners;
- the overview about the costs and the human resources consumed by the project; and
- some statistical data, e.g. timestamps of all modifications.

The costs and information about human resources consumed can be kept hidden from non-authorized people, such as sub-contractors, e.g. the designer or the engraver, and as well from the eyes of the customer. The owner of the DesignTexNet system, in this example the printing company, has the possibility to check the cost-effectiveness of the whole development project, which can be a crucial factor.

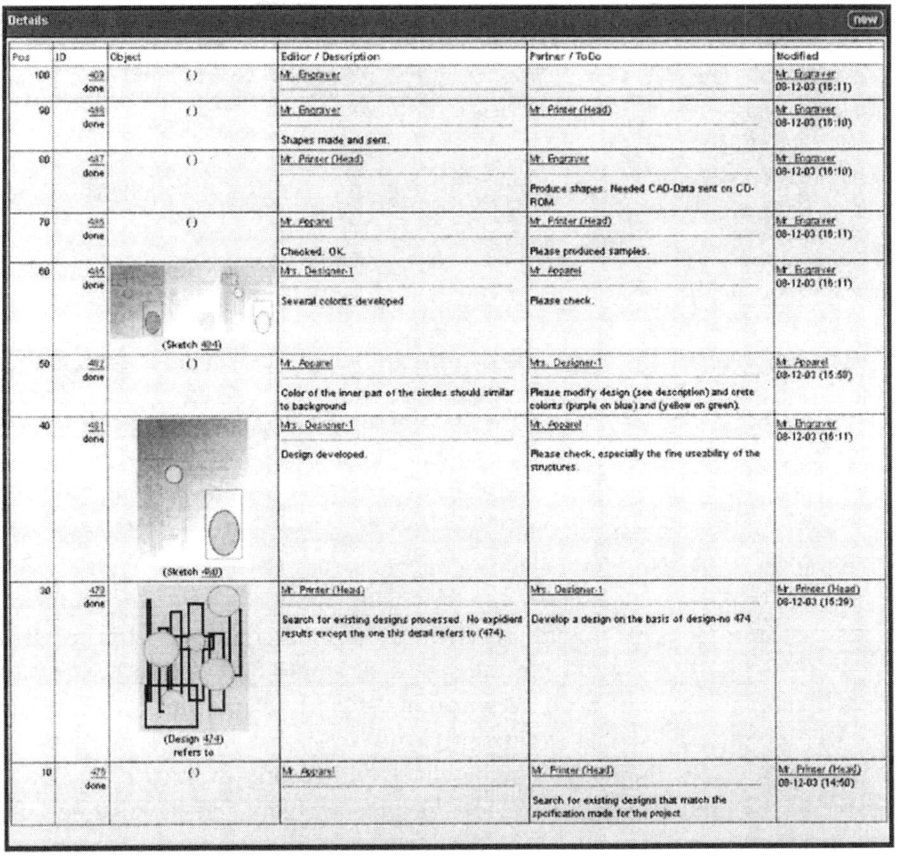

Figure 7-9. Screenshot – Status of the Project after 10 Working Steps.

3.2 Requirements of the Implementation

The platform DesignTexNet has been proved to be a user-friendly web-base system that can be customised to a partner's requirements and is useable with a variety of colour separation and colouration systems. Table 7-1 illustrates the impact when using DesignTexNet in a design-network of textile and apparel companies. The most important result was the increased speed of product development as well as the faster retrieval of data and information by using a central database. In addition a reduction of cost for coordinating the network and the cost of logistics (exchanging samples and information) is reduced. Moreover, an increase 3% market share, by better fulfilment of customer need, is estimated.

Table 7-1. Advantages using DesignTexNet (according to Boudon/Robinet, p. 17 et seq.)

Affected Areas	Improvement (in %)
Speeding up the process of product development	80
Acceleration of retrieval of needed data and information	90
Cost reduction	
in coordinating the network	60
in product development	20
Increase of market share	3 (estimated)

In the wake of the implementation of DesignTexNet new problematic issues arose that deal with trust and security. Trust is definitely needed between the partners of a network in order to guarantee a successful cooperation. If not there must be sanctions applied against those, which have e.g. sold designs to other companies. It is mandatory to agree on property and exploitation rights in the network dealing with the development (making) and the exploitation of a design, sketch or colourit. For example there can be an agreement on simple rights (one-off usage), or exclusive rights for designs, or exploitation rights, which are limited to territories, a specific period of time etc. Consequently a customer only contracts and pays for a defined package of exploitation rights.

It is recommended that the main partner in such a network is keeps the rights and defines the r oles, competences and r esponsibilities of the o ther partners. All inputs must be free of third-party rights and have to be held, managed and maintained after the input by the leading company. A copyright protection for the developed sketch, design or colouration may be obtained through deposit of electronic server data with local and international authority.

4. CONCLUSIONS

The implementation of e-Business services like the platform DesignTexNet strongly supports the competitiveness of textile and apparel companies. All dimensions regarding time, flexibility, costs and quality can be improved. Furthermore business partnerships are strengthened, processes in the context of product development re-designed and improvements to customer satisfaction made easier and much faster to achieve.

Finally companies have the chance to work not only in regional clusters but also in global clusters. The capability to work in value networks is one of the main competences that especially SMEs have acquired in order to stay competitive.

NOTES

1. cf. IKB, p.2 et seq.
2. cf. EPPA, p. 4.
3. cf. Euratex.
4. cf. Porter, p. 77.
5. CREATIV - Commercial Retrieval for Fabrics and Design Patterns in Virtual Structures, founded by the European Commission, IST-1999-20534.
6. Because of protection of trademarks the original name, "DesignNet" had to be replaced by "DesignTexNet".
7. cf. Winkler/Grau, p. 668.

REFERENCES

1. Boudon, E., and Robinet, P.,, *Yardsticks Report – Deliverable D2.2, confidential report of project CREATIV*, Paris: IFTH, 2001
2. EPPA, *Die Auswirkungen der ATC-Liberalisierung auf die deutsche Textilwirtschaft – Die Auswirkungen der Liberalisierung im Rahmen des ATC (Agreement on Textiles and Clothing) auf die deutsche Textilwirtschaft und die von ihr beschäftigten Arbeitnehmer/innen unter besonderer Berücksichtigung des Beitritts Chinas zur WTO,* retrieved December 2002 from: http://shop.ceps.be/BookDetail.php?item_id=43
3. Euratex, *Memorandum on preferential rules of origin*, 9th of November, Brussels 1999
4. IKB Branchenbericht, *Technische Textilien – Impulsgeber für die Textilindustrie*, retrieved November 2002 from: http://www.ikb.de/frames/Services/bp_analysen.html
5. Winkler, M., Grau, G., "CREATIV: Co-operative Product Development in a Virtual Network". In: *E-work and E-commerce: Novel Solutions and Practices for a Global Networked Economy, Vol. 2,* Stanford-Smith, Brian and Chiozza, Enrica, ed. Amsterdam, Berlin, Oxford, Tokyo and Washington: IOS Press, 2001

Chapter 8

REVERSE MARKETING, CONSUMER VALUE NETWORKS AND THE NEW BRAND INTERMEDIARIES

Christopher Lawer, Simon Knox
Cranfield School of Management, Cranfield University, England

Abstract: This conceptual paper presents an approach to consumer brand management that challenges current practice in the light of sustained criticism of marketing. We propose a reverse market concept that defines markets and marketing activities from an explicitly consumer value perspective. We describe how reverse market brands are able to create new forms of mutual value by organizing consumer demand and by supporting their lifestyles, need sets and complete consumption experiences. A simple framework is proposed through which brand management can review and respond to the unmet, complex and increasingly non-material drivers of consumer value. It outlines the benefits and the barriers that await brand management wishing to develop reverse marketing practices and concludes by evaluating the validity of the reverse market logic.

Key words: Reverse marketing, Consumer value networks, Brand transformation

1. INTRODUCTION

In many markets, there is a growing friction between marketing strategy and practice and the expectations and experiences of the newly empowered consumer. As power shifts to individuals, enabling them to seek out and extract value from sellers rather than the other way around (Pitt et al. 2002,

Hagel and Singer 1999, Wind and Rangaswamy 2001, Prahalad and Ramaswamy 2000), some writers have called for a radical overhaul of marketing thought and practice (Achrol and Kotler, 1999, Sawhney and Kotler 2001, Mitchell 2001, McKenna 2002). The same writers also identify the potential for a new marketing logic: one capable of building superior brand equity through mutual loyalty from networked customer and partner relationships. This logic has been termed the reverse market concept and the new brand asset, the consumer value network.

Based on case studies, interviews and our own experiences engaging with reverse market pioneers, we explore the logic, principles and form of the concept in this paper. We define and contrast it with transactional and relational marketing perspectives in the context of growing criticism of current marketing axioms. We also propose a framework of emerging consumer value drivers to identify sources of reverse market brand and customer value. Finally, we discuss the benefits and the barriers that await brand management wishing to develop reverse marketing practices.

2. MARKETING IN TRANSITION

Over the last decade, there has been widespread acknowledgment of how relationship marketing can create superior customer value (Christopher et al. 1991, Gummesson 1987, Payne and Holt 2001). Despite these advances, marketing managers remain under a lot of pressure to improve their performance further. For example, Day and Montgomery (1999) argue that there are still many pointed questions about the productivity of marketing expenditures, the appropriate position and influence of the marketing department and the contribution of marketing activities to overall financial performance.

McKenna (2002) points to the paradoxical impact of new technology upon the influence of the traditional marketing function. He argues that because of rapid technological change, marketing is now beginning to lose control over its very reason for existence. The irony is, however, that as technological advances drive marketing into obscurity, they also offer it the only hope for regaining a prominent, central place in today's organization.

Mitchell (2001) broadens the criticism of marketing in an attack at the systemic level. He suggests that marketers are suffering from an introspective seller-centric perspective and that marketing is in danger of retrenching into a hit-and-miss, poorly-informed tactical activity. The more marketing is practiced in its current form, he argues, the poorer the return and the less chance of the whole system reinventing itself. In other words, there is a negative system effect, where the sum of many individually rational decisions creates collective irrationality and diminishing returns.

Some writers predict problems arising from the ascendancy of the relationship marketing perspective in its own right. Fisk (2001) argues that if each company in every sector continues to pursue relationships with their customers – in particular high net-worth individuals - then the approach is doomed. Fournier, Dobscha and Mick (1998) take a similar view. Their research shows that each new successive consumer relationship-building initiative and invitation appears increasingly trivial and useless, rather than unique and valuable to today's consumer.

Other authors suggest that relationship marketing is having a growing negative impact on consumer perception and experience of brands, even though it claims to resolve many of the limitations of transactional, mass marketing. Zuboff and Maxmin (2002) argue that despite the valid discovery of the individual consumer in relational marketing thinking, in practice many brand managers use the approach to pursue transaction cost-efficiencies rather than fundamentally reinvent the means to create and deliver customer value.

Firat, Dholakia and Venkatesh (1995) critically describe the failure of marketing to adapt to the post-modern age of fickle consumers, unstable markets and the attendant loss of predictive capability, particularly as the traditional variables used in consumer behavior models are now less helpful in guiding marketing decision-making (Webber 1999). They suggest that there is a need for practitioners to prepare for a reversal of many long-held marketing axioms and to find new ways to close the growing gap between marketing strategy and practice on the one hand, and consumer value on the other. Before exploring this gap in more depth, we first examine current marketing practice and its influence on consumer decision-making and outcomes.

3. THE TWO FACES OF MARKETING

Recent empirical evidence provided by Coviello et al. (2002) shows that contemporary marketing practice within individual firms is characterized by a hybrid approach, partly transactional and partly relationship building. McKenna (2002) concurs and suggests that marketing today is a descendant of past practice, which can no longer be viewed from a single perspective, just as it no longer operates through only one medium. Although Coviello et al. were able to conceptualize the pluralistic nature of modern marketing practice, they acknowledge that their research does not explain why marketers choose to implement a hybrid or even a predominantly transactional approach. We regard this pluralistic pattern of marketing practice as illuminating in its own right for three reasons. First, it suggests that many marketers are engaged in competitive leapfrogging (Hunt 2001), where the focus is on the incremental, trial and error improvement of marketing mix practice rather than the achievement of a fundamental breakthrough in the role and value of marketing itself. Second, it indicates that marketers use the relationship marketing language, techniques and managerial practices (interaction, database and network marketing as defined in Coviello's framework 2002) to pursue tactical, promotion-intensive, selling goals. Third, the findings confirm that marketing management has an identity problem, which underlines why many writers continue to question its content, emphasis, boundaries and even its very essence (viz. Day and Montgomery 1999, Achrol and Kotler 1999, Mitchell 2001, Zuboff and Maxmin 2002).

We label the transactional and relationship marketing perspectives the two faces of marketing. Their objectives, orientation, tools and techniques are summarized in figure 8-1.

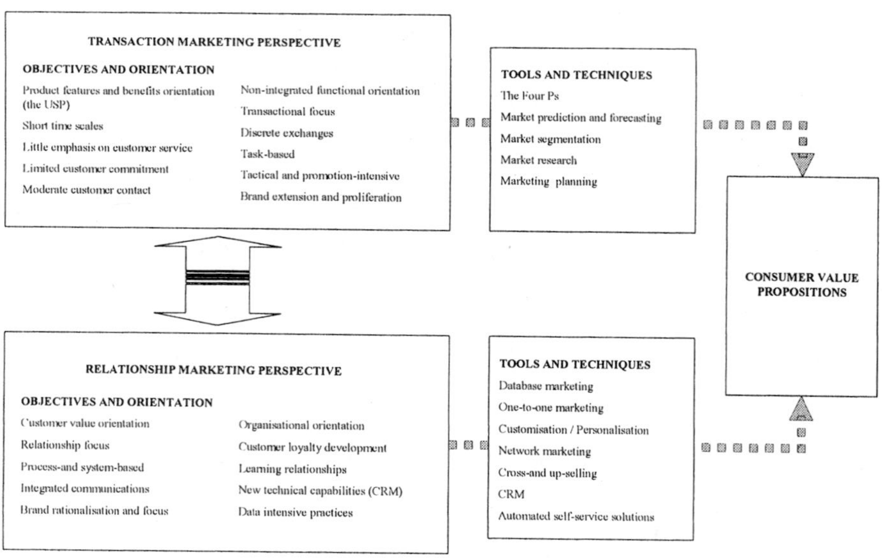

Source: Adapted from Christopher et al., 1991; Webster, 1992; Knox and Maklan, 1998; Coviello et al., 2002; Mitchell and Papavassiliou, 1999; Fournier, Dobscha and Mick, 1999

Figure 8-1. The Two Faces of Marketing

As the two faces of marketing are subject to growing uncertainty and criticism, evidence suggests there is also an increase in consumer uncertainty and confusion.

4. THE RISE OF CONSUMER CONFUSION

In a review of previous studies, Mitchell and Papavassiliou (1999) identify consumer confusion as an outcome of marketing, as individuals are provided with increasing amounts of information in their decision-making environments. With a growing number of product categories and high levels of portfolio purchasing in many consumer markets, consumer overload can result in stress, frustration and sub-optimal decisions. Mitchell and Papavassiliou isolate the three most important marketing-generated influences on consumer confusion. These are (1) a surplus of brand and store choice, (2) product similarity and (3) ambiguous, misleading or inadequate information conveyed through marketing communications.

Fournier, Dobscha and Mick (1999) have also researched the connection between the confused consumer and relationship marketing. They were concerned that in many instances, the very techniques used to create relationships are often the ones that are destroying those relationships. They describe the negative influence on consumer outcomes due to the proliferation in computer-generated personal communications, the one-way exchange of personal data and the problems of gaining access to human operators via call-center routing systems.

5. CHANGING DRIVERS OF CONSUMER VALUE

We argue that to prevent further diminishing returns for consumers and marketers alike, firms must begin to ask some fundamental and searching questions about how brands need to be renewed to create customer value in the future – especially as the business model shifts from "make, sell and tell" to "listen, serve and customize" (Knox 2000). At the heart of this shift in consumer marketing is a much greater understanding of consumer buying patterns, attitudes and perceptions of value. Consumer marketing without these insights is like flying blind (Child 2002). The future goal must be to build the business working back from the consumer by adopting radically new thinking which translates into radically new solutions. Solutions which are constructed around the changing drivers of customer value are unlikely to be just the products and services that have been offered in the past. To begin to address these issues and to develop a more explicit consumer orientation, we propose a simple framework consisting of four emergent and overlapping drivers of consumer value (Figure 8-2). We examine each in turn.

5.1 Value-for-Time

In a 1988 Harvard Business Review article, George Stalk wrote, "Today, time is on the cutting edge. In fact, as a strategic weapon, time is the equivalent of money, productivity, quality, even innovation". Since then, most businesses have overhauled their supply chain, reengineered their innovation process and, in consumer markets, introduced efficient consumer response (ECR) to take time out of their business system in order to remain competitive. However, what most consumer businesses have failed to do is to recognize the growing importance of consumer "value-for-time" as a critical driver of future brand value.

Despite significant progress in personal time- and labor-saving technology, many people still report that they now have less time. In a survey conducted by the Henley Centre (2000) for example, the majority of UK adults questioned (66 per cent) agreed with the statement that they "never seem to have enough time to get things done". Southern and Johnson (1999) suggest that one consequence of this time famine is that time is rapidly becoming a currency with many consumers being more aware of how they spend it and how they save it. This finding is supported by the Henley Centre study. It found that 37 per cent of respondents indicated that they would be willing to spend money to save time.

Mitchell (2001) identifies that "value for time" is not just about saving time and just-in-time delivery. Rather, it is about time enrichment and the provision of valuable brand experiences too.

In other words, brands now have an opportunity to define their customer value propositions across one, some or all of the value-for-time factors depicted in figure 8-2.

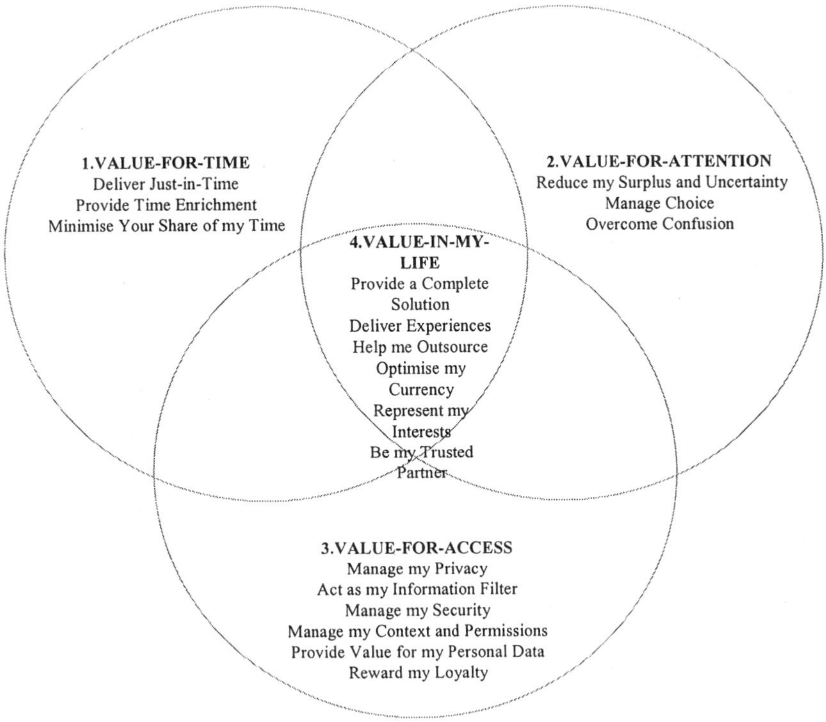

Figure 8-2. The Four Consumer Value Drivers

5.2 Value-for-Attention

Falling costs and lower barriers of market entry have led to a proliferation of marketing communications in the past few years, both in volume and pervasiveness. In the UK, for example, total direct marketing expenditure grew by 80 per cent in the period 1996 to 2001, rising from £6.1 billion to £11.14 billion (Direct Marketing Association UK 2002). In fact, it was recently estimated by Western International Media (quoted in Southern and Johnson 1999) that every UK adult is subjected to over a thousand commercial messages in an average week. In a study of US markets, Godin (1999) demonstrates that US consumers are exposed to roughly five thousand advertising messages every single day. Although information overload was a phrase coined as long ago as the 1950s by Miller (1956) (in his study of the psychological limits on the human capacity to process information during a given unit of time), the term has particular resonance in a consumer context today.

To help individuals cope with the marketing surplus of choice and accompanying consumer uncertainty, many writers argue that some brands are now ideally placed to perform a new mediating role that effectively manages consumer value-for-attention (Hagel and Singer 1999, Achrol and Kotler 1999, Sawhney and Kotler 2001). Others suggest that as technology finally catches up with consumer demand, future marketing practice and customer value-creation will be increasingly defined by individual permission and preference and by enabling consumers to manage brand attention (Coots 2003). The introduction of TiVo, a personal television programming and recording system, is an example of how one new technology now provides people with the means to control their exposure to marketing and advertising. Although user take-up in the US has been slow to date[1], a study by CNW Marketing Research (2002) identified that 72 per cent of TiVo users program their recorders to skip or remove all television advertisements, a much higher rate than those watching live television or those using standard VHS videotape recorders.

5.3 Value-for-Access

Time and attention are joined by personal information and privacy as the new consumer currencies of the information age. Currently, companies attempt to gather extensive data on consumer socio–demographics, identification and behavior often via loyalty cards and online internet registration and tracking systems. Rapidly proliferating and interconnecting new technologies such as Radio Frequency Identification (RFID), global positioning systems (GPS), sensors, smart cards and biometrics provide additional means to gather personal data and track people's behavior. As such devices continue to multiply, it will become possible for organizations to collect data about individual consumers almost anytime and anywhere. Whilst there is great potential for such technologies to transform marketing practice and the consumer experience in terms of improved targeting accuracy, reduced cost and superior personalization, there are serious concerns about possible abuses of collected personal information. As a result, the opportunities for personal data gathering and use are reducing owing to a growing consumer privacy backlash and new data protection legislation.

In a recent IBM-Harris survey (IBM-Harris 1999), 78 per cent of UK and 94 per cent of US consumers expressed concern about the "possible misuse" of their personal information by companies. Prabhaker (2000) suggests that such privacy worries are the defining element in the competition for online consumers. He argues that businesses have a choice in how they respond to this matter. They can see it as a threat and simply react defensively or they can treat it as an opportunity and be proactive in maximizing the gains. O'Malley and Tynan (2000) also highlight a number of issues regarding invasions of consumer privacy, particularly as a by-product of the growth in direct marketing communications, concerns that were largely denied or overlooked in the rush to develop a relationship marketing approach using database technologies. Hagel and Singer (1997) argue that as more and more managers begin to build strategies based on capturing personal data, a major change is underway that may undermine their efforts. They speculate that consumers themselves are going to take ownership of their personal information and demand new forms of value in exchange for it.

We distil issues of privacy, security, data ownership and information exchange into a consumer value dimension which we call value-for-access. Again, we support the view of Hagel and Singer that certain brands are capable of capturing this important and emergent driver.

5.4 Value-in-My-Life

Value-in-My-Life, proposed by Mitchell, Bauer and Hausruckinger (2003), resides at the core of the consumer value framework. As Mitchell et al. suggest, brand managers who recognize the long-maturing shift in what consumers value and understand how to create this value will be able to develop a new role for their brands in the life of their customers. As studies indicate that consumers are now spending more on services and solutions than on material goods (Henley Centre 2000), this brand role would be to optimize an individual's productivity and to provide support for their lifestyle by seeking to become a trusted partner. Lewis and Bridger (1999) suggest that this shift in expenditure is an important trend because the majority of consumers in mature markets have largely exhausted the things they need to purchase and are focusing instead on seeking opportunities and experiences which make their lives happier, more balanced and rewarding. Recent research into patterns of work-life balance in the UK provides parallel evidence of this shift in personal objectives. A joint survey by the employment web site Reed.co.uk and the Department of Trade and Industry (2003) found that of nearly 5,000 UK workers and job seekers interviewed, 31 per cent would prefer flexible working hours over a £1000 pay rise. Other studies have identified high levels of consumer dissatisfaction and unhappiness with existing products and services, notwithstanding the proliferation of choice and personalization (Chandy 2001).

Despite such evidence, Firat, Dholakia and Venkatesh (1995) suggest that marketers continue to emphasize the role of material conditions in shaping consumer needs and demand for products rather than the broader context in which consumers select, buy, and use products and services. Seybold (2001) extends this idea by introducing the concept of Customer Scenarios. She argues that by thinking broadly about the challenges consumers face, rather than narrowly about what firms can sell them, new ways to make their lives easier can always be found. She concludes that many existing brands are able to deliver new forms of value by providing solutions that seek to maximize customer utility. Nevertheless, for brand management to begin to capture and distribute the value locked in these emergent needs, much will depend on the growing willingness of consumers to outsource their preferences, specify their requirements and engage in new forms of brand relationship.

6. THE CONSUMER AS SPECIFIER

Today, consumers have a powerful new set of information and access media. They have the means to acquire objective information to compare brand pricing, features and performance. They can also achieve new levels of convenience and direct communications with companies, designing and submitting personal specifications for products and services (Sawhney and Kotler 2001). Indeed, Dupuy (1999) refers to this as the "victory of consumers" over sellers. Even so, Moynagh and Worsley (2002) suggest that consumer attitudes to choice and negotiation are still evolving as the world becomes increasingly tailor-made. In the age of mass-consumption and standardized products, the idea of having more customized choice had a strong appeal. Choice enabled consumers to jump out of the standardized box to become the specifier. In the years ahead though, as mass-customization becomes a conduit to infinite choice, people will take this for granted. What they will then want is help in knowing how to choose. We argue that marketing will increasingly focus on enabling people to manage such choices, which becomes the new currency derived from the four emergent drivers of consumer value.

Bovet and Roucolle (2000) argue that brand, product and communication proliferation is not the sole cause of consumers' changing attitudes to choice. They suggest that many companies are just poor at giving customers what they want and that the issue lies more with outmoded supply chains and a lack of commitment to place customization on the business agenda. Decades after demand-pull manufacturing was first introduced, the standard supply chain still pushes products down the line and out the factory gate, in the hope that someone will want them. Customized goods and services under this system require a long wait and premium pricing, and customers often complain that it was not what they really wanted in the first place!

Instead of trying to adapt existing supply chain and brand management systems, we suggest that there exists a significant opportunity for brands to configure, create and operate a new type of networked consumer system organized around individual consumer demand rather than supply. Such networks have been given many names, from infomediary (Hagel and Singer 1999) to intermarket and opportunity networks (Achrol and Kotler 1999). We term them consumer value networks to express how they are able to realize the four drivers of consumer value described above.

Because of the distinctive positioning and roles of brands that appropriate and orchestrate consumer value networks, we propose a new market concept to illustrate the unique and necessary transformations in brand management mindsets and capabilities. We term this the reverse market concept. In the next section, we define the key elements of a reverse market brand orientation, position and role and we explore the concept in practice with reference to examples of emergent reverse market organizations.

7. THE REVERSE MARKET CONCEPT

In the above discussion, we suggest that both the transactional and relationship marketing perspectives are primarily focused on value generation, efficiency and effectiveness from the point of view of the firm. The reverse market concept[2] fundamentally challenges this orientation. Whilst still acknowledging the importance of mutual relationship exchange and value-creation, it views markets and marketing activities from an explicitly consumer viewpoint. Importantly, the reverse market concept offers new modes of customer transaction and relationship; modes that capture and respond to the emerging patterns of behavior which reflect the consumer as specifier. Importantly, it defines how brand management can support individual consumers by helping them to achieve their best personal solution and by reducing their risk of encountering market uncertainty, complexity and confusion.

To characterize the distinction between reverse marketing and the established market concept, we illustrate differences in market orientation and position, marketing processes and brand role.

7.1 Market Orientation

From a macro perspective, the reverse market orientation is explicitly buyer-centric, prioritizing the needs, wants and resources of individual buyers and enabling individual customers to gain maximum value from their transactions and relationships with multiple sellers. This contrasts with the established market orientation which is often directed towards improving the productivity and profitability of the firm (Mitchell, Bauer and Hausruckinger 2003), despite claiming the fundamental goal of satisfying the needs and wants of its customers (Shapiro 1988, Ruekert 1992, Kohli and Jaworski 1990, Narver and Slater 1990, Deshpande et al. 1993).

7.2 Market Position

Reverse market brands act as buyer-seller intermediaries which Vandermerwe (1999a) describes as "go-between service providers". They mediate across all channels by orchestrating consumer value networks consisting of multiple online and offline sellers. Such networks are designed to fulfill complete consumer need sets by delivering the solution that individual buyers seek (Vandermerwe 2000). Their market position departs from tradition where sellers compete to offer part-solutions to consumer need sets.

7.3 Marketing Processes

In many instances, consumers now lead the marketing exchange process. In response, marketing processes are reversing as consumers become empowered knowledge partners and active agents in value creation and value-sharing (Gibbert Leibold and Probst 2002). Kotler (2002) anticipates a number of changes in the direction and emphasis of operational marketing as a result of this shift. He argues that there will be a gradual reversal of product design, pricing, advertising promotions, distribution and segmentation practices (from "push" to "pull"); the sum effect of which will be to conceptualize and personalize the marketing process (Kenny and Marshall 2000).

7.4 Brand Role

The reverse market brand performs as a consumer consultant (Achrol and Kotler 1999) and market specialist (Sheth, Sisodia and Sharma 2000) through an advocacy and support role (Zuboff and Maxmin 2002). They seek to optimize their client's interest and personal productivity as characterized by the four value drivers shown in figure 8-2. Reverse market value-exchanges are largely initiated and controlled by customers and are driven by deep levels of customer commitment, involvement and brand trust.

Bailey and Bakos (1997) identify four brand roles based upon empirical research of electronic intermediaries:

1. Trust - Reverse market brands perform as trusted agents by acting for and on behalf of buyers and by maximizing their benefits, protecting their interests (such as personal information) and minimizing their risk and uncertainty when dealing with the market;
2. Matching - They match specific buyer needs to appropriate sellers. When combined in a value network, multiple sellers are able to provide complete solutions and experiences to individual consumers;
3. Aggregation - They aggregate knowledge of customer demand in order to lower buyer and seller transaction costs and take advantage of economies of scale by reducing asymmetries in the bargaining power of buyers and sellers;
4. Facilitating - By functioning in these multiple roles, reverse market brands become facilitators of consumer buyer motivations, preferences, permissions and requests as well as orchestrators of seller capabilities and responses.

Reverse market buyer and seller exchanges can be turned into brand equity based on the development of new knowledge and network management competencies. These are leveraged across the whole value network to sustain brand commitment and growth from the acquisition and retention of new customers and partners. Figure 8-3 illustrates the essential characteristics of the reverse marketing concept and the central role of the brand in the consumer value network.

We now discuss applications of the reverse market concept with reference to FreelanceZone[3], a brand seeking to support the self-employment needs of the independent, professional homeworker, and myciti.com, Citibank's online portal.

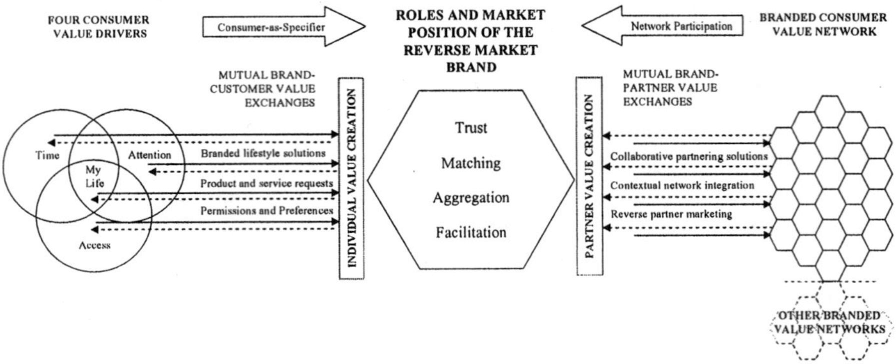

Figure 8-3. The Reverse Market Concept

8. REVERSE MARKETING: CASES

FreelanceZone: A reverse market brand model

Over the next decade, forecasts indicate that the number of self-employed professionals in the UK will double to 3.2 million, making up 12 per cent of the total working population (Mori 2001). However, despite the growing size and importance of this community, the same research reveals that many independent workers are underserved and under-represented by government, financial services and many other businesses. FreelanceZone recognized this shortcoming and the opportunity for building a branded value network that supports a number of home, work and lifestyle support and solution services for the professional self-employed. It began by modeling a number of value network scenarios derived from the explicit involvement of future users rather than an internal process or operational perspective. In this way, the active, early involvement of potential customers in the design process led to the discovery of a powerful set of future service differentiators. By aligning with the key value drivers and core interests of its customers and by intermediating between its users and multiple network partners, FreelanceZone hopes to create and deliver enhanced forms of sustainable relationship value.

The FreelanceZone value network consists of commercial partners who have been carefully selected for their fit with distinct elements of the total needs of the self-employed professional community. They include office equipment, telecommunications, home services, travel, personal assistant, financial services and small business advice providers. Each has been chosen on the basis of their ability to provide superior levels of user support and service as well as their alignment with the core values of the reverse market brand proposition. Through the deployment of shared customer insight, interactive technology and a commitment to the network, FreelanceZone hopes to support the fluid and dynamic exchange of information, knowledge and guidance, as well as products and services. Its revenue streams consist of a mix of client and partner payment mechanisms including user subscriptions, partner response fees, permission to marketing access charges and aggregated knowledge services income. Critically, the value-generating process is controlled by FreelanceZone members who have access to a number of customized, intelligent and interactive reverse marketing interfaces and systems. These enable them to specify their needs, permissions and preferences and to develop their relationship with FreelanceZone, the self-employed community and its network partners. FreelanceZone recognizes that its success depends on distributing value to and within its core community as well as democratizing its role, responsibilities and influence within the network.

Freelance Zone has four main functions which correspond to the roles of the reverse market brand shown in figure 8-3:

a) Trust – FreelanceZone builds relationship value by providing expert levels of individual consumer advocacy, protection and support. Not only is it seeking to represent its community's interests to public sector and industry bodies, but it also acts as a trusted, accountable and transparent value-in-my-life partner, responsible for resolving client problems, supporting their lifestyles and monitoring the performance of partner capabilities and support services;
b) Matching – At the heart of the FreelanceZone value network are a series of processes that effectively reverse and conceptualize the flow of marketing communications from users to network partners and back again. By enabling individuals to specify their preferences to the network and by providing multiple partners with the means to evaluate requests and customize responses using a mix of both online and offline channels, FreelanceZone is "customerizing" (Wind and Rangaswamy 2001) network marketing activities. It is hoped that by matching efficiencies in market demand and supply, it will be able to command greater buying power and deliver superior value-for-attention for its installed customer base as well as reduced customer acquisition and other marketing costs for its network partners;
c) Aggregation – FreelanceZone provides tools for its customers to protect and manage their appointments, contacts, newsletter subscriptions, client

engagements and personal information. Uniquely, an important by-product of the FreelanceZone role is the generation of user knowledge in two forms. First, knowledge about customers and partners, their characteristics and preferences and second, the vested knowledge held by customers and partners. Through a continuous process of customer knowledge management, FreelanceZone distributes its knowledge competencies to co-create mutual value-for-access within the community;

d) Facilitation – FreelanceZone is devising new modes of customer and partner transaction and interaction and aims to support its users throughout their entire consumption cycle, whether online or offline. It delivers independent product and service recommendations derived from its own knowledgebase and its customer's previous experiences. It also provides price comparison tools, personal decision-making guides, reverse auctions, payment management services and fulfillment and delivery support. This is backed-up with a peer-to-peer network that enables its community to rate and review products, services and trade partners as well as to post and respond to freelance assignments.

Myciti.com

FreelanceZone has a bold strategy to position and grow its brand by fulfilling four reverse market roles. Other brands are focusing on just one or two roles. For instance, Citibank's online portal, *myciti.com*, is positioned primarily as an "aggregation" brand[4]. The site provides a full range of Citigroup consumer products and services, including credit and charge cards, banking services, investments, mortgages, loans and insurance. By allowing their customers access to all their supplier's email, financial and loyalty accounts (such as frequent flier miles), as well as to store personal data and passwords, Citibank is creating new opportunities for customer differentiation and service in a highly competitive, commodity market (Trombly 2000). Despite privacy concerns some consumers have about the service, the myciti.com site attracted more than 300,000 users in its first year of operation and adds some 3,000 new users a day (Thomas 2001). The myciti.com portal does not stop at aggregation and access however. It also provides financial planning support linked to key customer life stages or events, such as becoming a student, getting married, starting a family, buying a home and planning for retirement.

Both Freelance Zone and myciti.com illustrate how early-stage reverse market brands are beginning to unlock the consumer value that is located beyond the provision of products and services by performing a new market mediation role. They understand that there are mutual benefits to be realized by improving on their customer's time, attention and access to markets. However, aspiring reverse marketers face a number of barriers in the development of successful brands. We now consider these benefits – and the potential barriers – from both a consumer and a managerial perspective.

9. REVERSE MARKETING: BENEFITS AND BARRIERS

9.1 Consumer Perspective

There is presently little empirical evidence to help our understanding of consumer need, participation and behavior in a reverse market context. Indeed, most studies are either anecdotal or are limited to consumers' attitudes towards agent-supported shopping in complex, high involvement product categories. One such survey conducted by the Consumers' Association (2001) found that 41 per cent of people interviewed were either "very likely" or "quite likely" to use a brand that advised them of the best possible product or service for their particular needs and which helped them get the best deal. For these individuals, the primary benefits cited were convenience, the reassurance gained via the provision of expert and trusted advice, and the opportunity to save time, hassle and money.

However, such benefits form only a limited part of the total set of consumer outcomes that could be engineered and delivered through the reverse market brand. Provided the customer interface is intuitively designed and the value network is representative of the lifestyle choices of a particular market segment, reverse market brands should be able to act as a single, trusted destination for consumers.

Much more research is necessary to validate the proposition that substantial benefits can be realized by existing brands from unlocking more consumer value. Equally, studies are required to empirically establish the relationship between appropriate levels of brand trust, privacy concerns and the barriers which may prevent individuals from subscribing to and using reverse market brands.

9.2 Managerial Perspective

To become a successful reverse market organisation, brand managers need to address a new set of questions and overcome a number of significant internal barriers. The reverse marketing concept requires bold leaps in managerial imagination and attitudes towards customers as they become the specifiers of value. It requires managers to acknowledge that the boundaries between organizations, their partners and individual customers are disappearing. The reverse market concept also necessitates new marketing processes defined by collaboration with active, empowered individuals and value network partners rather than by the organized control of passive consumers. This is particularly the case regarding the ownership and distribution of customer and network knowledge. Here, brand management must develop ethical, transparent and accessible information systems and expert customer knowledge management capabilities in order to maintain sufficient levels of trust and collaboration within the network.

In addition, an aspiring reverse marketer must assess whether their existing brands have the necessarily levels of trust and reputation to support a substantive shift in market orientation in the development and mediation of a consumer value network. The brand owner must also weigh up the potential negative effects of distributing the ownership of its accrued brand trust across a network of partners. In fact, for many brand managers, this risk of brand exposure represents the greatest barrier to the adoption of the reverse market concept.

A further barrier is the difficulty of achieving coordinated network innovation, collaboration and critical mass. As Chesbrough and Teece (1996) describe, delivering a systemic innovation of this nature is particularly difficult when industry standards and operational infrastructure do not exist and must be pioneered. Although the advent of web services promises to enable a company to connect its applications to any number of trading partners relatively inexpensively and easily, many organizations remain islands of technology, operating its own assortment of systems, applications, databases, and communications technologies (Hagel 2002). Thus, a critical challenge also lies in discovering how and when to build the bridge to the consumer, which customer communities to select, who to share the costs of developing the standards for constructing network value.

However, for firms choosing to adopt a reverse marketing strategy, there are evidently some important benefits, not least of which is the opportunity to develop deep levels of customer commitment leading to greater levels of loyalty and expenditure. By focusing on overcoming consumer's perceived risk, particularly in situations they may not have previously experienced, studies show that greater brand trust can be earned and new brand extensions achieved more easily (Delgado-Ballester and Munuera-Alemán 2001). It is possible, therefore, that reverse marketing value networks, founded on clear principles of consumer dialogue, support and advocacy, might realize a more authentic means of building long-term, mutual dependency and brand loyalty. It would also provide new roles and skills for front-line employees many of whom will be empowered to resolve customer problems.

From an economic perspective, reverse market brands have the potential to deliver greater economies of scope through network convergence (from the removal of functional duplication), knowledge-sharing and reductions in network customer acquisition and retention costs. Indeed, cost improvements in the matching of demand and supply are fundamental to the participation of value network partners.

Because the management of reverse market brands can lead to superior market-sensing and market-relating capabilities (Day 2000), reverse marketers are likely to continually challenge established practices, sense new market opportunities and create economic value more quickly for customers, partners and shareholders (Gibbert, Leibold and Probst 2002).

10. SUMMARY

Whilst acknowledging the complexity of the reverse market concept and the potential barriers for consumers and brand management alike, we concur with Achrol and Kotler (1999) that the new competitive landscape of reverse market brands could provide the most dramatic scenario for change in marketing this decade. Recently, Zuboff and Maxmin (2002) predicted the rise of what they term, "distributed capitalism" founded on the provision of deep support to individual consumers. They conclude that a "copernican inversion for commerce will occur as brands utilize new capabilities to liberate the vast, but suppressed, reserves of relationship value with the individual (p.317/318)".

Even if Zuboff and Maxmin's predictions of radical marketing and brand transformation are overstated, there is already some evidence of the gradual emergence of the reverse market concept. In this paper, we describe the embryonic examples of FreelanceZone and myciti.com. However, if their predictions do prove to be right, the implications for marketing theory, research and practice are revolutionary. By engaging in this fundamental transition by aligning with the needs of empowered consumers, the reverse market concept might represent a way for marketing management to silence some of their more outspoken critics.

NOTES

1. TiVo was introduced in the US in 1999. By July 2002, 500,000 users were owners of the system. The TiVo digital-video recorder is a set-top box that can perform functions similar to those of a VCR, but instead of using a videotape, shows are stored on a hard disk drive. The set-top boxes can also perform other functions, pause live programming, skip or remove ads and automatically schedule the recording of future shows. Set-top boxes that use TiVo's recording service connect via a phone line to a server to download a schedule of shows and times. The service can also suggest which shows the viewer would like based on previous selections.
2. The term "reverse marketing" was originally coined by Leenders and Blenkhorn (1988) in the context of B2B purchasing and negotiation. It is derived from the idea that the purchaser adopts a marketing perspective and persuades the seller to proffer a solution to the buyer's problem (Plank and Francis, 2001). To date, the concept has had little discussion in the academic literature, particularly from a consumer to business (C2B) perspective.
3. We have changed the name of the organisation for reasons of commercial confidentiality.
4. Citibank's MyCiti portal available at http://www.myciti.com

REFERENCES

1. Achrol, R., Kotler, P. Marketing in the network economy. Journal of Marketing 1999; 63: 146-163
2. Bailey, J.P., Bakos, J.Y. An exploratory study of the emerging role of electronic intermediaries. International Journal of Electronic Commerce 1997; 1(3): 7-20
3. Bovet, D., Roucolle, G. Unlocking the rusty supply chain. Ivey Business Journal 2000; 65:1 (Sep/Oct): 31-35
4. Chandy, C. (2001) quoted in Carter, M., "Dancing to very different tunes: Advertising campaigns are failing to reflect the growing dissatisfaction in British society", Financial Times, May 1st, p. 15
5. Chesbrough, H.W., Teece, D.J. When Is Virtual Virtuous? Organizing for Innovation. Harvard Business Review 1996; 74: 1: 65-72
6. Christopher, M., Payne, A. and Ballantyne, D., *Relationship Marketing*, Oxford: Butterworth-Heinemann, 1991.
7. Child, P.N. Taking Tesco global. The McKinsey Quarterly 2002: 3: 134-144
8. CNW Marketing Research (2002) quoted in Friedman, W., (2002) "72.3% of PVR Viewers Skip Commercials", Retrieved January 26th 2003 from: http://www.adage.com/news.cms?newsId=35293
9. Consumers' Association, Right Side Up Research Report", unpublished research commissioned for The Marketing Forum, 2001.
10. Coots, L., quoted in anonymous, "IdeaFest – 23 Bright Ideas for a Stellar 2003", Fast Company, 2003; 66(Jan): 104
11. Coviello, N.E., Brodie, R.J., Danaher, P.J., Johnston, W.J. How firms relate to their markets: An empirical examination of contemporary marketing practice. Journal of Marketing 2002; 66(3): 33-46

12. Day, G.S. Managing market relationships. Journal of the Academy of Marketing Science 2000; 28: 1: 24-30.
13. Day, G.S., Montgomery, D.B. Charting new directions for marketing. Journal of Marketing 1999; 63 (Special Issue): 3-13
14. Delgado-Ballester, E., Munuera-Alemán, J.L. Brand trust in the context of consumer loyalty. European Journal of Marketing 2001; 35(11/12):1238-1258
15. Direct Marketing Association (UK), DMA Census of the Direct Marketing Industry, 2000-2001, London, The Direct Marketing Association (UK) Ltd., 2002.
16. Dupuy, F., *The Customer's Victory: From Corporation to Co-operation*, London: Macmillan Business, 1999.
17. Firat, A.F., Dholakia, N., Venkatesh, A. Marketing in a postmodern world. European Journal of Marketing 1995; 29(1): 40-56
18. Fisk, P. (2001) "The rise of the customer gateway", Retrieved November 2002 from: http://www.allaboutbranding.com/printhis.lasso?print=157
19. Fournier, S., Dobscha, S., Mick, D. Preventing the premature death of relationship marketing. Harvard Business Review 1998; 76(1): 42-51.
20. Gibbert, M. , Leibold, M., Probst, G. Five styles of customer knowledge management, and how smart companies use them to create value. European Journal of Management 2002; 20(5): 459-469
21. Godin, S., *Permission Marketing*, New York: Simon and Schuster, 1999.
22. Gummesson, E. The new marketing – developing long-term interactive relationships. Long Range Planning 1987; 20(4); 10-20.
23. Hagel, J., (2002), "Edging into web services", The McKinsey Quarterly, No 4, Retrieved January 2003 from: http://www.mckinseyquarterly.com/article_page.asp?ar=1245&L2=21&L3=37&srid=21&gp=1
24. Hagel, J., Singer, M. The coming battle for customer information. Harvard Business Review 1997; .75: 1: 53-65
25. Hagel, J. and Singer, M., *Net Worth*, Boston: Harvard Business School Press,1999.
26. Henley Centre (2000), Planning for Consumer Change 1991-2000
27. Hunt, S. Commentary: A general theory of competition: Issues, answers and an invitation. European Journal of Marketing 2001; 35: 5/6: 524
28. IBM-Harris (1999), IBM Multi-National Consumer Privacy Survey. Retrieved December 2002 from: http://www-1.ibm.com/services/files/privacy_survey_oct991.pdf
29. Johnson, C (2001) "e-commerce brokers arrive" quoted in Gehman, J. (2001) "The rise of e-brokers". Retrieved January 2003 from: http://www.clickz.com/aff_mkt/aff_mkt/article.php/840351
30. Kenny, D., Marshall, J.F. Contextual marketing: The real business of the internet. Harvard Business Review 2000; Nov-Dec: 119 - 125
31. Knox, S.D. (2000) "The Marketing and Branding of Intangible Assets: Whose role is it anyway?", Paper No 266, Proceedings on Managing across Boundaries, R. Kerley (ed.), CD ROM for BAM 2000, Edinburgh.
32. Kohli, A.K., Jaworski, B.J. Market orientation: the construct, research propositions, and managerial implications. Journal of Marketing 1990; 54: April: 1-18.
33. Leenders, M.R. and Blenkhorn, D.L., *Reverse Marketing: The New Buyer-Supplier Relationship*, New York: Free Press1998.
34. Lewis, D. and Bridger, D., *The Soul of the New Consumer*, London: Nicholas Brearley Publishing, 1999.

35. McKenna, R., *Total Access – Giving customers what they want in an anytime, anywhere world*, Boston: Harvard Business School Publishing, 2002.
36. Miller, G.A. The magical number seven, plus or minus two: some limits on our capacity for processing information. Psychological Review 1956; 63(March): 81-92.
37. Mitchell, A., Right Side Up, London: Harper Collins Business Books, 2001.
38. Mitchell, A., Bauer, A. and Hausruckinger, G., *The New Bottom Line: Bridging the value gaps that are undermining your business*, London: Capstone Wiley, 2003.
39. Mitchell, V-W., Papavassiliou, V. Marketing causes and implications of consumer confusion. The Journal of Product and Brand Management 1999; 8(4): 319-339
40. Mori (2001) "Professionals shifting to self-employment - but with no support from bank managers or politicians", Retrieved January 2003 from: http://www.mori.com/polls/2001/alodis.shtml
41. Moynagh, M., Worsley, R. Tomorrow's consumer-the shifting balance of power. Journal of Consumer Behaviour 2002; 1(3): 293-302
42. Narver, J.D., Slater, S.F. The effect of a market orientation on business profitability. Journal of Marketing 1990; 5(October): 20-35.
43. O'Malley, L., Tynan, C. Relationship marketing in consumer markets - Rhetoric or reality? European Journal of Marketing 2000; 34(7): 797-815
44. Payne, A., Holt, S. Diagnosing customer value creation in relationship marketing. British Journal of Management 2001; 12(2); 159-182.
45. Peck, H. (2002), "Tesco Clubcard Forever?", Case study, Cranfield School of Management.
46. Pitt, L.F, Berthon, P.R., Watson, R.T., Zinkhan, G.M. The Internet and the birth of real consumer power. Business Horizons 2002; 45(4):7-14
47. Plank, R.E., Francis, D. Does reverse marketing reduce conflict in buyer-seller relationships. American Business Review 2001; 19(1): 76-83
48. Prabhaker, P. Who owns the online consumer. Journal of Marketing 2000; 17(2): 158-171
49. Prahalad, C., Ramaswamy, V. Co-opting customer competence. Harvard Business Review 2000; Jan-Feb: 79-87
50. Reed.co.uk and Department of Trade and Industry (2003), The Reed / DTI Work Life Balance Report, January 2003. Retrieved August 2003 from: http://www.reed.co.uk/surveys.asp
51. Sawhney, M., Kotler, P., "Marketing in the Age of Information Democracy". In *Kellogg on Marketing*. Iacobuuci, D., Ed. New York, John Wiley Ltd, 386-408, 2001.
52. Sawhney, M., Prandelli, E. Communites of creation: managing distributed knowledge in turbulent markets. California Management Review 2000; 42(4):24-54
53. Seybold, P. Getting inside the lives of your customers. Harvard Business Review 2001; 79(5): 80-89
54. Shapiro, B. What the hell is 'market-oriented?, Harvard Business Review 1998; November-December: 119-25.
55. Southern, H., and Johnson, M., *Consumer Value in the Future*, London: WPP, The Store, 1999.
56. Stalk, G. Time – the next source of competitive advantage. Harvard Business Review 1988; 66(4): 41-52
57. Sveiby, K.E., *New Organizational Wealth: Managing and Measuring Knowledge-based Assets*, San Francisco: Berrett-Koehler Publishers Inc., 1997

58. Thomas, O., (2001) "Anatomy of a Website: One-Stop Account Management. MyCiti.com tracks all your bank, brokerage, and other online accounts." Business 2.0, April, Retrieved August 2003 from: http://www.business2.com/articles/mag/0,1640,9561,FF.html
59. Trombly, M. Citibank's aggregation portal a big draw. Computerworld, 2000; September 18: 40
60. Vandermerwe, S. (1999a), The electronic 'Go-between Service Provider': A new 'middle' role taking centre stage. European Management Journal 1999; 17: 6 (December)
61. Vandermerwe, S. (1999b), Customer capitalism: Increasing returns in new market spaces, London: Nicholas Brealey, 1999.
62. Vandermerwe, S. How increasing value to customers improves business results. Sloan Management Review 2000; Fall: 27-37
63. Varey, R.J., *Relationship Marketing - Dialogue and Networks in the E-Commerce Era*, John Wiley and Sons, 2002
64. Webber, A. Are you on digital time? Fast Company 1999;22(February): 114-118
65. Wind, J. Ramaswamy, A. customerization: The next revolution in mass customization. Journal of Interactive Marketing 2001; 15(1):13-32.
66. Zuboff, S. and Maxmin, J., *The Support Economy – Why Corporations are Failing Individuals and the Next Episode of Capitalism*, New York Viking Penguin, 2002.

Chapter 9

EXPERIENCES OF WWW SITES AS A DECISION SUPPORT SYSTEM IN DIFFERENT ACQUISITIONS

Rainer Breite, Hannu Vanharanta
Tampere University of Technology/ Pori Industrial Management, Finland

Abstract: In this chapter we examine how hypertext based Internet applications support decision making in different purchasing situations. Our research methods are based on hyperknowledge framework, which is characterized to create an ideal computer based decision support system environment. We have performed three different purchasing case tests. In the two cases test subjects acquired a product or service from Internet without a defined www address or www page. In one case, test subjects used a definite product and www address. The test results of our study are based on comparisons between the above mentioned three cases. The comparisons suggest that if decision-makers use Internet based applications in complex purchasing situations, hypertext based Internet applications seem not to sufficiently support the purchaser's decision making.

Key words: Application, Hyperknowledge, Hypertext, Internet, Purchasing, Validity

1. INTRODUCTION

Purchasing and supply chain management many times lead to unstructured problems and issues and therefore purchasing professionals have to acquire information and new knowledge from several different places and sources. Nowadays different Internet applications (e.g. e mail, search engines, portals, etc.) are used to make this work easier. With the help of Internet applications purchasing professionals have the possibility to get rapidly into different sources of information and knowledge and they can use this information and knowledge interactively during their purchasing decision making with an Internet application.

In this chapter we have conducted a study that looks into how Internet applications support purchasing professionals for two different purchasing situations. The Internet applications with hypertext functions contained only www addresses and www pages with search engines. Three empirical test runs (laboratory experiments) were conducted, compared and evaluated against the computer based hyperknowledge environment, described as an extended hypertext environment.

During the empirical tests, the test subjects made purchasing decisions based on information on www sites. In the first, ("Specialty Service/Product", conducted in the spring of 1999) and third ("Virtual Enterprise", conducted in the autumn of 2000) tests, the subjects looked for the product or service that they especially needed. In those cases the test subjects did not have any particular www address to contact. In the second test ("Specialty www address", conducted in the autumn of 1999), the test subjects did have a particular www address and a particular product to be purchased. The results of our research revealed that the hyperknowledge framework used is well suited for analyzing Internet applications and that in purchasing situations different amounts and types of support are needed during the decision making process. In the following we briefly present the hyperknowledge environment used as well as the underlying decision support system configuration. We then describe validity and utility evaluation principles, after which we describe our test cases and research results.

2. METHODOLOGY

The objectives of this chapter are to examine the utilization of hyperknowledge frameworks and decision support systems in the Internet context and how the Internet supports decision-making in different acquisition situations. With the help of literature, we formed the theoretical framework, where the usability and utility of the Internet were described in the context of hyperknowledge and decision support systems. Based on this framework we made three separate empirical tests, with the help of which we evaluated the ability of the Internet to support decision-making in different acquisition situations.

3. THEORETICAL FRAMEWORK

3.1 Hyperknowledge and Decision Support System (DSS)

Hyperknowledge is strongly related to the technologies of hypertext and hypermedia (Bush 1945, Engelbart 1963), and to the architecture of the hypertext systems (Nelson 1987, Nielsen 1990). Hypertext consists of pieces of text or other information, which can be interlinked. Users can read this interlinked text network in a non-sequential manner. This technology gives a reader different options from which to choose. Users determine what to read, in other words, they determine what kind of information to pick and read. Hypertext is part of hypermedia. Hypermedia in turn means the free combination of information using media such as text, data, pictures, voice and video (Woodhead 1991).

The concept of hyperknowledge is wider than those of hypertext or hypermedia, although it follows the same general principles. Hyperknowledge is an ideal working and learning environment that holds knowledge and, at the same time, defines the nature of hypertext and hypermedia. The user can navigate freely in this environment (as on the Internet), and widen his or her own knowledge (Chang et al. 1989, 1993, 1994). The basic goal of this framework is to serve active decision support that enables the decision-maker to participate actively throughout the decision process. The framework takes into account that decision maker cognitively possesses and processes many diverse and interrelated pieces of knowledge e.g. procedural knowledge, descriptive knowledge, reasoning knowledge, etc. The user (i.e. user's mind) is able to freely deal with and control these different pieces of knowledge, thus the decision support systems (DSS) (cf. Dos Santos et al. 1989) should be a natural extension of the decision maker's internal activities (Chang et al. 1989, 1993, 1994).

A DSS (Figure 9-1) consists of three main components:

- The Language and the Presentation System -mediate messages to and from the decision support system;
- The Problem Processing System -handles all the user requests or responses to and from the various knowledge sources in the system;
- The Knowledge System -contains all the decision support system's knowledge and it stores, in groups, concepts that are related to each other by definition and/or by association.

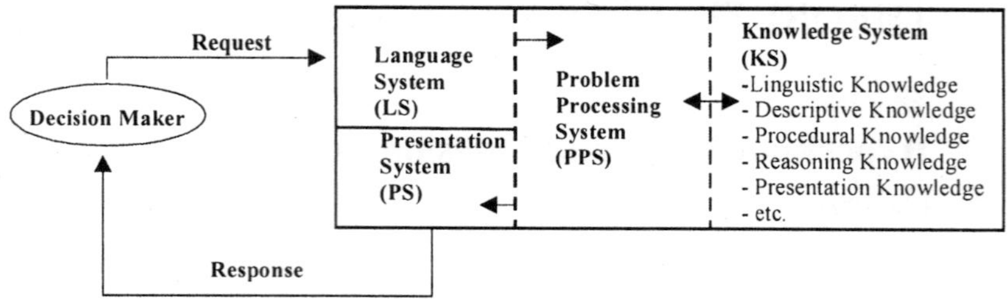

Figure 9-1. Configuration of Decision Support System (Dos Santos et al. 1989)

The environment (Figure 9-1) makes it possible for the user to navigate logically in the hyperknowledge environment. In the ideal hyperknowledge environment the pieces of the knowledge (i.e. concepts) and the user are united (Figure 9-2). The user receives an impression or several impressions as the system presents an image or images (visualization) on a graphical user interface (GUI), i.e. computer screen. The negative is an internal characterization of the image created by the problem processor (Chang et al. 1989, Vanharanta et al. 1995).

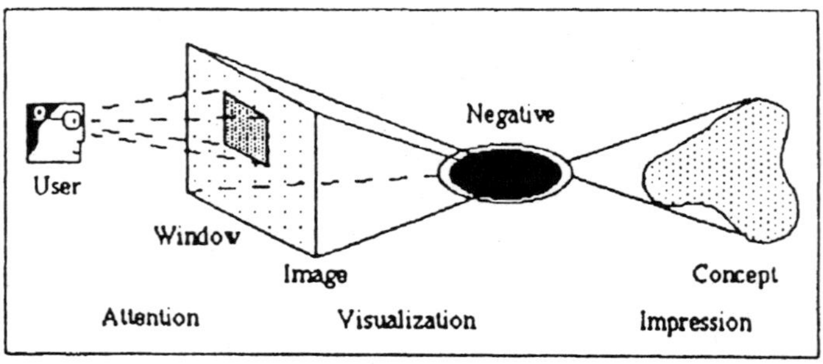

Figure 9-2. User Connection to Hyperknowledge Environment (Vanharanta et al. 1995, p. 223)

3.2 Internet Applications and the Hyperknowledge Environment

In this chapter we have utilized Internet applications so that this practice contains only www addresses, search engines and www pages. This means that when user or decision-maker "goes" to some www address, so he or she goes to a certain specific www page, which is also a specific page on information. Thus we can include the following characteristics of Internet applications into our purchasing decision support system:

- Knowledge System (KS) on the Internet applications means all available purchasing information, knowledge and procedures in www addresses, sites, and pages;
- Problem Processing System (PPS) on the Internet means search functions and processes by the Internet application;
- Using hypertext transfer protocol (http) to control the language System (LS) on the Internet. The user activates a www –page and www address by using a mouse and keyboard;
- Presentation System (PS) on the Internet application is everything presented in www page format and which can be printed in paper form.

4. VALIDITY AND UTILITY EVALUATION

Verification and validation are two important steps in any software evaluation process. Adrion et al. (1982) defined verification as "the demonstration of consistency, completeness, and correctness of the software at any stage and between each stage of the development life cycle" and validation as " the determination of the correctness of the final program of the software produced from a project with respect to user needs and requirements." Usually, verification concentrates on software aspects and validation on modeling aspects. Within expert systems research, verification and validation have often been translated into popular slogans: "building the system right" and "building the right system" (Fox 1990, O'Keefe et al. 1987). Usability and utility are two equally important factors in the evaluation process. Newell et al. (1972) defines utility as "the question of whether the functionality of the system in principle can do what is needed". Usability is usually described in terms of criteria like 'learnability', 'efficiency of use', 'memorability', 'a small number of errors', and 'subjective satisfaction' (Nielsen 1990 and 1993). We assume that verification of the tested applications has been conducted and therefore it is important to know how these applications meet user requirements and the targeted overall functionality in a purchasing situation.

4.1 Evaluation Methods

The result of previously executed validity and utility tests of on hyperknowledge applications (Vanharanta et al. 1995), constitutes the point of departure for our validity and utility study. The present study focuses on the two most critical components of the hyperknowledge environment: the user and the contents of the application. We try to find out if the Internet applications used are the right system for purchasing decision support activities, and what are the advantages and disadvantages of Internet applications from the user's point of view. The validation methods are the same, which have been used in expert systems (O'Keefe et al. 1987 and O'Leary 1988) and in the hyperknowledge system evaluations, i.e. form of performance validation combined with a questionnaire.

4.2 Test Subjects

In the "specialty service/product" test cases the test subjects consisted of twenty M.Sc. students (Industrial Management), all of whom have a B.Sc. degree in engineering. All the test subjects were familiar using Internet applications and moreover, most of them knew what an important role of purchasing and supply management plays in a company. In the "specialty www –address" and "Virtual Enterprise" test cases the sample consisted of M.Sc. students (Industrial Management) and M.Sc. students (Economics and Business Administration). All of these test subjects have a Bachelor of Science degree and are active daily users of the Internet. Also for them the importance of purchasing and supply management in a company was understood. (Table 9-1 ~ Table 9-6) So background of our test subjects suited the targeted test cases very well, furthermore the test subject at least partially reply to Holleran's (1991) criteria of test subjects.

Table 9-1. Age (years) of Test Subjects

Test experiment	N	Minimum	Maximum	Mean
"Specialty Service/Product"	16	24	50	31,2
"Specialty www address"	27	25	48	29,1
"Virtual Enterprise"	30	24	39	28

Table 9-2. Gender of Test Subjects

Test experiment	Male	Female	Total
"Specialty Service/Product"	15 (75 %)	5 (25 %)	20 (100 %)
"Specialty www address"	18 (62 %)	11 (38 %)	29 (100 %)
"Virtual Enterprise"	19 (63 %)	11 (37 %)	30 (100 %)

Table 9-3. Understanding the Purchasing Management Concept

Test experiment	Agree	Partially agree	Neutral	Partially disagree	Disagree
"Specialty Service/Product"	11 (55 %)	9 (45 %)			
"Specialty www address"	14 (48 %)	13 (45 %)	1 (3.5 %)	1 (3.5 %)	

Table 9-4. Previous Use of the Internet

Test experiment	Very much used	Much used	Somewhat used	Little used	Not used
"Specialty Service/Product"	3 (15 %)	6 (30 %)	9 (45 %)	2 (10 %)	
"Specialty www address"	9 (31 %)	16 (55 %)	2 (14 %)		

Table 9-5. Demographic Data of the "Virtual Enterprise" Test Subjects

	N	Minimum	Maximum	Mean
Participation in Internet projects	29	0	7	2,3
Proficient with the Internet	29	1	5	3,2
Years of experience	28	0	11	3,7
Valid N (listwise)	28			

Table 9-6. Frequency of Use of the Internet by Test Subjects

Test experiment	Every day	Several times a week	Once a week	Total
"Virtual Enterprise"	12 (40 %)	15 (50 %)	3 (10 %)	30 (100 %)

4.3 Test Situation, Assignments and Questionnaires

Every test case was arranged as exercises. In the "specialty service/product" case the test subjects were in the role of a purchasing manager, who was obliged to look for some product, service or partner via the Internet, for example from the address: http://www.manufacturing.net. In this exercise, the purchasing manager was in a situation, where the product's supply chain and company's value chain were changed or affected by the Internet application. In these test situations the Internet application(s) used behaved like a decision support system. This means that data is retrieved from the Internet by a web browser, which activates for example, descriptive knowledge in a DSS (Figure 9-1 and Figure 9-3). In the "specialty www address" case the test subjects were in the role of a purchasing manager, who had to purchase a product through the address: http://www.titus.co.uk. In this test exercise, the purchasing manager's company was one part of a virtual supply chain and the manager affected the supply chain by choosing and using the Internet address. Also in this test case, the purchasing manager had to make decisions by using information, which was obtained only from the Internet application's www address. In the "virtual enterprise" case the test subjects were in the role of a purchasing manager, who was obliged to look for some product, service or partner via the Internet. In this exercise, the manager was in a situation, where the product's supply chain and company's value chain were changed or affected by the Internet application. In these test situations the Internet application(s) used behaved like a decision support system. Also in this case the meaning is that data is retrieved from the Internet by a web browser that activates for example, descriptive knowledge in a DSS (Figure 9-1 and Figure 9-3).

In every test case, the principles and objectives of the studies were explained to the test subjects at the beginning of the tests, moreover, every question of the charting questionnaire was presented to them. After that, the test subjects performed their tasks. After c ompleting the test, the subjects answered a questionnaire charting their subjective satisfaction with the Internet application in acquisition activities as well as various other issues related to the validity and utility of the internet application in the purchasing situation.

In the "specialty service/product" and "virtual enterprise" test the assignment was designed so that each subject was obliged to give a report, where test subject described his or her company's industry, what kind of service, product, or partner company was needed. Moreover, all the www addresses needed were illustrated in the report. In the "specialty www address" test case the assignment was designed so that each subject was obliged to give a report, where all the www addresses that belonged to the supply chain were illustrated.

In every case the same questionnaire was used. The questionnaire was divided as follows: background information about the test subject (9 questions); test subject's own attitude and estimation about the Internet and Internet environment (29 questions); test subject's own attitude and estimation about suitability of the internet to purchasing (30 questions); test subject's satisfaction with the Internet application in purchasing (7 questions).

5. RESULTS

Our validity and utility analysis in both test cases focuses on eleven different research constructs, which are based on the hyperknowledge framework, and the validity and utility model for the hyperknowledge environment as implemented in the Vanharanta's benchmarking application (c.f. Vanharanta 1995). The model is illustrated in figure 9-3. In this model the knowledge of the system has been divided into five different types of knowledge: linguistic knowledge (e.g. computer explanations), descriptive knowledge (e.g. fact data etc.), procedural knowledge, reasoning knowledge and presentation knowledge (These areas form a flexible user interface e.g. audio and visual systems. See, Chang et al. 1994). Our paper is concerned with the descriptive knowledge only, thus we examine how the information, which is transmitted via the www pages, supports the user's purchasing decision making in the different purchasing situations.

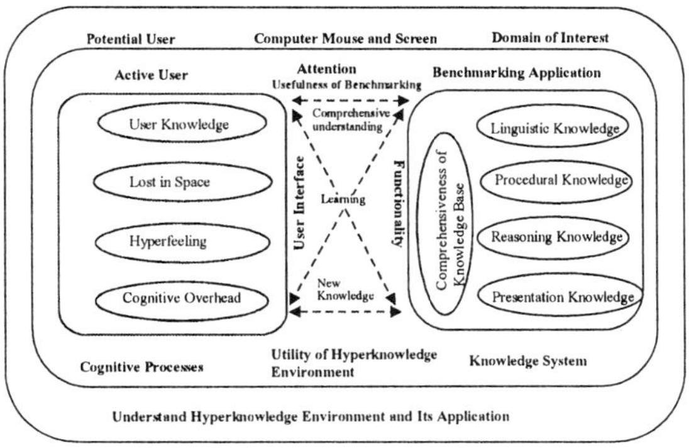

Figure 9-3. Validity and Utility Model for the Hyperknowledge Environment as Implemented in the Benchmarking Application (Vanharanta et al. 1995, p. 227)

The different constructs as used and mentioned above were divided into four different categories:

- The first category describes traits and feelings perceived by the test subjects themselves during and after the tests. These constructs in our study are "lost in space", "hyperfeeling", and "cognitive overhead" (See Conklin 1987);
- The second category describes perceived outcomes for the test subjects interacting with the Internet application, e.g. what was the contribution of Internet for the user. These constructions in our study are: "learning", the "creation of comprehensive understanding", "understanding the Internet and its applications", "usefulness in acquisition", and the "utility of the Internet";
- The third category describes the perceived "comprehensiveness of the Internet" as a whole;
- The fourth category (satisfaction) describes "satisfaction of the customer"

5.1 Statistical Results

Tables 9-7, 9-8, and 9-9 review statistical results for the eleven constructs. The results are very similar in the both test cases and the mean values for all constructs below are close to neutral three, which we regard as very satisfactory.. The result for the "usefulness in acquisition" constructs is good in both cases and also the results for the "lost in space" and the "understanding the Internet and its application" are rather good in the specialty www address case. In the specialty product/service case, however, the result for the "comprehensiveness of the Internet" construction is below satisfactory level. The detailed results are as follows (Table 9-4, 9-5, and 9-6).

Table 9-7. The Result of Subjective Validity and Utility Assessment of "Specialty www Address" (N = 29).

Constructs	Mean	St dev
User Knowledge	2,33	0,62
Lost in Space	3,59	0,91
Hyperfeeling	2,97	0,76
Cognitive overhead	3,01	0,89
Learning	2,76	0,62
Creation of comprehensive understanding	2,41	0,58
Comprehensiveness of the Internet	3,07	0,64
Understanding Internet and the applications	1,98	0,57
Usefulness of acquisition	1,31	0,59
Utility of the Internet	2,52	0,61

Table 9-8. The Result of Subjective Validity and Utility Assessment of "Specialty Product/Service" (N = 20)

Constructs	Mean	St dev
User Knowledge	2,35	0,69
Lost in Space	3,13	1,07
Hypefreeling	3,27	0,62
Cognitive overhead	2,62	0,95
Learning	3	0,61
Creation of comprehensive understanding	2,94	0,88
Comprehensiveness of the Internet	3,6	0,52
Understanding Internet and the applications	2,34	0,70
Usefulness of acquisition	1,08	0,24
Utility of the Internet	2,85	0,72

Table 9-9. The Result of Subjective Validity and Utility Assessment of "Virtual Enterprise" (N = 30)

Constructs	Mean	St dev
User Knowledge	2,40	0,44
Lost in Space	3,55	0,96
Hyperfeeling	2,68	0,75
Cognitive overhead	2,79	0,60
Learning	2,62	0,68
Creation of comprehensive understanding	2,56	0,75
Comprehensiveness of the Internet	2,84	0,67
Understanding Internet and the applications	2,84	0,60
Usefulness of acquisition	1,25	0,39
Utility of the Internet	2,68	0,70

The differences between the three groups were analyzed statistically. Specifically for each of the ten items used the Kruskal-Wallis method (Mendel and Sincich 1995, pp. 945-948) to test the null hypothesis.

> H_o: the distributions of the three groups are the same against the alternatives hypothesis. H_a: the distributes of the three groups have different locations

The alternative hypothesis essentially means that the three groups differ. The significance level of or p-value from the Kruskal-Wallis test is shown in table 9-10. (The closer p is to zero, the more easily we can reject H_o in favor of H_a.) Only the "Comprehensiveness of the Internet" item manifest a significant difference between the three groups ($p = 0.0003$).

Table 9-10. The p-values of the Constructs

Constructs	P
User Knowledge	0,7731
Lost in Space	0,2747
Hyperfeeling	0,2826
Cognitive overhead	0,0450
Learning	0,1970
Creation of comprehensive understanding	0,1212
Comprehensiveness of the Internet	0,0849
Understanding Internet and the applications	0,0003
Usefulness of acquisition	0,1315
Utility of the Internet	0,3291

5.2 Validity and Utility Assessment

Because the above mentioned results in every case are very similar in the following we handle both cases simultaneously.

User Knowledge: The level of existing knowledge in purchasing issues is mapped using the questions concerning the background of each test subject. The test subjects are not experts in purchasing management but all are quite well grounded in the theories of purchasing management.

Lost in Space: According to Vanharanta et al. (1995), one of the problems in hypertext and hypermedia applications is the feeling of being "lost in space". This means that application users loose their "coordinates". In Internet applications, especially www sites this means that the user is unable to find "the right page" and she or he is not sure, where they exactly lie and also what the context of the www page is. So the user does not have any familiar framework. In the questionnaire we asked for the subject's opinion. In the "specialty service/product" test case 25 % and "virtual enterprise" 13 % the subjects replied that they felt "lost in space" and in the "specialty www address" test case 14 % the subjects replied that they felt "lost in space". Our test result revealed that if the www address has not been clearly defined the user could feel several times "lost in space".

Hyperfeeling: The hyperfeeling construct tries to discover how the user has internalized the available information knowledge, and how deep into the knowledge the user has gone. Moreover, whether the user felt that, via the system (the Internet application) functionality and its interface, the Knowledge System became an integral part of the user, actively supporting his or her work, decision making and learning (c.f. Vanharanta et al. 1995). In the "specialty service/product" test case 10 % and "virtual enterprise" 33 % of the subjects experienced "hyperfeeling" and in the "specialty www address" test case 31 % of the subjects experienced "hyperfeeling".

Cognitive Overhead: In this construct the user feels that there is too much information available and that she or he cannot digest it all, e.g. the user does not have the time to try out and check all the available addresses. Moreover, the user's cognitive capacity is not large enough to handle vast amounts of information and the user is therefore reluctant to use it or cannot complete the assignment (c.f. Vanharanta et al. 1995). In the "specialty service/product" test case 55% and "virtual enterprise" test case 43 % of the subjects and in the "specialty www –address" test case 28 % of the subjects experienced "cognitive overhead", e.g. a www page included too much information and too many addresses.

Learning: In the "specialty service/product" test case 45 % of the subjects agreed that the Internet at least partially widened their understanding about the product/service that they needed and 50 % of the subjects agreed, at least partially, to being encouraged to acquire the information from the Internet. However, 20 % of the subjects experienced, at least partially, "holistic" learning. In the "specialty www address" test case, 69%) of the subjects agreed that the Internet application at least partially widened their understanding about the product that they needed and 72 % of the subjects agreed, at least partially, to being encouraged to acquire the information from the Internet application. However, at least partially, 38 % of the subjects experienced "holistic" learning. In the "virtual enterprise" test case 70 % of the subjects agreed that the Internet, at least partially, widened their understanding about the product/service that they needed and 43 % of the subjects experienced, at least partially, "holistic" learning.

Creation of Comprehensive Understanding: In the "specialty service/product" test case, 55 % of the subjects agreed that the Internet at least partially helped to create understanding about the product or the supplier but 35 % of the subjects got sufficient information about the product or the supplier. In general 35% of the subjects experienced "creation of comprehensive understanding." In the "specialty www address" test case, 62 % of the subjects agreed that the Internet at least partially helped to create understanding about the product or the supplier and 69 % of the subjects got sufficient information about the product or the supplier. In general, 62 % of the subjects experienced "creation of comprehensive understanding." In the "virtual enterprise" test case, 63 % of the subjects agreed that the Internet at least partially helped to create understanding about the product or the supplier but only 43 % of the subjects got sufficient information about the product or the supplier. In general 35% of the subjects experienced "creation of comprehensive understanding". It is also significant that in every test case a minority of test subjects did not experience "creation of comprehensive understanding".

Comprehensiveness of the Internet: This construct is concerned with the questions that lead to the evaluation of the knowledge structure and content of the Internet applications. The goal is to reveal important information concerning the utility and future development of Internet applications and their use in supply chain management. In the "specialty service/product" test case, 20 % of the subjects agreed that the Internet, at least partially, makes it easy to compare different products or services and none of the test subjects agreed that the Internet is sufficient for the acquisition function. In general, 5 % of the subjects experienced "comprehensiveness of the Internet". In the "specialty www address" test case, 41 % of the subjects agreed that the Internet, at least partially, makes it easy to compare different products or services and 17 % of the subjects agreed that the Internet is sufficient for the acquisition function. In general only 21 % of the subjects experienced "comprehensiveness of the Internet". In the "virtual enterprise" test case, 30 % of the subjects agreed that the Internet, at least partially, makes it easy to compare different products or services and 43 % of the test subjects agreed that the Internet is sufficient for the acquisition function. In general 30 % of the subjects experienced "comprehensiveness of the Internet".

Understanding the Internet and the Applications. It is important that every test subject understood the test environment and the target of the test before starting the test assignments. In the "specialty service/product" test case 70 % of the subjects understood the goals and possibilities of the Internet in the acquisition function and 75 % of the subjects knew what the Internet in generally contains. In general, 60 % of the subjects experienced "understanding the Internet and the applications". In the "specialty www address" case twenty-five, 86 % of the subjects, understood the objects and possibilities of the Internet in the acquisition function and 93 % of the subjects knew what the Internet generally contains. In general 79 % of the subjects experienced "understanding the Internet and the applications". Correspondingly, in the virtual enterprise 77 % experienced "understanding the Internet and the applications".

Usefulness in Acquisition: In the "specialty service/product" and "virtual enterprise" test cases all of the test subjects agreed that purchasing management is, at least partially, useful for a company. In the "specialty www address" test case 96 % of the subjects agreed that purchasing management is, at least partially, useful for a company.

Utility of the Internet: This group of questions handled the Internet like a hyperknowledge construct. Our purpose was to indicate how well the user can handle the knowledge, which the Internet contains, how aware she or he is of that knowledge, and how close the Internet can be to the ideal hyperknowledge environment. In the "specialty service/product" test case 30 % of the subjects agreed that the Internet, at least partially, can be utilized to support their acquisition decisions. In the "specialty www address" test case the corresponding figure was 59 %. In the "virtual enterprise" test case the figure was 43 %.

6. CONCLUSION AND DISCUSSION

This chapter is based on the idea that the Internet partially determines an ideal hyperknolwledge environment and an assumption that the hyperknowledge environment can be suited to both of our cases. Even though the constructs of hyperknowledge are clearly overlapping, we believe that these more descriptive than normative constructs help us to determine the user's or decision-maker's decision situations in the Internet environment. We also believe that even though these constructs are more unclear than clear, they take at least partially into account the user's body, mind, and situation. As a result of these assumptions, we constructed a questionnaire by which a test of the validity of Internet applications was performed.

In this chapter we have described and evaluated how well some Internet applications, mainly www addresses and www pages, support purchasing decision making and how they meet purchasing requirements in different acquisition situations. In the "specialty service/product" and "virtual enterprise" cases the test subjects needed a product or service but did not have any exact www address or www page. In the second case, the test subject had a particular www address and www page.

In every test case, users generally experienced that Internet applications suited their needs in a purchasing situation. However, experiences varied and in the "specialty service/product" case they perceived some dissatisfaction. 25 % of the test subjects in this case experienced that they were "lost in space", only 10 % of the test subjects experienced "hyperfeeling", and 55 % of the test subjects experienced "cognitive overhead". The corresponding figures of the "specialty www address" test case were "lost in space" 14 %, "hyperfeeling" 31 %, and "cognitive overhead" 28 %. In the "virtual enterprise" case users also perceived some dissatisfaction. 47 % of the test subjects in this case did not experience that they were "lost in space", only 33 % of the test subjects experienced "hyperfeeling", and 13 % of the test subjects did not experience "cognitive overhead". In our test cases the purchasing situations were relatively simple, in more complex situations Internet applications would give less support and would be more inadequate. We believe that these results indicate that Internet applications do not meet the demand for the ideal hyperknowledge environment and thus users/decision makers do not get sufficient decision support from Internet www pages for their decision making.

The contribution of the Internet for the user's purchasing activities was also relatively low. In the "specialty service/product" case zero and in the both "specialty www address" and "virtual enterprise" 17 % of the test subjects agreed that the Internet is sufficient for real purchasing functions. As a result of this, we assume that Internet applications are still inadequate in real purchasing situations, where an application user needs much more specific information for his or her decision making. The available information and the form of the information on www pages or different www' sites seem not to fully support, or satisfy the basic needs of decision makers'.

It is evident that the research results of our test cases are in conflict with the trend in supply chain management, where development is leading to more and more virtuality. In a virtual business world, information management in the supply chain is important and the roles of Internet applications are essential parts of the information acquisition and supply. Our experiences, which are based on our test cases, indicate a great deal of potential new research for example, how to add characteristics of hyperknolwledge to Internet applications. However, we feel that the most important future research will concern the relationships among: decision-maker – supply chain management – the Internet application. There seems to be a need for a concept, which describes these relationships. Especially we need more information about what kind of support from Internet applications the decision-maker will get in purchasing situations. Moreover, we need information about what kind of roles Internet applications play in a company's future supply chain management. All of this means that we need more prototyping and laboratory testing as well as extensive field testing in real management situations. We also need more information for how and in which way the Internet affects in supply chain.

REFERENCES

1. Adrion, W. Brandstad, M. Cherniavsky, J. Validation, verification and testing of computer software. ACM Computing Surveys 1982; 14(2): 159-182
2. Bush, V. As we may think. The Atlantic Monthly 1945; 176(1): 28-32
3. Chang, A-L. Holsapple, C.W. Whinston, A.B., A decision support system theory. Working Paper, 1989, University of Arizona Tucson, USA, 53p.
4. Chang, A-L. Holsapple, C.W. Whinston, A.B. Model management issues and directions. Decision_Support System 1993: 9(1): 19-37
5. Chang, A-L. Holsapple, C.W. Whinston, A.B. A hyperknowledge framework of decision support systems. Information Processing and Management 1994; 30(4): 473-498
6. Conklin, J. Hypertext: An introduction and survey. IEEE Computer 1987; 20(9): 17-41
7. Dos Santos, B.L. Holsapple, C.W. A framework for designing adaptive DSS Interfaces. Decision Support Systems 1989; 5: 1-11
8. Engelbart, D. "A Conceptual Framework for The Augmentation of Man's Intellect". In: *Vistas in Information Handling*, Howerton, P.W. and Weeks, D.C., ed. Cleaver-Hume Press London 1, 1963, pp. 1-29
9. Fox, M. AI and Expert Systems, myths, legends, and facts. IEEE Expert 1990; 5:1: 8-20
10. Mendelhall, W. and Sincich, T., *Statistics for Engineering and the Sciences*. 4th ed., Prentice-Hall, 1995.
11. Nelson, T. Litterary Machines. The Distributors, Indiana 0/1-1/21, 1987

12. Holleran, P.A. A methodological note on pitfalls in usability testing. Behaviour and Information Technology, 1991; 10: 5: 345-357.
13. Newell, A. and Simon, H., *Human Problem Solving*. New Jersey: Prentice Hall, 1972
14. Nielsen, J., *Hypertext and Hypermedia* Academic Press Inc., San Diego, 1990
15. Nielsen, J. *Usability engineering*. Boston: Academic Press Inc1993
16. O'Keefe, R. Balci, O. Smith, E. Validating Expert System performance. IEEE Expert 1987; 2(4): 81-89
17. O'Leary, D. Methods of validating Expert Systems. Interfaces; 1988:18(6):72-79
18. Vanharanta, H. Käkölä, T. Back, B. Validity and utility of Hyperknowledge–based financial benchmarking system, Proceedings of the 28th Annual Hawaii International Conference on System Sciences, 1995, pp. 221-230
19. Woodhead, N. *Hypertext and Hypermedia, theory and applications* Addison-Wesley Publishing Company, 1991

3

PART 3. CONFRONTING NEW WAYS OF WORKING

"Manufacturing businesses are not loosely knit confederations of marketing, engineering, finance and manufacturing activities but are with their suppliers and customers, a single entity needing a unified, integrated effort with all functions working closely together utilizing common strategies." G.W.Plossl, 1991, 'Managing in the New World of Manufacturing', Prentice Hall

"To administer a social organization according to purely technical criteria of rationality is irrational, because it ignores the non rational aspects of social conduct" Peter M. Blau, 1956, 'Bureaucracy in Modern Society', New York: Random House

We see around us a state of upheaval in industry, as businesses face intensive global competition and new consumer demands, and in combating this they have to become more proficient and resourceful, as well as exploring new market opportunities. Large companies have been very active; they have stripped to core competencies; they are in ongoing processes to form new strategic alliances; they have new regional centers for manufacturing to serve local markets and they are very energetic in strengthening their supply chains. Moreover, at the same time they are very large users of ICT. Whereas, the less powerful companies have to increase their strengths in local markets and seek extended market horizons by making much better use of ICT together with organizational changes to gain a competitive advantage. All companies belong to a number of supply chains and any change has to take account of all relationships with external companies. And new communication channels such as the Internet emphasize the linkages between communication, human activity, human behavior and the social structure. The role of national governments is to ensure that not only the right technologies and communication services exist but that they are available to everyone and will serve all on an equitable basis. New technologies create a different potential with different possibilities for suppliers and customers, and in general companies to enjoy and to raise their expectations for the future.

New organizational structures have emerged naturally as business networks to serve a particular objective or to serve a longer-term use with closer partnerships. The dynamics of networks is concerned with new balances between structures and processes that are strategically orientated.

Confronting new ways of working means understanding and finding solutions for network strategies and stimulating collaborative interactive relationships between companies, as well as people, together with the employment of a judicious blend of new tools and ICT. This helps to form and evolve virtual and adaptive value networks with a collective behavior that provides an ability to be innovative, and provide growth and value in conjunction with others to compete in today's highly dynamic business environment and international markets.

Part three of the book deals with analysis of network models and provides guidelines for methodologies and the care and attention necessary to create new organizational networks to serve the consumer, customers, stakeholders and the purpose of future business through interplay with both new ICT and the market.

Chapter 10, written by Rainer Breitz and Hannu Vanharanta, explores ways to examine a value chain to meet the demands of a dynamic environment. They submit that companies must be able to manage and hold their own in several value chains at once in a responsive and flexible manner. But the properties of value chains depend largely on business processes and systems as well as their activities. Their analysis and methodology creates four classifications for the value network, the static value chain, the virtual value chain, the knowledge focused value chain and the dynamic value chain. The properties of how a dynamic value chain may be achieved in the context of the management of technology deployed and the knowledge level of the value chain is explained.

Chapter 11, written by Antonio Lucas Soares and Jorge Pinho de Sousa, has a wide ranging look at social-technological networks and explore some of the issues leading to greater collaboration in business networks. They set out theoretical approaches, methodologies and analyses. They look at how far these networks need to evolve and how new technology, tools and standards are assisting the process. A plea for a greater understanding, through new research, of inter-nodal relationships in networks and ontologies is made. They see in future intelligent agents and peer-peer grid computing as a powerful means of processing network management data making true value networks a reality.

Chapter 12, written by Patrizia Fariselli, takes a wide-ranging view from a number of perspectives of the relationship between ICT and value networks. Concepts of value are discussed within highly dynamic environments. Rigorous cross analyses, with support of technical literature surveys and case studies, provides substantive evidence for symbiotic triggers between ICT and value networks and the catalytic nature of information exchange and knowledge sharing. Conclusions are drawn about the main drivers for change across a range of business and public services and offer a number of pointers to ease the process by considering the barriers, enablers, preferences, choices and investment returns.

Chapter 13, written by Hugh Wilson, Matt Hobbs, Chris Holder, and Malcolm McDonald, deals with the marketing side of demand management. They explore the nuances of the many channels that may be used for customers to communicate with suppliers/vendors. They look at behavior, choice, and the value of customers to an organization. The authors suggest various frameworks for suppliers/vendors to consider when evolving a strategy for investment with the right mix of channels.

Chapter 14, written by Yongjiang Shi, starts from the point that classical manufacturing strategy no longer applies and gives explanations for this. He then goes on with a thorough analysis based on a wealth of knowledge from case studies, involving many industrial areas, over an eight-year period. Multi National Corporations are explored in some depth and compared to collaborative inter-clusters involving SMEs against the background of globalization. A number of different network configurations and capabilities are identified. New conceptual three-dimensional frameworks are described. The reader is guided through the difficult choices posed for industry in their evolutionary path to Global Manufacturing Virtual Networks and the extra work that remains to be done.

Chapter 15, written by Yoon Chang and Duncan McFarlane, reviews the basics of Auto-ID technology and presents some highlights of current research. A number of ICT difficulties in current supply networks are exposed, from company and customer perspectives. The range of manufacturing and business systems used in industry are discussed and problems with identifying, querying and tracking precisely customer orders are illustrated with simple cases for a better understanding of the challenges. The issues for implementation of Auto-ID and RF-ID are raised and simple examples given to illustrate points. The chapter ended with guidelines for future work in this area.

Chapter 10

VALUE CHAIN METHODOLOGY FOR DYNAMIC BUSINESS ENVIRONMENTS

Rainer Breite, Hannu Vanharanta
Tampere University of Technology/ Pori Industrial Management, Finland

Abstract: The requirements of the dynamic business environment have made companies' value chains unstable, which means that companies have more and more been forced to make decisions concerning the value chain and its management. This, in turn, means that companies are changing their value chain management into more external network management, information management and communication management. The dynamic business environment seems to affect both internal as well as external business processes. Its recent leap from internal business processes to external business processes has been very quick, due to different IT-applications. These applications now strongly affect the form and content of the whole value chain, which also sets new demands on company strategies, tactics and daily operations. Managers have to understand how to manage these new structural changes and they also have to understand many new behavioral features in the new type of value chain. In this chapter we have created a method for managers and other decision-makers to examine and manage the value chain in order to meet the demands of the dynamic environment.

Key words: Dynamic business environment, Value chain

1. INTRODUCTION

The concepts of the company, with the aid of which the company creates and adds value, are normally described by supply and value chains. Today the business environments are becoming more and more complex, dynamic, and uncertain, which sets new demands on companies' ability to add value and manage their value chains.

The drivers of a dynamic business situation are supposed to be complex and differ. Moreover, these drivers are suggested to consist of several different elements (Introna, 2001). For example, Stewart (1993) emphasizes elements like the globalization of markets, the development of information technology, the information flow itself, and changes in organizational structure, which are supposed to be elements of the main drivers of a dynamic environment. Drucker (1994) and Ashkenas (1995) emphasize the role of the information technology and organization, as the drivers of the dynamic business environment. Huey (1994) has shown that companies' good capabilities to form different networks are also suggested to increase the dynamics of the business environment. Generally, it can be seen that utilization of information and communication technology (ICT) and remote, virtual working methods, are also to be found in a dynamic business environment. The dynamic business environment affects business processes in the form of rapidly changing technologies, shorter product life cycles, increased competitive pressure, and organizational changes (Davenport, 1993). This also means that a business environment, where customers and working groups vary strongly and where demands on quantity and quality change all the time, and where the production methods change quickly, requires new management methods.

It can also be supposed that the phenomena of New Economy are strongly related to the dynamic environment. The New Economy and its relation to the utilization of the Internet and new computer applications are strong in the dynamic business environment. Moreover, the following new phenomena (see the list below) on both the mental and practical levels clearly increase the dynamics of the business environment:

- A more globally focused company, which aims at a global presence;
- Rampant entrepreneurialism, which means high product/service; innovation and short product and business model life cycles;
- A shift to Service Economy, which means disruptive business systems, outsourcing of all functions, networked structures (in- and outside the company);
- Accelerating creation and diffusion of technology;
- Pervasive connectivity, which means near-zero access costs and plummeting transaction costs.

The New Economy also seems to be driving this change in the value chain thinking towards a new level where information, and information and communication technology (e.g. computers, CD-ROMS, and the Internet), plays the major role.

1.1 Dynamics in Companies

On the company level, the trends of dynamics have meant that companies utilize ICT, especially the Internet, more and more in their daily tasks. This in turn gives companies possibilities to develop more activities which are based on the concepts of virtuality and networking. The business processes of these companies are then becoming more and more visible and transparent. This also means new connectivity and interactivity mechanisms between companies and customers (Dutta and Segev, 2001, Johnston et al. 2001).

The dynamics of the business environment has also strongly affected manufacturing/production philosophies, or so-called "isms", and their development have strongly affected the development of new business processes. "Modern" manufacturing/production philosophies like "lean production" (Womack et al. 1990), "agile manufacturing" (Kidd 1994), and "target costing" (Ansari and Bell 1997, Cooper and Slagmulder 1997) have caused major changes in companies' business processes (also Hammer 1990 and Hall et al. 1993). For example, the "make or buy" strategies of companies have been weighted heavily towards buying; the economic scopes of companies have changed; companies have invested in new Information and Communication Technology (ICT); organizations have been using the concept of networks, etc. It can be supposed that these different manufacturing/production philosophies increase companies' competitive capability in the dynamic business environment.

In this new dynamic management situation we need more understanding about the properties and characteristics of the value chain. The properties of the chain depend largely on the business systems, specific business processes and different activities inside the supply net. We can say that the company's added value activities form its real value chain, and the main cost drivers of these activities form the cost behavior of the value chain (Wilson 1995).

As a result of all this, the role of the management of information and knowledge in companies has become more important. Information glues together value chains, supply chains, and organizations across the entire business economy. The role of ICT is a very strong and important. It changes the properties of companies' value chains and enables new competitors to come out of nowhere (Hof, 2000; Junkkari, 2000; Malin, 2000; Ewans and Wurster, 2000).

How well a company succeeds in its dynamic business environment depends mainly on the company's business processes, or in other words, the company's main activities. How well the business processes suit the demands of the dynamic business environment and how capable the business processes are, depends on the company's abilities to manage its value chain. For these reasons the activities should be organized to produce services or products in the most effective way, so that customers receive the best possible added value, which, in turn means that the company has to manage its value chain in the best possible way.

This new development has forced decision-makers to change their companies' business processes, i.e. the supply and value chains of a product or service. Managers have to combine different knowledge sources to one solid entirety by using information and communication technology in the most innovative way (Hamel 2000 and Cox 1996). In such a dynamic working environment decision processes must be flexible. Thus the management team must able to handle conflict situation in a competent way (Drucker 1985, Vanharanta 1995). In order to secure the future of their companies they need new concepts and methods to support their decision making in the context of the value chain.

The nature of the participation and contribution of the members of the supply and value chains become very important and complex in the situation described above. It is therefore essential for companies to evaluate all their value chain activities as well as human resource allocation and company management throughout the whole supply chain. Moreover, it is important that companies understand new modern technology and its current possibilities (Porter 2001).

1.2 Methodology of the Chapter

We suppose that in the dynamic business environment, companies cannot be successful by managing only one type of value chain, but they need a combination of several different value chains. Based on this supposition, we focused on one main goal: creating a methodology, with the help of which the manager/decision-maker will be able to manage his or her company's value chain in a dynamic environment.

We created a new theoretical management methodology, which is based on several definitions and evaluations of the value chain and value chain management. Based on this theoretical literature review, we introduced four different management concepts, which are based on two clusters, in value chain management: "The Utilization Level of Technology in the Value Chain", and "The Knowledge Level of the Value Chain". These management concepts are related to a company's management situation. We also created fourteen different properties for value chains. By using the different properties of these value chains and four different management concepts, it is possible to form a methodology for value chain management that concentrates on the demands of the dynamic environment in different management situations.

2. THEORETICAL FRAMEWORK

Production philosophies

A new concept with which these requirements of dynamic business environment have been dealt with is agile manufacturing. In this concept companies' capabilities to adapt into dynamic business environment have been improved by adapting into short product life cycles, smaller and smaller repeat order and batch sizes, and an increased variety of products (Kidd 1995). This has meant that companies have been forced to create new concepts which quickly react and respond to fluctuations in demand, i.e. larger flexibility, higher quality, better customized products, etc. (Booth 1996, Kidd 1995). The concept consists of three basic elements: the learning organization, intelligent people, and high-tech technology (Kidd 1995). This strong change process, in turn, has led these companies to a new situation, in which they form virtual enterprises or extended enterprises (cf. Christopher 1998). The main point in virtual or extended enterprises is that each company has concentrated and focused on core competence (people), which supports the whole value chain (Cox 1996, Kidd 1995). Routines and business processes are more and more automated and the whole business system is becoming totally transparent (Sinha 2000). Communication between members of different organizations is carried out using Graphical User Interface (GUI) and keyboards. This in turn affects an active user's personal identity, commitments, social environment and his or her working functions(Flores 1998, Turkle 1995). Moreover, every modern company has to show that it is willing to learn, and they must also possess the necessary high-tech technology for the virtual or extended enterprise. In other words, the company must belong to different networks and supply chain which means that organizations are forced firstly to adapt to the virtual enterprise organization, and then the personnel in different companies have to be capable of working in teams, supporting transparency and crossing the boundaries of companies, with the help of the latest technology.

Supply Chain and Network Management

If a company wants to be a member of the supply chain or value network, it must add value either directly or indirectly to the end customers. If the company adds value directly to the end customer, it is essential to know how to create value for the customer and what things affect and form the customer's value thinking (Porter and Kramer 1999, Kim and Mauborne 1999). Day

(1990) emphasizes value superiority, where customers take into account perceived costs and perceived value.

When the company's added value is examined from the supply chain management point of view, a company's ability to add value to the whole supply chain is essential. Moreover, what is also essential is a single company's ability to acquire value from its suppliers. How value is delivered to customers is also important. Normally, a single company's value adding is described by using the value chain concept. The content of the value chain concept varies from case to case, but a very popular value chain concept and thinking is based on Porter's (1985) value chain. However, during the last few years this value chain thinking has been changed. The utilization of IT for value chain management has concentrated on the development of communication between members of an organization and also between different organizations. This development concentrates on connecting the various members in the value chain to some common network by using EDI, Intranet, Extranet, and Internet applications. The objective seems to be to create an agile manufacturing concept, a virtual network, or a virtual factory (Kidd 1995, Booth 1996, Upton and McAfee 1996).

Information and Communication Technology

The modern way of using information technology in these new applications is to use the Internet. Utilization of the Internet has increased dramatically (see e.g. http://www.isc.org), and so that the Internet's World Wide Web provides a relatively easy, inexpensive way for companies to communicate in a global network. Also small companies can join any network. This removes the distance between a company's suppliers and the distributors of a company's products. It is easy to leave or enter the network, because there is an open protocol (TCP/IP) for all members of the network (Upton and McAfee 1996, Cronin 1996, Martin 1999). Companies' marketing strategies are also affected in many different ways, for example the managing of advertising and collaboration between business partners (Hoffman and Novak 2000). The Internet affects a company so that the company's location becomes less and less of a competitive advantage, while customers' and suppliers' commitments grow weaker. On the other hand, the Internet provides a company with virtual characteristics so that they can extend their own ability to do business. Not only can a business find customers from all over the world, but also a business can find suppliers, distributors, and allies who also add value to its products all over the world. According to Flores (1998) the World Wide Web (WWW) affects a company's identity so that this identity turns into something more like today's brand-name identities and reputations. It is evident that the power of positioning does not depend on a strictly conceived value chain as much as it does on an interlocking set of negotiated relationships in which everyone has at least their minimal conditions of satisfaction met. The Web transforms company positions in two ways (Flores 1998):

- It enhances a company's ability to find suppliers, customers, and value-enhancing associates around the globe;
- On the Web, the best-positioned company is the one that has a site through which the greatest numbers of people pass.

The above mentioned results of the changes in companies mean that decision-making is based more and more on information which the decision-maker gets from a computer's GUI (Graphical User Interface). It is therefore important for a company to know how relationships between people, a company, and technology affect a single company's business processes, that is, its value chain. A new type of value chain management is therefore needed. This trend has led to several new value chain concepts. In the following we will present fourteen characteristics of different value chain concepts, which are based on existing management literature.

2.1 Managing Features of the Value Chains

Our literature study reveals that the general value chain concepts have been described in different ways. The value chain concepts used are based on the model of these authors: Samuelson (1981); Porter (1985, 1998); Day (1990); Hines (1994); Cronin (1995); Rayport and Sviokla (1995); Cokins (1996); Cox (1996); Ansari and Bell (1997); and Christopher (1998). The properties mentioned above cover all the main features of current value chain concepts. In this taxonomy some of the properties of the elements are clearly overlapping and the differences between properties are not fully clear. However, using these fourteen properties, a single type of value chain can be analyzed to find its specific characteristics. The fourteen properties are:

1. Ability to Manage Resource Allocation – Resource allocation in the value chain means that we use resources in an optimal way. The goal of the value chain is that a company's resources are used in a way that adds value to the customer. Thus, in the value chain it is taken into account that all of a company's resource consumption is concentrated on adding value;
2. Effectiveness of Material Fluidity - This property means that right flow of material is an important source of value creation;
3. Information Fluidity - This property means that the appropriate information flow is an important source of value creation inside an

organization (Tapscott 1999). It also means that we need the relevant information for other partners in the network;

4. Utilization of Cost Management – Normally, the term cost management is used to describe the performance of executives and others in the cost implications of their short-term and long-term planning and control functions (Horngren and Foster 1991). Cost accounting, in turn, handles added value in a product oriented way (Horngren and Foster 1991). It examines how to allocate different costs to a product and how different activities add value to the product. The cost management perspective means that the definition of the value chain concept is taken into account to show how efficient resource utilization adds value, and how the value chain is controlled by effective resource utilization;
5. Ability to Sustain Profit Margin – The profit margin is the essential part of the effectiveness of funds fluidity. It is the difference between the cost of value and the price that is paid for added value. In the definition of the value chain concept "profit margin" means that all activities in the value chain should strive to reduce costs and increase prices;
6. Ability to Sustain Value Margin – The value margin is strongly based on Porter's idea that a company's value-making is related to a customer's willingness to pay for products or services continuously;
7. Networking Capability – This property is concerned with how the value chain takes into account a company's external environment. It is important that the definition of the value chain contains elements, which describe relationships within the industry specific network. (Easton 1992);
8. Adaptability to External Changes – This property means that the value chain contains proactive elements to external changes. The value chain should be more "market driven";
9. Integration Degree with the Supplier's Value Chain – The material and the information flows from suppliers are stressed in this property. The idea of this property is based on the lean concept and JIT philosophy, so value is believed to be increased by managing the information and material flows to and from the suppliers (Hines 1994);
10. Integration Degree with the Customer's Value Chain – The material and the information flows from company to its customer are emphasized in this property. The idea of this property is based on the lean concept and JIT philosophy, so value is believed to be increased by managing the information and material flows (Hines 1994);
11. Utilization of Information Technology – In general, technology is an essential element which supports the formation of a technically well-working value chain. Especially understanding the role of information technology in a company's daily operations and improving and managing

these operations with the help of IT are essential sources of efficiency. (Porter 1985, Drucker 1985);

12. Utilization of the Internet –This property means that the construction of the value chain is utilizing modern Internet applications. The Internet is an essential part of electronic commerce (e-commerce), which, in turn, creates very much new added value to an organization (e.g. Tuunainen 1999, Kalakota and Whinston 1997, Martin 1999);
13. Capability of Adapting the Extended Enterprise Concept – Christopher (1998) describes extended enterprise as: "... extended enterprise is a common information "highway". It is the use of shared information that enables cross-functional, horizontal management to become reality. Even more importantly it is information shared between partners in the supply chain that makes possible the responsive flow of product from one end of the pipeline to another.";
14. Accessibility to Virtual Company Structure – Properties of virtuality means that the value chain is based on virtual concepts. Value in an organization is thought to form if the activities of the organization utilize virtuality principles or a concept of virtual enterprises. This also means that the construction of the value chain is essentially based on virtuality. Virtuality is here defined as a virtual organization or virtual enterprise. Therefore, the concept of virtuality includes the definition of remote work (Stables and Ratnasinghamn 1998), virtual organization (e.g. O'Leary 1998, Park et al. 1993) and virtual enterprise concepts (e.g. Goldman et al. 1995). These different definitions of "virtual" are not completely separated from the concepts of network, IT, Internet and extended enterprise, but it is more useful to handle these concepts separately, because features of the value chain are easier to discern when these categories are presented separately.

Most of the properties mentioned above contain some characteristics of technology, either directly or indirectly. Especially information technology (more often the Internet) affects many properties. It seems to directly affect the following properties: *Accessibility to Virtual Company Structure; Utilization of Information Technology; Utilization of the Internet; Adaptability to External Changes; and Capability to Adapt Extended Enterprise.* The technological (ICT) part is notable also in the following properties: *Networking Capability; Effectiveness of Information Fluidity; Integration Degree with the Supplier's Value Chain;* and *Integration Degree with the Customer's Value Chain.* The result of these perceptions is that information technology strongly affects the value chain concept and a company's ability to add value to its organization. This also supports the general understanding that information, knowledge, and information technology affect a company's ability to add value. They form an essential part of a company's functions (Earl 1989, Porter 1998 and 1985, Drucker, 1985, 1998 and 1988).

Figure 10-1 illustrates how the different value chain concepts are comprised of the above properties. It can be seen from the figure that most of the different value concepts are marked either fully (•) or partially (◐) with these properties. There are ten different value chain concepts and in each of them information technology is playing a key role. Four value chains have this property fully in use and five value chains have this property at least partially. Only one property is not used at all. This analysis reveals that information technology is very well embedded in value chains. Figure 10-1 also illustrates how a company can create a methodology for its e business strategy (follow line --o--). Companies have to stress characteristics, which promote virtuality and the use of IT. Moreover, it also illustrates an example of an ideal value chain in the context of the Internet environment (follow line --□--).

Number	Properties of Value Chain / Authors and Type of Value Chain	Samuelson	Porter	Day	Hines	Cronin	Rayport & Sviokla	Cokins	Cox	Ansari & Bell	Cristopher
1	Ability to Manage Resource Allocation	●	◐	◐	◐	○	[●]	◐	○	◐	◐
2	Material Fluidity	●	◐	◐	◐	●	[●]	◐	○	◐	●
3	Information Fluidity	○	○	○	●	(●)	◐	○	[●]	●	◐
4	Utilization of Cost Management	◐	○	○	○	●	●	[●]	○	○	(●)
5	Ability to Sustain Profit Margin	●	◐	◐	◐	◐	[●]	◐	○	◐	◐
6	Ability to Sustain Value Margin	○	●	●	[●]	●	○	●	●	(●)	◐
7	Networking Capability	○	◐	◐	(●)	○	○	○	◐	◐	[●]
8	Adaptability to External Changes	○	●	●	(●)	●	◐	●	[●]	●	◐
9	Integration Degree with Supplier's VC	○	(●)	[●]	◐	◐	○	●	●	●	◐
10	Integration Degree with Customer's VC	○	(●)	●	◐	◐	[●]	○	◐	◐	◐
11	Utilization of Information Technology	○	○	○	◐	○	○	○	◐	●	[●]
12	Utilization of the Internet	◐	○	○	◐	[●]	(●)	○	◐	○	●
13	Capability to Adapt the Extended Enterprise Concept	◐	◐	(●)	◐	[●]	◐	○	◐	◐	◐
14	Accessibility to Virtual Company Structure	○	◐	◐	(●)	○	○	○	◐	◐	[●]

● Value chain contains this property
◐ Value chain contains partially this property
○ Value chain does not contain this property

—○— A road map for a hypothetical company utilizing e- business concepts

—□— A theoretical example of ideal value chain in the context of the New Economy

Figure 10-1. Features and the Different Value Chains

2.2 Classification of Value Chain Management

In order to understand value chain management better, we have divided those different value chain management concepts into four different categories, which are based on the properties of the value chains presented. In figure 10-2 the categories are arranged so that they describe how the value chain concept supports the two main factors already presented in the supply chain analysis, i.e. *Utility Level of Technology in Value Chain,* and *Knowledge Level of Value Chain.* From these dimensions we have created four classification categories: the "*Static Value Chain*"; the "*Virtual Value Chain*"; the "*Knowledge Focused Value Chain*"; and the "*Dynamic Value Chain*".

Utility Level of Technology in Value Chain indicates how well the value described chain concept takes into account the technical connection to a supply chain. This dimension illustrates technical connections to suppliers and customers. If the degree of technical supply chain integration is high, the value chain concept allows, and indeed requires, us to follow the changes of product specifications and to take into account different suppliers and distributors. If the degree of technical supply chain integration is low, the value chain concept is usually part of the company's normal activities and reflects mainly a company's internal functions.

Knowledge Level of Value Chain indicates how sensitive the value chain is to changes in the market. If both the knowledge level of the supply chain and the ability to explain the market are high, the value chain pulls information into the supply chain. The value chain members manage information in a proactive manner so that the information indicates future changes in the market's demands. If the ability to explain the market is low means that the value chain is passive also to any changes in the market.

The first category, *Static Value Chain Management,* describes a company's state at a certain moment. The management methods in this kind of value chain are usually functional descriptions of a company's activities, where the position of suppliers and the relationship with retailers, distributors, and markets has not been taken into account. The value chains which lie in this area usually stress the following properties (Figure 10-1): utilization of cost management (7); ability to sustain profit margin (8); ability to manage resource allocation (9); and ability to sustain value margin (10).

The second category, *Knowledge Focused Value Chain Management,* contains a value chain management method that usually concentrates only on a company but does not take into account the technology dependent relationship to suppliers. The market driven value chain strongly indicates market needs and thus this management method takes into account the changes in the market and connects retailers, distributors and end customers to the company. The value chains in this area usually consist of the following properties (Figure 10-1): utilization of cost management (7); ability to sustain profit margin (8); ability to manage resource allocation (9); ability to sustain value margin (10); adaptability to external changes (11); and integration degree with customer's value chain (13).

The third category, *Virtual Value Chain Management*, has a network viewpoint along with strong utilization of information technology (IT). In this category the main idea is that the company's value chain is tightly connected to the supply chain. The goal of the strong relationship is to produce material, information and knowledge for the market. Relationships between all members of the chain are maintained with IT. Usually the concept does not indicate changes in the market, or how in the future a customer's needs change. In this area the value chains have the following properties (Figure 10-1): accessibility to virtual company structure (1) is strongly supported; utilization of technology (3) is high; utilization of cost management (7) is applied; the ability to sustain profit margin (8) is strong; the ability to manage resource allocation (9) is good; and the ability to sustain the value margin (10) is possible.

The fourth category, *Dynamic Value Chain Management*, which the industry leaders utilize, is a management method that is usually based on a network viewpoint and strong utilization of IT. The main idea is that the value chain should be an inseparable part of the supply chain and that the value chain has strong connections to other parts of the supply chain. Moreover, the whole concept contains structures that immediately reveal changes in the market. All agreements are based on the win-win principle with suppliers and customers and on a transparent information flow from markets down to suppliers. This type of value chain fulfils most of the important properties mentioned above. By applying these four areas, we have now created a new emphasis for value chain management (Figure 10-2). The main factors, which affect this development, are companies' utilization of IT and the knowledge level of the value chain. Behind these concepts are the properties with which we can achieve a dynamic value chain management. It seems that a high degree of supply chain integration is not enough; companies need a high ability to explain markets. Technological skills alone are not enough either. Instead, managerial skills are needed to control technology and to understand the relationships between people, company and technology.

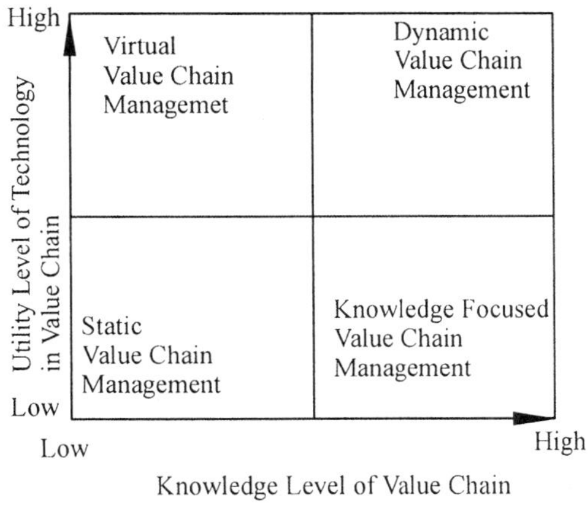

Figure 10-2. The Fourfold Matrix for Value Chain Management

3. CONCLUSIONS

During the last few years the trends mentioned above have changed the demands of the value chain concepts towards virtual and dynamic directions. IT has enabled new technological networks, which can connect several companies virtually and create extended enterprise models, where information flow and transactions are fast and reliable. New value structures reflect business processes and functions that respond to changes in the dynamic business environment. In addition, these new technical networks connect suppliers and customers to the company more tightly.

In this situation, organizations and members of organizations have to adapt to the demands of the new era. The everyday functions of people consist of more and more information handling and communications, which are strongly related to information technology, especially computers: the Internet, the Extranet and the Intranet. We believe that one single value chain does not sufficiently support managers' decision-making in a complex and dynamic business environment. There are also many behavioral as well as decision-making problems that are related to value chain management. The outcome seems to be that individuals and management in an organization feel that they are powerless to control and handle the value chain. It is therefore important that a manager as a decision-maker possesses the methodology to estimate and develop the company's value chain.

Through our research we have shown that the value chain concept and the role of IT in the value chain concept require a new supporting framework for management. This new framework, in turn, defines and describes a new type of management methodology, which concentrates on controlling the relationship between people (the manager as a decision-maker), company, and technology in the value chain concept.

REFERENCES

1. Ansari, S. L. and Bell, J. E., *Target Costing- The Next Frontier in Strategic Cost Management*. Irwin, 250 p, 1997.
2. Ashkenas, R. Capability: strategic tool for a competitive edge. Journal of Business Strategy 1995; 16: 13-14.
3. Booth, R. Agile manufacturing. Engineering Management Journal 1996; April: 105-112.
4. Cokins, G., *Activity-Based Cost Management: Making It Work: A Manager's Guide to Implementing and Sustaining an Effective ABC system*. Irwin, 1996.
5. Cooper, R. and Slagmulder, R., *Target Costing and Value Engineering*. Portland, Oregon: Productivity Press, 1997.
6. Cox, A. Relational competence and strategic procurement management. European Journal of Purchasing and Supply Management 1996; 2(1): 57-70
7. Cristopher, M., Logistics and Supply Chain Management – Strategies for Reducing Cost and Improving Service. Financial Times Management, 1998
8. Cronin, M. J., *Doing More Business on the Internet*. 2 nd edition. New York: Van Nostrand Reinhold Co 1995.
9. Davenport, T.H., *Process innovation: re-engineering work through information technology*. Harvard Business School Press, 1993.
10. Day, D. S., *Market Driven Strategy: Processes for Creating Value*. The Free Press, 1990.
11. Drucker, P., *Management: Tasks Responsibilities Practices*. New York: Harper and Row Publishers Inc., 1985
12. Drucker, P. The coming of the new organization. Harvard Business Review 1988; January-February: 45-53.
13. Drucker, P. The theory of business. Harvard Business Review 1994; 72: 95-104
14. Drucker, P., *On the Profession of Management*, A Harvard Business Review Book, 1998.
15. Dutta, S., Segev, A., "Business transformation on the Internet". In *E-commerce and V-Business: Business Models for Global Success*, Stuart Barnes and Brian Hunt, ed. Butterworth-Heineman, 2001
16. Earl, M. J., *Management Strategies for Information Technology*. Prentice Hall International (UK) Ltd, 1989.
17. Easton, G., "Industrial networks: a review" In *Industrial networks: A new view of reality* Axelsson, B., and Easton. G., ed. London: Routledge, 1992.
18. Evans, P., Wurster, T. Getting Blown to Bits. The Australian Financial Review 2000; Easter Special 20-25 April: 3-8
19. Flores, F. Information technology and the institution of identity. Information Technology and People 1998; 11(4): 351-372
20. Goldman, S., Nagel, R. and Preiss, K., *Agile Competitors and Virtual Organizations*. New York: Van Nostrand Reinhold, 1995
21. Hall, G., Rosenthal, J., Wade, J. How to make reengineering really work. Harvard Business Review 1993; May-June 1993: 119-131.
22. Hamel, G., *Leading the revolution*. Boston, Massachusesetts, Harvard Business School Press, 2000.

23. Hammer, M. Reengineering work: Don't automate, obliterate. Harvard Business Review 1990; May-June: 104-112.
24. Hines, P., *Creating World Class Suppliers*. London: Pitman Publishing, 1994
25. Hofman, D. L., Novak, T.P., How to acquire customers on the Web. Harvard Business Review, 2000; May-June 2000: 179-188.
26. Hof, R.D. Who will profit from the Internet agora? Business Week, European Edition, 2000; June 5: 20-30.
27. Horngren, C. T., and F oster, G., *Cost Accounting – A Managerial Emphasis*. Seventh edition. Prentice-Hall International, 1992.
28. Huey, J.F. The new post-heroic leadership. Fortune 1994; 129: 42-50
29. Introna, L., "Defining the virtual organizations". In *E-commerce and V-Business: Business Models for Global Success*. Stuart Barnes and Brian Hunt, ed. Butterworth-Heineman, 2001.
30. Johnston, R.B., Mak, H.C., Kurnia, S., "The contribution of Internet electronic commerce to advanced supply chain reform – a case study". In *E-commerce and V-Business: Business Models for Global Success*. Stuart Barnes and Brian Hunt, ed. Butterworth-Heineman, 2001
31. Junkkari, M. Uuden talouden supertähdet ovat vapaallakin töissä. Helsingin Sanomat 2000; 16:.3.
32. Kalakota, R. and Whinston, A. B., *Electronic Commerce – A Managers Guide*. Addison-Wesley, 1997.
33. Kidd, P.T., *Agile Manufacturing – Forging New Frontiers*. Addison-Wesley Publishing Company, 1994.
35. Kim, W.C., Mauborngne, R. Creating new market space. Harvard Business Review 1999; January-February: 83-93.
36. Malin, R. (2000) Uusi talous tulee sisältä päin. Talouselämä 20/2000 pp. 92-94.
37. Martin, C., *Net Future*. McGraw-Hill Inc., 1999
38. O'Leary, D. E., Virtual Organizations: Two choice problems. Proceedings of the Nineteenth International Conference on Information Systems. Helsinki, Finland pp. 145-154, 1998.
39. Park, H., Tenenbaum, J., Dove, R. Agile infrastructure for manufacturing systems (AIMS). Proceedings of Defence Manufacturing Conference, 1993.
40. Porter, M. E., *Competitive Advantage. Creating and Sustaining Superior Performance*. New York: Free Press, 1985.
41. Porter, M. E., Millar, V.E. How information gives you competitive advantage. Harvard Business Review 1985; July-August: 149-160.
42. Porter, M. E., *On Competition*. A Harvard Business Review Book,1998.
43. Porter, M.E., Kramer, M.R. Philanthropy's new agenda: creating value. Harvard Business Review 1999; November-December: 121-130.
44. Porter, M.E. Strategy and the internet. Harvard Business Review 2001; March 2001
45. Rayport, J.F., Sviokla, J.J. Exploiting the virtual value chain. Harvard Business Review 1995; November-December: 75 – 85.
46. Samuelson, K., InformatiCom and Multiway Video Communication as a Cybernetic Design and General Systems Technology. 25th Annual Conf. of SGSR Society for General Systems Research, 5-10 Jan. 1981, Toronto, Canada. pp. 207-214, 1981.
47. Sinha, I. Cost Transparency, The Net's real threat to prices and brands. Harvard Business Review 2000; March-April: 43-50

48. Staples, D.S., Ratnasingham, P., Thrust: The panacea of virtual management?, Proceedings of the Nineteenth International Conference on Information Systems. Helsinki, Finland, pp. 128-144, 1998.
49. Stewart, T.A. Welcome to the revolution. Fortune 1993; 28: 32-38
50. Tapscott, D., *Creating Value in the Network Economy.* A Harvard Business Review Book, 1999
51. Turkle, S., *Life on the Screen: Identity in the Age of the Internet.* Simon and Schuster, New York, NY, 1995
52. Tuunainen, V.K., Different Models of Electronic Commerce - Integration of Value Chains and Business Processes. Helsinki School of Economics and Business Administration, A-153, 1999.
53. Upton, D. M., McAfee, A. The real virtual factory. Harvard Business Review 1996; July-August: 123-133.
54. Vanharanta, H., Hyperknowledge and Continuous Strategy in Executive Support Systems. Abå Akademi University Press, 1995
55. Wilson, R.M.S. (1995) "Strategic Management Accounting (pp 159-190)", In *Issues in Management Accounting.* Ashton, D. Hopper, T. and Scapens R., ed. Prentice Hall, 1995.
56. Womack, J.P., Jones, D.T., and Roos, D. *The Machine That Changed The World.* HarperCollins; Reprint edition, November 1991
57. Internet Software Consortium, http://www.isc.org

Chapter 11

MODELS AND SYSTEMS TO MANAGE HIGH VALUE SOCIO-TECHNICAL NETWORKS

António Lucas Soares, Jorge Pinho de Sousa
INESC Porto and Faculty of Engineering, University of Porto

Abstract: This paper presents concepts and approaches to build models and systems for the management of High Value Socio-technical networks. A justification of the social networking perspective as applied to business organizations is made, focusing on methods and theoretical approaches. Then, it is speculated how social networking can be a paradigm for the management of networks of business organizations. The management of relationships (fostering collaboration and cooperation) and their support through new information system architectures are the key issues. Socio-technical networks are analyzed from this point of view, and focusing on the aspects of coordination and collaboration. Some research directions using the networking paradigm are discussed, focusing on the modeling tools for inter-organizational systems analysis and implementation.

Keywords: Networking theories, Socio-technical networks, Inter-organizational systems

1. INTRODUCTION

In the last years there has been an interest toward the networking concept as applied to management and technology studies. The general theoretical framework of the networked society (Castells 1996) strongly supported by a set of comprehensive empirical studies, settled up a new analysis perspective triggering several research programs developing and using network-centric theories and methodologies. "Networks are the fundamental stuff of which new organizations are and will be made" says Castells adding further that the network will become the basic unit of economic organization. The networking phenomena are not exclusive of the area of economics but spread to all areas of social activity, from solidarity to culture. It is within this context that socio-technical networks are nowadays developing. Today we find supplier networks, producer networks, customer networks, coalitions for standards, technology cooperation networks, research networks, collaborative heterogeneous networks, etc. which become more and more complex, and more and more demanding in terms of coordination, collaboration and general management to become and remain high valued. These requirements call for new perspectives and paradigms in the engineering and use of information technologies and systems, in particular in the design of inter-organizational systems as technological backbone in socio-technical networks. High value business networks can only be adequately studied through a socio-technical perspective (Lamb et al. 2001). Therefore, we use in this c hapter the term socio-technical network as the main unit of analysis.

In the next section the networking paradigm is presented and its importance in the s tudy o f s ocio-technical n etworks is justified. S ection 3 summarizes some of the key issues in the collaboration between businesses, presenting an outline of the state-of-the art in the frameworks to link business processes in different companies and presenting some of the more important conceptualizations of inter-organizational systems. Section 4 points to research directions in applying networking theories to the development of business networks. It ends with a speculation of how management and decision support tools should be from a strict networking perspective.

2. RELATIONSHIPS, NETWORKS AND NETWORKING

2.1 From Chains to Networks

In the manufacturing world, supply chains have been the most scattered type of network of organizations. In the last years, t he concept of supply chain has gone beyond the focus on logistic processes and is nowadays advancing steadily towards the integration of key business processes, from suppliers to final customers, encompassing products, services and information (Lambert and Coper 2000). More and more, companies interact with the clear purpose of cooperation. Some authors call this *networking* (Magnusson and Nilsson 2003, Triekenens 2002).

Generalizing, we can consider the socio-technical network concept as a physically decentralized social network made up of individuals who form a community but are not members of the same formal organization (Howard 2002). These socio-technical networks are called for example 'communities of practice' in sociology (Bijker et al. 1987) or 'knowledge networks' in management (Podolny and Page 1998, Uzzi 1996). We can identify socio-technical networks in each of the specific types of networks enumerated above.

For example within the manufacturing area, knowledge networks are becoming particularly important. One key factor in such networks is the ability to integrate the knowledge managed by each of the organizations. This happens in supply chains, where the level of knowledge integration is low, in business networks, built upon a central actor (hub-firm) that identifies a business opportunity and creates a network to satisfy that need, and have a medium knowledge integration level, and in research networks where knowledge creation is the main goal and a high level of knowledge integration is achieved (Magnusson and Nilsson 2003). A range of methodological approaches have been used to study these social phenomena (depending of course on the aim of the study), but every approach is built upon a set of models used to explain or just describe relevant aspects of those phenomena (Howard 2002).

Networks can be seen as an evolution of chains in the sense that they possess a stronger social structure, which helps to generate trust, to stimulate networked learning and to internalize external issues such as technologies or standards (Jonkers et al. 2001).

2.2 Studying Business Collaboration and Cooperation as Networking

The study of socio-technical networks is not new and has been within the interests of disciplines such as economy, management, sociology and information systems for a long time. Nevertheless, in the last few years, there was a renewed interest in this area mainly because of the synergies created between information technology and new business models based on networks of companies. With respect to new business models, Trienekens (2002) analyses comprehensively the visions and conceptualizations of inter-company relationships, focusing on two complementary approaches for the design of relationships: (1) vertical relationships between companies (value chain management, supply chain management) considering the transaction costs theory and activity theory as fundamental conceptualizations, and (2) the position of the company within its business and within the socio-economic environment (strategic management) being conceptualized by networking and social capital theories. The first approach has been extensively dealt with in the literature (see e.g. Lazzarini et al. 2001). For us more promising work is the work reported by Trienekens (2002) which is focused on the relationship between companies with the other businesses, organizations and institutions.

In the theories of networking the forms of collaboration are not only based in economic motivations, but also power and trust are key concepts (Uzzi 1997). Social capital theory is centered on the capitalization of relationships, which is becoming a branch more and more important within the network approaches. "Social capital metaphor means that people do better when they are connected" (Burt 2001).

The objects of study of networks and chains encompass, besides the actors and their relationships, the patterns of activities creating and transferring value objects. These patterns are referred as the network business processes and the pattern of relationships associated as network structures (Jonkers et al. 2001). Nevertheless, there is a complex inter-relation between structure and processes in networks of organizations (Soares et al. 2003).

The objects of study of networks and chains can also be analyzed from three main viewpoints: (1) socio-economic, where networks and chains are characterized by the creation and exchange of value through specific functions and mechanisms, by actors having common or conflicting interests; (2) informational, where networks and chains are viewed as information processing systems required for control and management; (3) technical, where networks and chains are viewed as distributed production systems characterized by technical and logistics transformations.

2.3 New Paradigms and Models

In business and technology studies, and particularly in the topics of virtual organizations and collaborative networks development, network-centric approaches are being extensively used. Two of these approaches are currently been given some focus (Lamb et al. 2001): social network analysis and actor-network theory.

The concepts of Graph Theory have an enormous potential in modeling Collaborative Networks (CN) and Virtual Organizations (VO), as a natural framework for representing relationships, interactions and collaboration. This framework is supported by a graphical representation of entities and relationships, and enhances the understanding and sharing of network relationships, supporting network management methods and algorithms.

In general, a network is a set of nodes (entities) and channels linking pairs of nodes. Both nodes and channels may have several attributes that may, for example, quantify the average amount of exchanged information, the level of trust between partners, or the degree of ownership between companies. This "open" view of a network should allow us to model quite different situations and problems, such as client/supplier or co-operative planning environments, technical cooperation, informal information exchange, interest groups, etc.

Although sometimes not in a straightforward way, complex and non-tangible factors (i.e. factors which are crucial in the set-up and management of VO) can also be modeled by this type of framework

There are a lot of commercially available network optimization tools; packages for supporting network design; and software for drawing several types of networks. These tools are used for modeling interactions and quantitative relationships, but generally they have a very limited capacity to represent non-tangible, qualitative relationships (such as the level of trust or the exchange of tacit information); and to deal with multiple criteria in evaluating different operating alternatives.

Although Graph Theory has been extensively used in other domains for a long time, there are few practical applications in supply chains, in enterprise networks or in virtual organizations.

Social network analysis builds on a long tradition of social-structural perspectives. Nevertheless, there was recently a resurgence of this approach (Burt 2000, Zack 2000, Lamb et al. 2001). In networked or virtual organizations this approach is well-suited as social structure is not bound to the formal organization and the analysis may apply naturally to inter-organizational relationships. For example, the work of Wellman et al. (1996) on cooperative work and telework builds directly on social network concepts. Gitell's work (2000) also highlights the role of social networks and the use of various enabling technologies (including ICTs) to support forms of relational coordination (Lamb et al. 2001).

Another new research direction also draws on sociological theory to develop a better understanding of social interaction and information technology and systems. Latour's (1987) actor-network theory (ANT) combines the broad-scale thinking of the SST tradition with new conceptualizations that raise technologies (such as computers and networks) to an equal status with human actors (Lamb and Kling 2001). This perspective explores the intricate interrelationships that develop between people and the technologies they employ to interact with other individuals, organizations and institutions within complex, interconnected networks (Walsham 1997). By considering the connection between human and non-human actors, each of them with the same degree of importance (from an analytical point of view) ANT seems an important and refreshing framework for the analysis of socio-technical networks.

3. COLLABORATION IN SOCIO-TECHNICAL NETWORKS

The term "collaboration" has been used for a lot of business scenarios, with slightly different meanings, becoming therefore too much vague. An encompassing but still useful definition is proposed in (Hall 1999): "Collaboration is broadly defined as the interaction among two or more individuals and can encompass a variety of behaviors, including communication, information sharing, coordination, cooperation, problem solving, and negotiation".

3.1 How Do Businesses Collaborate Today

The streamlining of internal business processes has enabled a substantial part of this evolution, supported by projects of ERP implementation and integration, mainly aiming at automating and integrating internal business processes. The challenge, to go beyond this streamlining and to tackle inter-enterprise business processes, has emerged as a natural way to enable a new round of efficiency gains. In this business strategy, which has been called collaborative business (c-business), business partners in the whole value network collaborate, sharing information and processes, to provide a higher value product/service, with increased profitability.

C-business is based on the enhanced connection of participants in Internet-based value networks, and, as such, it may be regarded as an evolution of e-business. Naturally, c-business leverages the potential for value creation in e-business by enhancing opportunities for efficiency gains, exploitation of complementarities, customer and partner lock-in, and the introduction of novel products, services, processes and new business and revenue models (Amit and Zott 2001).

In very broad terms, c-business inhibitors may fit into three categories: financial bottlenecks; trust and organizational concerns; and technological issues. In the current context of very slow economic growth, with many enterprises still dealing with the aftermath of the dot.com bubble, an increasing erosion of margins, and doubts cast on the cost/benefit potential of c-business technologies, it has become increasingly difficult for providers and customers to invest in solutions in this area.

Alongside financial obstacles, trust and organizational issues firmly stand in the way: distrust between parties; functional "stovepipes"; difficulties in reaching goals and cost/benefit sharing agreements; difficulties in business and technology redesign; and inexperienced management (Those are some of the most frequently cited obstacles in surveys on c-business).

Perhaps surprisingly, technological issues do not appear to be a major concern, when compared to the previous categories of inhibitors. In this area, some of the major concerns are: the difficulties with systems integration; data protection and security; and the overall confidence of individuals and companies in electronic business systems.

3.2 Current Frameworks for Business Process Integration

Nowadays there is a strong activity in the area of integration toolsets that support both internal (Enterprise Application Integration) EAI and external B2B interactions. The main issue is to integrate inter-organizational business processes. Business networks need technologies and systems for managing inter-organizational business processes that are easy to use, flexible and capable of integrating systems across organizational boundaries and technology barriers. The goal is: to integrate systems; to automate routine activities; to manage all phases of processes; to deploy processes seamlessly and to provide end-to-end visibility and control.

Tools for inter-organizational process management should enable coordination and cooperation among distinct processes at many levels: applications; systems; business units; the individual corporation; and the extended enterprise. E-business in general has created a new set of requirements and challenges that are leading to the development of new advanced architectures and technologies for the design and deployment of information systems. Among these are the frameworks for business-to-business integration, which are the basic infra-structure to implement more complex systems supporting the management of high value networks from a socio-technical networking perspective. State-of-the art B2B frameworks are presented below.

ebXML (Electronic Business using eXtensible Markup Language) enables companies to exchange business messages, conduct trading relationships, communicate data in common terms and define and register business processes (http://www.ebXML.org). It aims at providing an open XML-based infrastructure enabling the global use of electronic business information in an interoperable, secure and consistent manner by all parties. The way each party can exchange information using ebXML, in the context of Business Collaboration, can be described by a Collaboration-Protocol Profile (CPP). The agreement between the parties can be expressed as a Collaboration-Protocol Agreement (CPA). The CPP describes the capabilities of an individual party and CPA describes the capabilities that two parties have agreed to use to perform a particular Business Collaboration. These CPAs define the terms and conditions that enable Business documents to be electronically interchanged between Parties. To enable parties wishing to do Business to find other parties that are suitable Business Partners, CPPs may be stored in a repository such as is provided by the ebXML Registry (ebRS).

RosettaNet is a non-profit consortium working to create, implement and promote open e-business process standards (http://www.rosettaNet.org). RosettaNet leverages existing open e-business standards, guidelines and specifications for cross-platform, -application and -network communication. RosettaNet develops and produces three elements, to support the partners in the network running their B2B processes: "Partner Interface Process" - PIPs define B2B processes agreed upon by the partners; RosettaNet Implementation Framework (RNIF) - provides common exchange protocols; Dictionaries (business and technical) - provide a common set of properties for business transactions and products.

The Trading Partner Agreement Markup Language (tpaML) is used to describe the terms and conditions of the electronic interaction between businesses and can be expressed as an electronic contract or a Trading Partner Agreement (TPA). The TPA expresses the rules of interaction between the parties while maintaining complete independence of the internal process at each party from the other parties (http://xml.coverpages.org/tpa.html). It uses XML to define and implement electronic contracts. This is already a critical issue in the domain of collaborative networks and "extended products". With the growing complexity of products and services, as well as the increasing sophistication in legislation and regulations, this will become an even more important research and development topic.

3.3 State-of-the Art of Inter-Organizational Systems

Inter-organizational information systems are built on top of the previously described frameworks. Major software providers already supply sophisticated solutions for managing supply-chains and customer-chains. Powerful groups are joining efforts for managing their businesses by recurring to comprehensive vertical portals. Nevertheless, there are not available yet neither the concepts, nor the applications, nor the tools to consider the management of relationships between companies as the fundamental way of managing modern businesses. We present now some fundamental concepts of inter-organizational systems in order to understand how far we are from the ideal concepts of socio-technical networking.

The term "Inter-Organisational Systems" (IOS) was introduced in the beginning of the 80's by Barrett and Konsynsky (1982) to refer to a computer-based information system shared by two or more organizations. An IOS may therefore be defined as an IS built upon a network, that goes beyond the borders of the traditional company. Research in the domain has led to results that emphasize various aspects of inter-organizational networking such as inter-organizational relationships, the strategic planning of IOS (McFarlan et al. 1983), and their network structure (Malone et al. 1999).

More recently some emphasis has been given to issues related to inter-organizational dependencies. Kumar and Dissel (1996) view IOS as a technology designed and implemented to simplify relations between organizations. Thompson (1967) proposed the well known framework based on three types of inter-dependency involving information resources, value/supply chains, and networks:

a) shared inter-dependency: information resources are shared and used as common. As examples you may have common databases, communication networks and applications, leading to scale economies and sharing of costs and risks;
b) sequential inter-dependency results directly from customer-supplier relationships along the value chain. These IOS reflect the sequential dependency among organizations. The output of a node is therefore the input of another node;
c) reciprocal inter-dependency is associated to networks where organizations exchange information with mutual advantages, thus eventually creating rather complex webs.

With the exponentially growing use of the Internet by companies to conduct commercial transactions, the research in the domain has naturally become closer to that pursued in the e-commerce field. For example, Sawney and Kaplan (1999) suggest that e-marketplaces linking suppliers and buyers may be viewed as IOS, that may be classified into two types (according to their purpose or strategic positioning): (1) horizontal, organized around a given industrial sector, e.g. as a way to link companies of a single industry, or promoting the cooperation among competitors; (2) vertical, focusing on specific functions or business processes, as it is the case of some heterogeneous value chains or when a company promotes some type of partnership with customers, suppliers, or even with organizations that provide complementary products, resources or services.

Another perspective on IOS comes from considering the level of decision support provided (Farbey et al. 1995): (1) operational, e.g. focus is put on the automation of processes and mainly aiming at increasing efficiency; or (2) strategic, e.g. global transformations of the business may take place. In this line, and assuming that the connections of the value adding activities are a good basis to understand the features of an IOS, Hong (2002) classifies IOS as providing one of the following functionality:

- resource sharing: among participants that perform activities of similar value (e.g., competitors), leading to sharing costs, risks and resources, and promoting coalitions to compete with large companies;
- complementary cooperation: co-operating firms perform different functions within the value chain of a given industry, and therefore they expand their business capacity beyond their own resources; In general they aim at accessing market by providing complementary advantages and services;
- operational cooperation: firms in the same value chain that aim at enhancing the quality of their services and to share information with competitors in order to improve common operations; This leads to some kind of "virtual organizations";
- operational coordination: an IOS may be configured to interlink different functions performed by different companies, in a given industrial value chain, with increased operational efficiency.

3.4 New Forms of Collaboration

A large number of these initiatives are taking place in the discrete manufacturing and distribution sectors. In the first case, these initiatives fundamentally target the most sensitive areas of collaborative engineering, (shortening) product life cycle management and outsourcing (supplier) management. For the distribution sector, specific targeted areas include order, inventory and transportation management. In the process and asset-intensive sectors, the number of initiatives is significantly smaller, focusing essentially on the management of capital assets and suppliers. Obviously, the greatest challenge faced today by technology providers is the unification of collaboration services with other technologies, in particular ERP, in order to offer solutions that simultaneously and coherently support business and work processes, and the integration and optimization of internal and external business process. This is a very important part of the requirements for a new generation of ERP solutions, which has been called ERP II.

Initiatives in the area of Collaborative Business have been taking place for a few years, supported by current technological solutions such as Customer Relationship Management (CRM), Supply Chain Management (SCM), Collaborative Planning, Forecasting and Replenishment (CPFR), Simple Extranets, e-Markets, Collaborative Engineering Systems, Direct Procurement, etc. A very significant number of these initiatives are taking place in the discrete manufacturing and distribution sectors. Significant challenges faced by technology providers in this area are arising from financial bottlenecks, trust and organizational concerns and other technological issues. Several aspects of C-Business technologies require critical consideration with respect to their cost/benefit trade-off, their ability to adequately support business redesign, and issues of systems integration, data protection and security.

One particularly important technological challenge is the unification of collaboration services with other technologies, in particular ERP, in order to offer solutions that simultaneously and coherently support business and work processes, and the integration and optimization of internal and external business processes. Collaborative Business process integration has also already been taken up by a few consulting companies, but success is quite limited in terms of coping with heterogeneous "legacy" processes and systems. Further research is clearly necessary to create an integrated and adaptable solution for a broader application.

SAP has designed a tool to create Collaborative Business Maps (C-Business Maps). These are a collection of industry-specific and cross-industry process blueprints. They explain proven state-of-the-art Internet business processes and show how to apply the mySAP Business Suite to support those processes. In addition, they help to understand what needs to be invested and what will be gained with respect to time, money, and competitive advantage. Again, this solution is only usable in the SAP context and has not at all been validated in relation to industrial needs.

One important aspect in adopting collaborative business is to prepare an organization for Collaborative Business Process (c-bp) management by developing a blueprint or "architecture" that lays out the process and technology design. This exercise includes an integrated vision of the individual participants' business processes, as well as a clear view of the process and technology connections that the firms must make to one another. Successful firms involve key external customers/partners in the process of creating the blueprint, and implement it jointly. By doing so, firms will increase the likelihood of success by recognizing and incorporating the requirements of each participant (Amit and Zott 2001). E-business frameworks have an important role in the planning and implementation of these issues as they establish the basic technical and management agreements that will guide the collaborative processes. In terms of IT requirements, this is reflected in the need to design applications that integrate business functionality (mainly at the management and operational levels) and collaborative functionality. This obviously requires a new design paradigm that is beyond the traditional approaches based on a single application, a single user, and a single business process (Ackoff et al. 2001).

Another factor becoming more and more important in c-bp management is knowledge sharing. Collaboration tends to be, and will surely be in the near future, based on accumulated knowledge encompassing the content and form of collaboration. This goes well beyond what current e-business frameworks can provide as they focus on the operational aspects of inter-organizational processes. Nevertheless, knowledge sharing must also be supported by more or less well defined agreements and structures.

The implications in individual and group work practices are twofold: first, the emergence of a more intense collaboration between companies requires different forms of work organization as well as new skills; second, responding effectively to these requirements is only possible if new concepts of information systems arise, integrating seamlessly collaboration features with operational management. If the knowledge sharing dimension is added here, we must also consider the requirements in terms of specific social skills. The implications of this, in terms of the effectiveness of collaboration in socio-technical networks, is that there is the need to consider new ways of understanding, modeling and formalization.

4. RESEARCH DIRECTIONS IN THE DEVELOPMENT OF HIGH VALUE NETWORKS

4.1 Social Actors' Networks

Social structure is related with regularities in relational patterns between concrete entities (Knoke and Kuklinsky 1990). To fully understand the structure and dynamics CN, for example in a virtual organization context, we need theoretical and conceptual tools that help analysts and the network members to manage the processes undertaken by those networks in the best way. Structural modeling of social interaction has been used for a long time in social sciences disciplines. We believe that the impressive body of knowledge created in this is of particular importance to understand and to manage collaborative networks and virtual organizations better.

The analysis based on social actors, is based primarily upon the assumption that a social structure does not organize itself in a random way but follows certain patterns (Freeman 2000). The social behavior of actors, at any aggregation level (individual, group) is manifested through patterns of interactions (latent or evident) that are dependent on established connections between actors. A network model is then adopted to represent the social structure and its characterization can be obtained by the use of statistic and mathematical methods.

4.1.1 Organizational Analysis Perspectives

SAN can be particularly useful in organizational (re)design, and reports of its effective use in that area are not new. For example, Tichy (1981) evaluates organizations using several network strategies after identify groups ("coalitions" and "cliques") as well as their connections. It is possible then through SAN based methods to determine which groups are dominant, how they agree or disagree and, consequently if it is required to adopt cooperation strategies or conversely it is more adequate to establish negotiation scenarios. In general terms it is possible to use SAN to analyze organizational views, e.g. social, political, cultural, technological, at different levels - inter-organizational, organizational, group, individual (Tichy 1981). Although a structural approach, SAN based methods enable organizational design to go well beyond formal, explicit structures: they can reveal the "informal" organizational chart (Mintzberg and Heyden 1999, Molina 2001).

4.1.2 SAN Analysis in the Context of Inter-Organizational Processes

As initially outlined in Soares and Sousa (2002), three analysis levels are possible with SAN:

a) Structural level tries to globally describe and analyze a given network, requiring complete data about the different actors and types of relationships. In the case of a complete supply chain, all the participating companies should be known, as well as all their client-supplier relations. Such analysis would for example be able to identify companies with common features (types of products, delivery performance, supply chain tier, etc.);

b) Relational level concerns the description of the relationships between the social actors. A large number of attributes may be used to characterize these relationships (e.g. distance, accessibility) allowing an analysis that may be both qualitative and quantitative. Examples of this are the set-up and study of indicators for the logistic performance in operations involving two supply chain partners (such as delivery lead times, quality performance), the degree of trust between partners, the level of information exchange (such as production plans, informal data on production status);
c) At the individual level, social actors are studied based on the relationships in which they participate (This characterization should probably be complemented by other analysis methods). Examples are issues such as the importance of an actor (e.g., measured by the centrality of his position in terms of information exchange), or his potential for establishing relationships in the network.

4.1.3 Combining Social Structure with Process Analysis

In the analysis and design of inter-organizational processes, the role of a SAN is to reveal how the structure of the social relations, between the actors performing the activities in the process, can influence the performance of the process either positively or negatively. One crucial aspect is the identification and characterization of the social actors. The approach followed in this work was to select the social actors based on the participation and/or influence in the inter-organizational processes. This implies that the first step in data collection is the processes identification, which is a complex task on its own. We will not detail the process identification and mapping here, as these methods are also extensively reported in the literature (see e.g. Eriksson and Penker 2000).

After the data collection phase, the next step was to select the measures, in this specific context, where to base the analysis of the network. Based on reported experiences (Ajuha and Carley 1998, Hagen et al. 1998), the following measures were chosen:

- *General measures* for characterization of the network as a whole - cohesion, density, transitivity;
- *Individual measures* that take as a referential an individual actor and the connecting categories with the rest of the set - adjacency, geodesic, connectivity and maximum flow;
- *Centrality measures* were used to determine the core actors in the network, the ones crucial for the development and support of the inter-

socio-technical network; *Subgroup identification measures* for a nClique analysis.

The results of the social structure analysis are used to identify relational problems and to redefine business processes in order to achieve an improved efficiency and effectiveness by considering appropriate relationships between the relevant actors in the process (see Soares et al. 2003 for a detailed description of this case study).

4.2 Actor-Network Theory

Because the implementation of information systems is inevitably intertwined with the innovation concept (Tatnall and Gilding 1999) and new technologies, business models, and social arrangements are under permanent and rapid transformation (Klischewski 2000), it is necessary to look into and exploit new methodologies (Neyland 2002, Tatnall and Gilding 1999), which enable the design of information systems that may consider both the human and technological aspects (Walsham 1997).

The Actor-Network Theory (ANT) considers the social process as an interconnection of human and non-human elements, each of them with the same degree of importance when considered from a social point of view. This approach is particularly important to the study of web based systems and collaborative work (Tatnall and Gilding 1999). The Actor-Network is a system of relationships, exchanges, alliances and negotiations among the actors (Underwood 2002). In the development of inter-organizational information systems, the application of ANT is its first steps, and there is not a single way, consensual or uniform, of using it in the domain of socio-technical development. There are a couple of attempts of using ANT to understand the introduction of an information system in an organization (Underwood 1998, Klishewsky 2001), with rather limited but promising results. According to ANT, the goal of developing an IOS in a socio-technical network is to establish a network of commitments between the actors involved. The development of the IOS is understood as a development process of a network in which the assumed commitments are closed in "black-boxes" and circulated in the sense of being perceived by the relevant actors, inducing the eventual change in previous social relations (Klishewsky 2000).

Gomes and Soares (2003) reported preliminary results of a study whose aim was to exploit the possibility of application of the ANT as a theoretic and/or methodological framework for analysis and design of socio-technological networks. They considered five conceptual stages, which are intertwined: Diagnosing and Problem Definition, Program Action, Construction, Evaluation and Learning Action. These stages lead to transformations in the system and the system will influence their implementation, originating a real network of interactions. The implementation of these stages is an interactive and dynamic process through the involvement, participation and cooperation of both the actors and the network. It is important to keep in mind that once the process is in movement, its various stages are boundless and they are not watertight. All the process stages are complementary and interact among each other.

4.3 Managing as "Managing Relationships"

The work on the application of networking theories to the development of socio-technical networks described in the previous section can be applied in the exploitation of new concepts for management, especially the management of networks of small and medium enterprises. For example, we can take a structural approach to the modeling of an SME network by considering two key concepts: social networking and hypertext. Here social networking refers to a range of approaches both from sociology and organization theory (as seen before), where the organization and their members are viewed as nodes in networks. Hypertext is a concept that underpins today's forms of interaction in the internet and information technology in general. From these concepts, we invent the concept of hyper-management with the goal of forming, to some extent, a unifying concept. Ideally this would mean: consideration of the local and the global; the detailed and the aggregate; the technical and the social; the structure and the behavior.

The management of an enterprise, or a network of enterprises, can be viewed as the management of relationships and interactions between the different actors, directly and indirectly involved in the activities. Relationships involve operations, processes, resources, knowledge, social interaction, trust, power, etc. Relationships can be related with other relationships. For example, an operational relationship between a procurement manager and one of its suppliers is related (or is influenced) by his relationship with the other suppliers, by the relationship of the supplier with other clients, etc. This can be considered as an evolution of the concept of meta-management (Mowshowitz 1997) in virtual organizations, going from the management of distributed tasks linked to some previously agreed overall goals, to the management of a web of relationships linked not only to a complex web of overall goals but also to individual goals and expectations.

The way a manager (which can be any node in a network) moves from one subject or issue to another, strongly depends on the linkages (relationships) he actually perceives and on his goals and objectives at the moment. Therefore, there can be several levels of relationships. For example, depending on the level of aggregation of the actors (e.g. nodes in a collaborative network) we can explore the relationships between companies, between teams, between roles, etc.

A holistic management view also requires managing the relationships and interactions of individuals and groups with technology, by adequately positioning the technological artifacts, such as software agents in information systems, in the social context of human action. In fact, today, information technology and systems have so strong an influence in human actors' relationships that it is interesting to conceptualize them as social actors on their own.

Through a "multi-perspective model" managed by a computer, each person or group is expected to be able to fully explore webs of relationships that are important in the scope of his responsibilities. Relationships can be analyzed, "designed" and managed at different levels and from several perspectives. Levels and perspectives are intertwined, linking the network of social actors in a hyper-web of relationships.

4.4 Information Systems for Managing Relationships

The functional/hierarchical foundational concepts of past management information systems are not compatible with the requirements of hyper-management. Rigid whole-part relationships (hierarchies) and functional dependencies are not appropriate to manage the complexity of today's webs of relationships in business networks. The need to freely navigate between management issues (e.g. operations, collaborations, negotiations, plans, etc.,) call for new systems' concepts and new information technology models away from the simple client/server architecture. Intelligent agents technology, peer-to-peer and grid computing are emerging in the business community as promising business information systems platforms. Nevertheless, the breakdown with the old models of management information systems has not yet occurred. Nowadays, new management concepts can only emerge in a complex interaction with new information systems concepts.

A new concept for management information systems in business networks should be based on three issues: networked collaboration support; networked decision support; and networked knowledge management. This requires multi-perspective (inter) enterprise models, including knowledge models and decision making models which are appropriate to enable the best management decisions by every individual and group. These models are inherently distributed and thus agent technology, peer-to-peer and grid computing naturally fit the implementation requirements of such models.

New reference architecture for management information systems should cope with two characteristics of future collaborative networks: heterogeneity and complexity. For that, highly flexible information technology infrastructures are needed. Intelligent agents, peer-to-peer and grid computing are promising technologies to answer these requirements. Agent based architectures provide a platform for the integration of heterogeneous services such as order distribution support in logistics management or skill capital optimization in human resources management. Software agents and in particular intelligent agents technology provide both the way to implement flexible/sophisticated business functionalities and the way to encapsulate the existing (legacy) ones in a way that reflects the networked nature of hyper-management. Peer-to-peer and grid computing provide a much more flexible and powerful way of processing management data at several levels than current data warehouse bases architectures.

5. CONCLUSIONS

Emerging standards and technologies are promoting integration and making high value business networks a reality, guaranteeing a fruitful cooperation among the participants in a socio-technical network. These standards and technologies are being fully integrated into applications for managing these networks. Nevertheless, we need to gain a better understanding of collaborative business processes in order to provide them with the required and adequate technological support by designing and developing powerful information systems that foster more effectiveness in the outcomes of collaboration. Most of the concepts, tools and approaches surveyed in this work have proved their potential in supporting collaborative business. However we actually believe that there is still a lot of research to be done for exploiting that potential in new, more powerful collaborative business models.

The ideas presented here point to one direction of research work that fits the needs of advances in networks of enterprises, answering to the four levels of complexity identified by (Moller and Halinen 1999): (1) industries as networks - vision of the networks; (2) industries in the network - management of the network; (3) relationships portfolio - portfolio management; and (4) exchange of relationships - management of relationships.

More specific research directions would include consideration of methods for the design of inter-organizational relationships including explicit models of the social-actors involved in such cooperation processes. This leads to a new inter-disciplinary engineering paradigm, where the network is both the development object and the means to undertake that development. For example, conceptual and computer based tools to be used in modeling for the analysis and management of inter-organizational processes hybrid modeling: qualitative/quantitative, technological/ sociological, conceptual/mathematical; reference modeling, "frameworking"; development methodologies: inter-disciplinary platforms, evolving and adaptable approaches balancing structure and process concerns.

Also other areas of research need to be more specialized as to be applied to e-manufacturing. Knowledge management is one of such areas, particularly research work on ontologies, domain and application ontologies, ontologies and information systems. Even in the field of management information systems, there is still work to do in inter-organizational information systems, distributed decision support systems, collaborative work support systems, negotiation support systems, specifically addressed to e-manufacturing.

REFERENCES

1. Ackoff, R. Chaddad, F. Cook, M. Integrating supply chain and network analyses: The study of netchains. Journal on Chain and Network Science 2001; 1(1):7-22.
2. Ahuja, K., M., Carley, M., K. Network structure in virtual organizations. Journal of Computer Mediated Communication, 1998:3:4: Retrieved August 2002 from: http://www.ascusc.org/jcmc/
3. Albert-László Barabási, Linked: The New Science of Networks, Perseus, Publishing, 2002, 2.parag CONTEXT Mark Buchanan, Small Worlds and the Groundbreaking Science of Networks, W.W. Norton and Company, 2002.
4. Amit, R., Zott, C. Value creation in e-Business. Strategic Management Journal 2001; 22: 493-520.
5. Barret, S. Konsynski, B. Inter-organizational information sharing systems. MIS Quarterly 1982: 93-105.
6. Bijker, W., *Of Bicycles, Bakelites, and Bulbs: Toward a Theory of Sociotechnical Change*, Cambridge, MA: the MIT Press. 1995.
7. Bijker, W.E., Hughes T.P., and P inch, T.J., *The Social Construction of Technological Systems: New Directions in the Sociology and History of Technology*, Cambridge, MA: MIT Press, 1987.
8. Burt, R., "The Network Structure of Social Capital in Research" In *Organizational Behavior*, Sutton R., and Staw B., ed. Greenwich, CT: JAI Press, 2000.

9. Burt, R., "The Social Capital of Structural Holes", Pre-print of a chapter In *New Directions in Economic Sociolog*", Guillen M. F., Collins R., England P., and Meyer M. ed. New York: Russel Sage Foundation. 2001.
10. Castells, Manuel, *The Rise of the Network Society, Volume* 1 of *The Information Age: Economy*, Society and Culture, Oxford: Blackwell., 1996.
11. Checkland, P. and Holwell, S., *Information, Systems and Information Systems: Making Sense of the Field.* Chichester: Wiley, 1998.
12. Cheshire Henbury, Retrieved July 2003 from: http://www.cheshirehenbury.com
13. Coleman J. S., *Foundations of Social Theory*, Cambridge, MA: Havard University Press. 1990.
14. Davenport, T., *Process Innovation: Reengineering Work Through Information Technology*, Harvard Business School Press, Boston, 1993.
15. Davidson, E., Lamb, R., *Examining Socio-Technical Networks in Scientific Academia/Industry Collaborations.* University of Hawaii. 2000.
16. ebXML, Retrieved July 2003 from: http://www.ebXML.org
17. Fahey, L Srivastava, R Sharon, J. Smith, D. Linking e-business and operating processes: the role of knowledge management. IBM Systems Journal 2001; 40:4 : 889-907.
18. Farbey, B., Land, F., Targett, D. A taxonomy of information systems applications: the benefits' evaluation ladder. European Journal of Information Systems 1995; 4(1):41-50.
19. Gitell, J. (2000a) Paradox of coordination and control. California Management Review 2000; 42:3: 1-17.
20. Gitell, J. (2000b). Organizing work to support relational coordination. International Journal of Human Resource Management 2000; 11(3): 517-534.
21. Gittell, J. Organizing work to support relational coordination. International Journal of Human Resource Management 2000, 11(3):517-534.
22. Gomes, P., Soares, A., Analysis and specification of information systems requirements based on the Actor-Network Theory. Proceedings of the 9th Intl. Conf. of European University Information Systems, 2003
23. Hanseth, O., e Monteiro, E., Socio-technical webs and actor-network theory., (on-line in 10-124-2002), Retrieved 1998 from: http://www.ifi.uio.no/~oleha/Publications/bok.6.html
24. Hong, I. A new framework for inter-organizational systems based on the linkage of participants' roles. Information and Management 2002: 39: 261-270.
25. Howard, P. Network ethnography and the hypermedia organization: new organizations, new media, new methods. New media and Society 2002: 4(4): 550-574.
26. Johnston, H., Vitale, M. Creating competitive advantage with interorganizational information systems. MIS Quarterly 1988; 12(2):153-165.
27. Jonkers, H. L., Donkers, H. W., Diederen, P. J. M., The Knowledge Domain of Chain and Network Studies. KLICT 2001: September.
28. Kaplan, S. Sawhney, M., E-Hubs: The new B2B marketplaces. Harvard Business Review 2000; May-June: 97-103.
29. Klischewski, R., "Systems development as networking", In: H. M. Chung ed.: Proceedings of the 2000 Americas Conference on Information Systems (August 10-13, Long Beach, CA). Association for Information Systems 2000, pp. 1638-1644, (2000).
30. Kumar, K., Dissel, H. Sustainable collaboration: managing conflict and cooperation in interorganizational systems. MIS Quarterly 1996; 20:3 (September):279-300.

31. Lamb, R., Sawyer, S., and Kling, R. "A Social Informatics Perspective On Socio-Technical Networks," in H. Michael Chung, ed. Proceedings of Americas Conference on Information Systems, Long Beach, CA, 2000.
32. Lambert, Ad. M., Coper, M. C. Issues in supply chain management. Industrial Marketing Management 2000; 29: 45-56.
33. Latour, B., *Science in Action: How to follow scientists and engineers through society.* Cambridge, MA: Harvard University Press, 1987.
34. Lazzarini, S., Chaddad, F., Cook, M. Integrating supply chain and networks analyses: The study of Netchains. Journal on Chain and Netchain Science 2001;1(1): 7-22.
35. Magnusson, J., Nilsson, A. Of knowledge management practice in SME networks. Journal of Knowledge Management Practice, February, 2003.
36. Magnusson, J., Nilsson, A. To facilitate or intervene – A study of knowledge management practice in SME networks. Journal of Knowledge Management Practice. Febrary, 2003.
37. Malone, T., Crowston, K., Lee J., Pentland, B. Tools for inventing organizations: Toward a handbook of organizational processes. Management Science 1999; 45(3):425-443.
38. McFarlan, F., McKinney, J., Pyburn, P. The information archipelago – plotting a course. Harvard Business Review 1983; January – February: 145-156.
39. Mintzberg, H., Heyden, L. V. D. – Organigraphs: Drawing how companies really work. Harvard Business Review 1999; September-October:87-94.
40. Molina, J. L. The Informal Organizational Chart in Organizations: An Approach from the Social Network Analysis. [on line] Connections 24(1) 2001. Also available at http://www.sfu.ca/~insna/Connections-Web/Volume24-1/8-Molina-24.1.pdf.
41. Moller, K. K., Halinen, A. business relationships and networks: Managerial challenge of network era. Industrial Marketing Management1999; 28: 413-427.
42. Mowshowitz, A. On the theory of virtual organization. Systems Research and Behavioral Science 1997; 14(6): 373-384.
43. Neyland, D., EVINCE Part One: Building a Methodology to Capture IT Culture, Proceedings of the 8th International Conference of European Information Systems, 2002; 217-221.
44. Omta S., Trienekers, J., Beers, G. Chain and network science: a research framework. Journal on Chain and Network Science 2001; 1(1): 1-6.
45. RosettaNet, Retrieved July 2003 from: http://www.rosettaNet.org
46. Pfeffer, J. and Salancik, G., *The external control of organizations.* New York: Harper and Row Publishers. 1978
47. Podolny, J., Page, K. Network forms of organization. Annual Review of Sociology 1998; 24: 57-76.
48. Sawhney, M., Kaplan, S., Let's Get Vertical, Business 2.0, Retrieved September 1999 from: http://www.business2.com
49. Sawhney, M., Parikh, D. Where value lives in a networked world. Harvard Business Review 2001: January: 79-86.
50. Soares, A., Sousa, J., 2002, "Multiple Perspective Configuration of Virtual Enterprises Using Social Actors Networks", in *Collaborative Business Ecosystems and Virtual Enterprises*, Luis M. Camarinha-Matos., ed. Boston: Kluwer Academic Publishers,
51. Soares, A., Sousa, J., Barbedo, F., 2003, "Modelling the structure of collaborative networks: some contributions". In *Processes and Foundations for Virtual Organizations*, Camarinha-Matos L. M. and Afsarmanesh H., ed. Boston: Kluwer Academic Publishers.

52. Steele, P. M., 1999, Modelling Frameworks: The Essential Link Between Models and Methodologies, 10th Australasian Conference on Information Systems.
53. Tatnall, A., Gilding, A., Actor-Network Theory and Information Systems Research, Proceedings of the 10th Australasian Conference on Information Systems, 2002; 955-966.
54. Thoben K., Jagdev, H. Typological issues in enterprise networks. Production Planning and Control 2001; 12(5): 421-436.
55. Thompson, J. D., *Organizations in Action: Social Science Bases of Administrative Theory*, New York, McGraw Hill, 1967.
56. Tichy, Noel, M. "Networks in Organizations". *In Handbook of Organizational Design. Vol.2: Remodeling Organizations and their Environments*. Nystrom,P. C. and Starbuck, W. H. ed. New York: Oxford University Press, 1981.
57. tpaML, Retrieved July 2003 from: http://xml.coverpages.org/tpa.html
58. Trienekens, Jaques H., Views on Inter-enterprise Relationships. 2002, Department of Management Research. Wageningen University. The Netherlands.
59. Underwood, J., "Not Another Methodology: what ANT tells us about systems development", Proceedings of the 6th Intl. Conf. on Information Systems Methodologies, British Computer Society, Also available from (http://www-staff.mcs.uts.edu.au/~jim/papers/ismeth.htm), 2002.
60. Uzzi, B. The sources and consequences of embeddedness for the economic performance of organizations: The network effect. American Sociological Review 1996; 61:674–98.
61. Uzzi, B. Social structure and competition in interfirm networks: the paradox of embeddedness, Administrative Science Quarterly 1997; 42: 35-67.
62. Veryard, 2002, Veryard Projects – "Enterprise Modelling", Retrieved February 2002 from: http://www.veryard.com/sebpc/enterprisemodel.htm
63. Walsham, G., Sahay, S. GIS for district-level administration in india: Problems and opportunities. MIS Quarterly 1999; 23(1): 39-66.
64. Walsham, G., "Actor-Network Theory and IS Research: Current Status and Future Prospects", In *Information Systems and Qualitative Research*, A. S. Lee, J. Liebenan, and J. I. DeGro ed. London: Chapman and Hall, 1997.
65. Wellman, B., Salaff, J., Dimitrova, D., Garton, L., Gulia, M., Haythornthwaite, C. Computer networks as social networks: Virtual community, computer supported cooperative work and telework, Annual Review of Sociology 1996; 22:213-38.
66. Zack, M., Researching Organizational Systems Using Social Network Analysis, 33rd Hawaii Intl. Conf. on System Sciences, 2002, Vol. 7.

Chapter 12

VALUE NETWORKS DYNAMICS IN ICT SYMBIOSIS

Patrizia Fariselli
Nomisma, Italy

Abstract: A wide notion of Value Networks (VNs) is adopted, considering value that is created both in and outside the market, and in the quasi market. An interplay between the value exchanged in market transaction and the information and knowledge exchanged in a non/pre-transactional dimension is detected. As ICT and Internet technologies alter the way in which information is organized, delivered, processed, the impact of ICT applications on VNs triggers dynamics which tend to alter the VNs' configuration and coordination models, enhancing the knowledge component in the value creation process. ICT symbiosis for increasing efficiency and/or reducing transaction cost cannot work as a self-contained approach; for moving towards the knowledge economy and society ICT research and applications for VNs have to address the demand for access to quality information and a knowledge creation relational approach.

Key words: Value networks, ICT, Dynamics, Hierarchy, Market, Partnership, Information, Knowledge creation

1. INTRODUCTION

While there is a general agreement on the definition and representation of the concept of network, the definition of Value Network (VN) is more equivocal, as it depends on the analytical perspective one chooses.

There are two major relevant perspectives, which are connected to different concepts of value:

a) The first one, which is dominant in the current literature about VNs and the corresponding impact of ICT, is the market perspective, where the exchange value is at the focal point. The exchange value is represented by two categories of price: market price, which is generated by the interaction of demand and offer; production price that is the value of the factors of production. The production price depends on the value of the direct and indirect labor that is employed in the production process[1].

If we look at the value as exchange value, it is measurable as the value of the goods and services exchanged at market price. In this case, VNs, which are the center of attention, are those aiming to increase this value, by using ICT (within this chapter ICT is a plural form meaning many possible technologies) with the purpose to optimize business processes in a way so as to reduce the production cost, and/or to increase the market price. The focus is on the processes, and on the ICT architectures and applications enabling the efficiency and productivity gains within VNs operating as market value chains.

b) The second one is a non-market - or pre-market - or quasi-market perspective, where the value is transactional, but not measurable through the market price. It can be interpreted in terms of the principle of reciprocity (Polanyi 1957) and it refers to intangible resources whose access, by individuals and organizations, makes possible the creation of market value. The relevant VNs are those networks aiming to increase the creation and exchange of these intangible resources across the knowledge creation value chain. The issue then is information and knowledge, which have to be considered essential factors of production, whose scarcity can severely limit the socio-economic growth, as much as the effects of natural resources, labor, capital and organization would do. In order to support the value (knowledge) creation and accumulation process the role of ICT is to ensure the widest access to VNs as well as to ensure exchange of structured and quality information both within and to VNs, which operate in between market and society.

The definition of VNs adopted in this chapter is the widest one that includes both perspectives. That is looking at the value that is created both in and outside the market, and in the quasi-market, because the particular interest is about the interactive dynamics that occur between these dimensions that are stimulated by the ICT. VNs encompass a broad range of cases, which can be categorized either by the field they belong to (business, policy, research), by the typology they fit (static analysis) or by the ways they adapt to change (dynamic analysis).

ICT impacts on configuration of VNs at various degrees between the two extremes: by strengthening their internal cohesion and by blurring their borders to the outside. In both cases ICT triggers a change, whose potential on the creation of additional value depends on the way the network controls - or does not control - the change itself. However, ICT is one among other drivers of change. The optimal balance to be achieved by VNs includes, beside technology and the market, a value creation process addressing growth.

Such a balance is very difficult to achieve, because it is situated at the crossroads of multiple interacting dynamics, started by a number of individual VNs (the market), by the administrative and regulatory institutions operating at local - national - international and global level (the policy environment), and also by the public and private organizations developing technology and socio-economic research (the research community).

Yet, the difficulties due to the increasing complexity cannot be overcome by short cuts, only zooming in on specific typologies of VNs, or extending the market rules and frameworks to any VN dimension. Policies are often pressed towards simplification, because of the weak link and time lag between research and the policy decision, limiting an integrated vision of the impact of multiple dynamics. On the other hand, the market tends to produce and diffuse more timely messages, and to impose a short-term horizon, reducing the potential of a strategy for complexity.

In this chapter we try to trace some ICT-driven dynamics in VNs, either in business or in society, through the cross-analysis of a relatively wide set of case studies, which have been surveyed in dedicated European research projects, and in the current literature on e-business and e-government. The lessons learnt from these broad but quite fragmented sources provides a solid basis for the analysis of policy and research programs' priorities addressed in the short but intense first phase of the Internet economy and society, and supports the formulation of recommendations for the next phase.

2. THE ANALYTICAL FRAMEWORK

Two major categories of VNs may be considered.

On the one hand, there are the VNs whose value can be eventually measured by means of variables such as cost, price, turnover, efficiency, productivity, profit. Value is basically exchange value. The network's glue is given by the ability of its members to contribute to the creation of additional measurable values. Membership depends - at different degrees - on a mix of trust and market selection, because of the interplay of the market, non-market and quasi-market value, which gives alternative bias to the VN configuration. The stability of the network is not ensured over time, as membership is dictated by the competitive environment. For example, in hierarchical supply chain networks in mature low- intensity knowledge industries, the position of subcontractors that produce pre-defined orders tends to be unstable. This is either because of delay in delivery time, or the mis-application of technical specifications may exclude the subcontractor from the network, and/or because the mis-management of the network's leader, or the market competition, may reduce the value dramatically and consequently lead to dissolution of the network. On the other hand, there are the VNs whose value is created in a non-transactional or pre-transactional dimension. The network's glue is the common interest in sharing information and knowledge, which can be subsequently used by partners to create measurable exchange value. Membership is normally open, as network economies increase proportionally with the size of the membership. However, over time the stability of the network is not ensured, and is dependent upon the balance between active and passive participation in the network. For example, a project-based network may be impoverished by the free rider behavior of members exploiting - without compensating - the information and knowledge made accessible by other members. Or, a cluster of firms in the same industry, networking across a portal or an exchange that provide information services, may unglue as soon as the quantity and the quality of the services turns out to be inadequate.

ICT, which have a fundamental network nature, play a powerful role in both VN categories. This is because they increment the communication channels dramatically within and between VNs, give visibility to data flowing across the digital channels, establish an environment where potentially an infinite quantity of information may be stored, accessed, downloaded, de-composed and re-composed at a very high speed. However, as it is commonly acknowledged, ICT does not build up any kind of VN, except the telecommunication ones, whose sustainability over time depends on the number of users. And users, including any kind of VN, must have reasons that are expectations to exchange or share value, to get connected: ICT provides the channels, not the motivations.

The distinction between the two categories of VNs is made exclusively for methodological reasons, to ensure that all the relevant phenomena are included. Indeed, there is a tight interplay between the two kinds of value, which can hardly be separated.

The outcomes from case analysis are controversial for three reasons, despite the general emphasis and concern expressed about the impact of ICT on productivity in non-ICT industries (Brynjolfsson 2002, van Ark 2002), where extra value is expected from digital technologies enabling the re-organization of the business processes as well as providing additional tools for marketing and for global trade.

Firstly, it is very difficult to measure the impact of ICT. This is either because the traditional analytical tools and statistical indicators are relatively inadequate, moreover not homogeneous across countries, and/or because the state of affairs of the digital economy is still too chaotic to provide convincing sets of comparable data.

Secondly, similar applications in different industries, firms, public administrations or agencies, provide different evidence. This reveals that the configuration of VNs is at times tenacious to any flexibility that may be introduced by ICT applications, which aim to rationalize the business process re-organization.

Thirdly, from the viewpoint of ICT users (citizens, workers, managers, researchers, firms, public and private agencies, and public-administrations) the highest demand is for access to useful information and communication through digital channels. Whatever the VN (Supply Chain, Extended or Virtual Enterprise, Project-based Network, Cluster, Community and so on) and whatever the technology (Information and Communication systems, Internet, TBP-Technologies for Business Processes, E-commerce platforms) the opportunity to access, store, and process information is a prevailing need. Its accomplishment often makes the difference in achieving a successful implementation of the technology, and improving the readiness of users to ICT. In other words, the communication and exchange of useful information, in an ICT environment, is in itself a powerful driver of change necessary for dynamic VNs'. Later these three points will be further developed during case analysis. In order to navigate through and comprehend the fragmented and heterogeneous material definitions and an analytical framework are needed.

2.1 Mapping Value Networks, Objectives and ICT

In table 12-1 and table 12-2 different typologies of VNs are identified as coordination mechanisms, interaction models, and the major objectives addressed by VNs when adopting different categories of ICT applications. Consistent with the broad concept of VN, we refer either to the Business or to the Society environments. The two tables have to be read by columns, where different possibilities are listed, which can be matched horizontally in different and multiple ways. This is done deliberately, because evidence from in-field analysis show that configuration of VNs and specific ICT applications are targeted at well identified objectives that often tend to change during the implementation process.

Table 12-1. Value Networks' Typology - Business

Value Networks	Coordination	Interaction	Objective	ICT
Supply Chain	Hierarchical	Market	to access,	Information &
		transactions	manage, process,	Communication
Extended	Horizontal	(cost-price)	exchange,	system
enterprise			disseminate	
	Mixed	Collaboration	information &	Intranet
Virtual		(functions)	data	Extranet
enterprise				Internet
		Participation	to control, share,	
Project-based		(competence)	convert, create	Technologies for
network			knowledge	Business
				Processes
Cluster			to optimize	(TBPs)
			business	
Community			processes	E-commerce
			organization &	platforms
			management	
			to innovate	
			marketing &	
			access new	
			markets	

Table 12-2. Value Networks' Typology - Society

Value Network	Coordination	Interaction	Objective	ICT
Public	Hierarchical	Market	to access,	Information &
Administration		transactions	manage, process,	Communication
	Horizontal	(cost-price)	exchange,	system
Policy Agencies			disseminate	
	Mixed	Collaboration	information &	Intranet
Social Groups		(functions)	data	Extranet
				Internet
Research		Participation	to control, share,	
Community		(competence)	convert, create	Technologies for
			knowledge	Business
				Processes
			to optimize	(TBPs)
			public services'	
			organization &	E-government
			delivery	services
			to ensure	
			accessibility,	
			transparency &	
			inclusion	

2.2 Definitions

2.2.1 Value Networks

The VNs, which have been identified, are the following:

- *Supply Chain*: a network of interdependent firms, inter-linking their activities along the processing and distribution channel of a product, from the sources of raw material to the delivery to an end consumer[2];
- *Extended Enterprise*: a network of "partner organizations, which combine their individual capabilities and business processes, in order to perform a joint business process" (Haake 2001);
- *Virtual Enterprise*: "a set of enterprises that in parts of their operations behave as one" (Wood and Milosevic 1998);
- *Project-based Network*: network of firms or firm's units working temporarily in a collaborative way to implement a complex project;
- *Cluster*: "a geographically proximate group of interconnected companies and associated institutions in a particular field, including product producers, service providers, suppliers, universities, and trade associations"[3];
- *Community (of Practice)*: "group of people in organizations that form to share what they know, to learn from each other regarding some aspects of their work and to provide social context for that work" (Nichols 2000);
- *Public Administration, Policy Agencies, Social Groups and the Research Community* are organizations, structured as networks and acting as VNs, as they process inputs (information, data, and knowledge) and produce outputs (policies, messages, research).

2.2.2 Models of Interaction and Coordination

The configurations for VNs span the spectrum between hierarchy and peer-to-peer settings, and depend on the type of interaction prevailing among the network's members.

The *hierarchical model* of coordination, from a static point of view, fits a typical market interaction within VNs. Price is a priority to partner selection, such as in a vertical (mono-industry) buyer-driven e-marketplace. The governing business takes the decision to procure from a range of suppliers competing on price, and although this kind of marketplace represents a market it is actually a vertical network. Such a network has borders, rules for admission, procedures to be adopted, and buyers exerting a strong market power on suppliers, who have limited choice whether to join or not to join the network.

The *hierarchical model* also may address Supply Chain Management requirements. This is especially true when the division of labor is based on functions and tasks whose specifications are imposed top down with a rollback mechanism from client to supplier to sub-supplier. Moreover, the Supply Chain network is often modeled on the pattern of a vertical enterprise in mass-production or retail industries, particularly for routine functions. From the Society side, the Public Administration (PA) can also be represented as a value chain, as its organizational processes fit the Supply Chain VN model. In the value creation process, and when interacting with Policy Agencies and Social Groups, the PA model gives birth to a number of VNs, that are very similar in purpose to those that have been listed in table 12-1. The Research Community also experiences hierarchical coordination, from the organizational point of view, but it is strongly marked from the value creation perspective by horizontal interactions and networking models. The collaborative and participatory models of networking prevail even though on the one hand market transactions occur in PA and in other policy and social organizations, especially in the form of consulting and advisory services to business, and on the other hand an increasing market orientation for large parts of the research community has occurred to cope with shortages in research funding. These collaborative and participatory models are essential to foster specialization, scientific and technical progress within and across a growing set of disciplines, as well as to provide evaluation mechanisms and centers of excellence.

The *peer-to-peer model* for coordination suits the types of interaction experienced in horizontal VNs, where value creation depends largely on collaboration and participation mechanisms among the partners that exchange and share intangible assets such as competence, know-how, and experience. A target for collaborative interactions is to perform jointly the functions of one or more specific business processes, as in the network of firms compounding an Extended Enterprise. In another setting it may be to achieve a goal, as for a Virtual Enterprise. In yet another setting it may be to successfully realize a complex project, as in the Project-based Network. However, a flexible configuration for these networks is not well suited for a hierarchical management approach. This is because the formation and the ownership characteristics of knowledge resources make them resistant to standard codification processes, so cannot be controlled by a vertical command line without losing a significant portion of their substance. Consensus and participation are indispensable to maximize knowledge transfer and management within competence-based VNs.

2.2.3 Objectives

The symbiosis between VN typology and ICT currently addresses six broad objectives:

- *To access, manage, process, exchange, disseminate information and data.* This is the core functionality of digital technology, which has been dramatically enhanced by the Internet, making ICT pervasive in society and in business. The demand for communication tools, online information, access, search and information management tools is very high, and diffused across all kinds of VNs;
- *To control, share, convert, and create knowledge.* These functions are usually grouped under the 'knowledge management' headline, often treated as a separate specialization, addressed by dedicated software and tools. Yet, knowledge management within complex organizations, such as VNs, is really a complex problem, and a confidence in complex technological solutions to solve it could be misleading. This is because, in the context of VNs, it may divert attention from the value creation process to technical procedures;
- *To optimize business process organization and management.* On the supply side, TBPs, originally based on information technologies, have enjoyed in the Internet environment great opportunities for development widening their scope of application and enhancing the possibilities of business process integration. On the demand side, the extremely fast and wide diffusion of Internet technologies across society and business has contributed to diffuse a network-oriented culture outside IT professional clubs. By connecting to the Internet everyone gets a perception of both actual and potential interconnectivity between individuals, organizations, activities, and processes in all fields, not only in telecommunication and information systems, but also in formalized settings such as enterprise or business networks;
- *To optimize the organization and delivery of public services.* TBPs by and large may be adapted to the context of PAs, as they may be seen as a large and complex value chain network. However, some caution should be taken, as the PA is not an enterprise or a VN operating in the market, and the efficiency gains have to be measured with respect to variables other than cost and price;
- *To innovate marketing and to access new markets.* ICT and Internet technology quickly penetrated information systems for the research community, military institutions, airline networks, and for large multi-national corporations operating in the global market. These were early users of TBPs and EDI, to assist business-to-consumer (B2C) and business-to-business (B2B) relations for all size companies aiming at

doing business online. In the late 1990s E-commerce solutions and platforms have developed at pace with the expectations of an inexorable fast move to the global digital economy. A significant effort has been paid by the ICT industry and by policy makers to sustain expectations with case studies, best practices, demonstration projects, pilot projects. However, the universe of the e-commerce practices has not significantly raised the market share for online sales (between 1-2% in Europe) with respect to total sales volume in the consumer market. Estimations for e-procurement account for about 10-15% of all purchases made by companies in Europe (e-business w@tch, 2003). The disappointing results for B2C e-commerce has shifted the attention to the B2B e-commerce, where its objectives are here included in the above TBPs category;

- *To ensure accessibility, transparency and inclusion.* The major goal addressed by the e-Government applications is to ensure accessibility to services for every citizen besides an objective of optimizing the management of the public service value chain. Online services are expected to improve their transparency, because software and information systems have pervaded PA organizations, and have virtually replaced the production of manual and paper-based documentation, its transfer, and control across the many PA units in between a public service and the citizen. However, two conditions would really increase the value creation along a PA digital value chain. They are: a) the safeguard of the public nature of a PA and its staff mission, which cannot be exchanged with automated procedures to be exclusively evaluated through efficiency and cost management; b) the opportunity to invest the public resources that are set free by the digitization process in the delivery of new services. The services that mostly benefit from information, communication and Internet technologies are those providing access to Public Sector Information usually through the control management of the PA agencies.

2.2.4 ICT

ICT have been grouped into four large categories:

- *Information, Communication and Internet technologies* span any type of VN, whatever the prevailing orientation of network interactivity. Digital technologies and the Internet fundamentally alter the way in which information can be stored, accessed, retrieved, delivered, and manipulated; the way of communicating; the way of doing research and sharing knowledge; and the way in which business processes are organized.

- *TBPs, E-commerce* technologies and *e-Government* applications are substantially based on information, communication and Internet technologies, but are mostly designed for market-oriented interactions in VNs. This broad category includes all the systems, platforms, techniques, applications, and solutions that are enabled by digital information and communication technologies that by and large address the business and management processes.

3. SYMBIOSIS OF THE DYNAMICS OF VALUE NETWORK WITH ICT

Over the last ten years interesting dynamics have come to our attention from a number of traced sources when surveying hundreds of VN cases[4]. Most of these dynamics are identified in literature by various authors who have analyzed individual or sets of typologies of VNs and ICT applications, either by theoretical means or through in-field research[5].

In tables 12-1 and 12-2 the analytical categories are represented in a static way. As we cautioned in section 2.1, the two tables must be read by column, there is no valid matrix effect by crossing horizontally 1:1 or 1:n the typologies listed in the columns. This is not simply a technical caveat, it is based on the in-field observation that the correlation between categories is not linear, therefore any static model does not help. The introduction of ICT systems and applications into a VN to address a specific objective does activate systemic effects and feedback that in many cases lead to the reconfiguration of a VN or parts of it, and/or to the reconfiguration of the ICT system or application. The dynamics that are generated deploy horizontally (implementation of the objectives targeted by the ICT application in the VN characterized by a certain coordination model) and vertically (change of VN typology, move to a different interaction and coordination model, modification of the ICT application, integration to other ICT applications). For example, if we take the following association as a linear one (Table 12-3),

Table 12-3. Example Association

Value Network	Coordination	Interaction	Objective	ICT
Supply Chain	Hierarchical	Market transactions (cost-price)	to optimize business processes organization & management	Technologies for Business Processes (TBPs)

we get a limited representation of the impact of the TBP applications on the supply chain, that is we only detect the impact on the interaction and coordination models of the network, as they were statically represented before the decision to adopt an ICT application, and before its implementation.

3.1 Supply Chain

The case analyses in this VN typology reveal various dynamics, either alternative or concurrent.

1. In some cases, particularly in low knowledge-intensive industries, the impact of TBPs (such as an ERP system) on a supply chain that is hierarchically co-coordinated does not lead to any reconfiguration of the network, but only to its downsizing. Yet, the integration of the network of suppliers to the new technological system normally raises problems that in a hierarchical networked organization may be addressed in the following alternative ways:

- A leading company takes the management and control, and is the center for the TBP system's key technical functionality (e.g. design and formulation of the technical specifications) and managerial activities with digital processing by the suppliers minimized. The TBP system is not extended across the network. The suppliers receive and fill in templates without any significant impact on their legacy system or learning process.
- An ASP server hosts the supply chain management system, and users (the client and the suppliers) access it via basic Internet technologies, with different degrees of control in the transfer of information to and from the ASP. In this case, additional ICT applications are necessary and have to be set up and operated for efficient operation of the TBP system. The ASP is additional partner in the network. The role of this partner is crucial, as it becomes the key node in the network, at the intersection of the information flow for the entire supply chain. It may happen that the management of the TBP system is split between the leading company, which would keep control of the strategic TBP applications for the knowledge intensive activities (such as design), and the ASP, which would manage and control the applications for the general management activities.

2. In supply chains in knowledge intensive industries the role of suppliers is differentiated by the quantity and quality of the knowledge they control in the value creation process. The leading company's objective to optimize the business processes may lead to a mixed strategy that aims loosely couple

with low quality suppliers, but strongly linking in a much tighter way with the high quality ones. The TBP applications become:

- a barrier for non key suppliers, who are excluded from accessing it;
- an opportunity for the key suppliers, who get access to the TBP system, and probably also incentives to integrate their legacy systems with the new one. In this latter context the likely dynamics that are triggered by a TBP application are the following:
    - a further investment in ICT technologies and applications is directed to information and knowledge management, as the value (knowledge) creation process is raised in priority after the investment in technologies to optimize the business processes;
    - a move from a hierarchical to more collaborative and participative interaction and coordination models with the high quality suppliers. To encourage their partnership, the leading company in the network is likely to provide incentives, such as some sharing of the added value that the key suppliers help to generate;
    - a change to other forms of VN could occur over time, for example changes to an extended enterprise, or virtual enterprise.

3.2 Extended and Virtual Enterprise

In the *Extended* and in the *Virtual Enterprise* types of VN, ICT plays a key role. Interestingly enough, most of the cases we have analyzed for enterprises that have developed an ICT infrastructure show that these enterprise networks have been relying on digital network technologies as a chance to recover and grow once more in the global markets after economic disasters that took place in the context of the old economy. Companies facing bad times in the market try to capitalize resources (knowledge), that are dispersed either through former suppliers or by technicians or managers try to use ICT to create a VN. In so doing they connect skills, expertise, experience without bearing the cost of establishing a new company in very competitive markets, where large size can be a necessary dimension. In these VNs the opportunities created by flexibility, tele-working, e-commerce, online business are opportunities for knowledge workers and managers alike not to miss, and to take commercial advantages from, the knowledge they control[6].

In these cases, ICT applications that are in use are not really sophisticated and the Internet technologies are more than sufficient. The starting position for these VNs is not how to deal with the complex problem of knowledge management, but not to lose it. The market success of these meta-companies sometimes leads to incremental use of ICT applications, extending their scope from knowledge management to business processes, organization and management.

There are also cases in which the small size of the individual firms makes it very difficult to operate in markets favored by mature industries. New knowledge-based VNs in the same business arena can be created and configured via the Internet in virtual networks that target niche markets[7]. Symbiosis with ICT here provides the glue to separate knowledge resources that alone would not be able to achieve critical mass for profitable business in a market. The e-commerce models then take shape in practice. Returning to the e-commerce language, in these networked enterprises online B2B relationships set up to pool knowledge and production resources, for creating or re-creating business opportunities, increasingly turn also to B2C and CRM facilities to help them to develop in a niche market.

3.3 E-marketplace

By way of introduction we will discuss the implications of the e-marketplace and its definition. In our analytical framework, most of the technologies and applications that are used for doing business online are grouped within the TBPs category. This is because either online B2C or B2B require, besides the Internet environment, systems and tools that impact in the business processes, organization and management of an individual firm as well as the network of firms participating in e-commerce practices.

By connecting on line buyers and sellers either in a final or in an intermediate, market E-commerce impacts on the various segments of the value chain. This impact is on the organizational setting and management of the VNs operating within a value chain, or across value chains. E-commerce, therefore, is a way of doing business that requires at different levels of sophistication with a judicious blend of both online and offline Information and Communication systems. These comprise Internet-Intranet-Extranet; and TBPs, to support and integrate front to back office operations as well as to manage relations within a firm, between firms in the network, and also with customer.

The key feature that is specific to E-commerce is e-trading, which is now available in an Internet environment. Trade over the Internet requires the development of additional (virtual) marketing tools, and expands the market that can be (virtually) reached by individual firms and by VNs.

Internet marketing is mostly based on Information, Communication, and Internet technologies but in addition online trading implies a networking platform, the *e-marketplace*. Since there are different models and definitions we use the term e-marketplace to encompass different typologies (portal, exchange, trading platform), "which all share one common denominator, in that they all bring together multiple buyers and suppliers through one exchange engine ... which enables companies to sell and/or to procure products"(EITO 2001).

For our analytical framework, an e-marketplace is a category in between a technology platform and a VN. Both are essential components, because, on the one hand, the e-marketplace exists only in an Internet environment, and, on the other hand, it helps create a network where exchange value is generated from e-trade transactions between partners and with the owner of the e-marketplace. As it has been pointed out in section 2.1.1, the e-marketplace is apparently a market, where interactivity is governed by the price formation mechanism, but it actually is a network, an entity in the open market.

Today, not so many e-marketplaces provide the means for fully-fledged commercial transactions, although all of them have this aspiration. E-marketplaces that work consistently with the transactional mission are mostly vertical, buyer-driven e-marketplaces, set up by major multinational companies in the commodity, materials, industrial components, or in the consumer products markets, to procure online from a large set of suppliers in the global market.

For example, in the auto industry, world-wide supply chains serving major buyers converge to an e-marketplace (Covisint), which Ford, Daimler-Chrysler, General Motors, Renault, Nissan, and PSA Peugeot Citroen have established, as an independent company. Although the Federal Trade Commission in the US has prohibited members to aggregate their purchasing, the e-platform aggregates individual suppliers by disconnecting them from the supply chains they served. So, a very large VN supplants large VNs working in parallel.

But, however large the number of suppliers attracted to this e-platform, and also however large the market share absorbed by this type of e-marketplace, it does not overlap the entire market. Prices are arrived at here in the same way as in the rest of the market. This raises some questions: (1) If the price interaction supports the market coordination model, are separate markets – in which each of them deliver the 'best' price - compatible with the global extension of the market? (2) Is there any contradiction between the global aspiration of transactional e-marketplaces and the concept of market? The free market competition has been conceived as the environment in which the price-interaction model legitimates the entry/exit of the competitors to/from the market. When in the 'global' market separate e-marketplaces plus non e-marketplaces concur to the formation of different prices for the same products, what happens to the freedom to entry/exit to/from the market(s)?

This is not simply an intellectual concern. In the vertical e-marketplaces suppliers are less happy than buyers are, because they get lower prices. The disappointing rate of success of e-marketplaces, compared to the expectation that was fostered during their take-up at the end of the 1990s, is generally ascribed to failure of achieving a critical mass. A critical mass is deemed necessary for the e-marketplace, as a business organization, to survive and make profits in the market. The success of a network depends on the network economy that is on the number of users, whose subscription fees or the commissions on the e -trade they activate provide the resources to the owner company, or consortium of companies. Many users/suppliers, if they have a choice, prefer not to join a transactional e-marketplace, whilst all of them are very interested in the complementary services that the e-marketplace provides.

These services comprise information about the market, on companies, and about the business. It is not surprising that this information is much more attractive to users or potential users of the e-marketplace. The information-related services, which were originally conceived to support commercial transactions in the e-marketplace, have in many cases become the real glue that aggregates the suppliers. The bias towards pre-market information services induces adaptive dynamics that enhances the portal or exchange dimensions of e-marketplaces that were conceived as fully-fledged e-trading platforms.

Resistance to online trade in e-marketplaces that are dominated by big buyers is to be ascribed more to business considerations rather than to ICT readiness, technological infrastructure, or security barriers. E-marketplaces create some instability, when they liberate network relationships that had been built up over time in order to reduce transaction cost that is generated in the market; and also re-establish a market context, where price is the only tool for coordinating business interactions.

In less developed markets, in which suppliers and buyers are very small and fragmented and where there is no network in place, a horizontal e-trading platform (many sellers-to-many buyers) can contribute to market expansion. But, in these cases, access and exchange of information drives a VN building process and the performance of fully-fledged e-trading remains a possible step.

3.4 Project-based Network

A project-based network is a temporary but dedicated network that is set up to harness knowledge and expertise from different companies and/or business units to meet a specific target. This may be a solution to a problem, or the realization of a complex project, where execution requires high quality knowledge resources, team building and management capabilities. These cannot be obtained through a make-or-buy alternative, nor is it necessary to create a permanent firm, be it a regular company, or an extended or a virtual enterprise. A project-based network may take a legal identity for management reasons and for interfacing to an institutional environment, nevertheless it normally works on a temporary base.

A Project-based Network may be set up for the realization of complex products in industries such as the engineering, construction, aeronautics, energy, telecommunications, and software. Different companies, business units, or expert staffs join to share competencies and knowledge in a task-oriented network. The separate organizations they belong to sign any contracts and provide the guidelines for their team members, but in a Project-based Network each contractual partner organization's internal hierarchy cannot be extended to the peer-to-peer network. The partners themselves decide upon the coordination model and work-plan used. The decision to set up this type of network is taken in a business environment that aims to achieve a business objective.

Project-based Networks are currently set up for the realization of research projects, where academic researchers, and/or private research institutions, and/or industry join to design, carry out, and evaluate research projects. This setting is normally one of the requirements of public or private sponsoring institutions, which assign public funds, or invest in research. Decisions that drive this are taken in the research policy environment, sometimes in coordination with other PA departments, and with relevant industry and business communities.

The same type of VN is often initiated in PA, in Policy Agencies as well as in Social Groups, in order to address complex problems that take advantage of the expertise which is available within and outside the individual organizations. Decisions are taken at the policy level, and team building may address either professional or political requirements. So, the project-based network is oriented to solve problems, and ICT does have a significant impact, as communication, information and data management applications are essential tools in any knowledge-based activity.

ICT and the Internet significantly contribute to foster the circulation of ideas and opportunities by dramatically enhancing the connectivity among people and documents. When people with knowledge are working together in a Project-based Network they get used to each other and if they are satisfied with the project results, the partnership or at least some of its members, may decide to join together in other projects, or business ideas, which may in turn give birth to spin-offs that could take shape as an individual, or an extended or a virtual enterprise.

Thus technologies targeted to information and knowledge management operating within Project-based Networks in research or in the PA can create an environment that produces new VNs, such as a business-oriented network that may need, in subsequent steps, TBP or E-commerce applications. Collaborative interactions can become a background to a market-oriented model of coordination for a VN that integrates knowledge sharing practices into a business.

Conversely, Project-based Networks set up in a market, bridging industrial experts to experts from the research and policy environment, may foster and consolidate professional relationships beyond the duration of a project that can further develop in partnerships on new research projects.

The digital exchange of information, data, knowledge (including know-how), the access to content produced by colleagues, together with the trust that develops by working together for the same objectives, stimulates the flow of experts across the boundaries of different organizations, communities, and networks.

3.5 Community

The Community model has been introduced to identify practices of information and knowledge exchange, as well as sharing that take place spontaneously in a firm, in society, or in a knowledge management dedicated program. The Community interaction model has a non-market transactional nature. It relies on a peer-to-peer interaction model, where ICT play a role, but TBPs do not. The Community represents the way by which workers or citizens self-organize themselves, or are driven to do so, by using basic Internet technologies and accessing common Internet channels, such as Portals or Intranets, in order to access and exchange information and knowledge they control.

In many business cases these Community networks are an alternative to TBP applications that aim at mining, storing, channeling, and transferring information and knowledge through codification processes and information system architectures. These applications are often top-down designed and managed and are controlled by managers and Information System professionals.

These practices show that people prefer to use basic Internet technologies to communicate and to exchange information, or knowledge components if they have some control. Also if there is no direct involvement of the knowledge stakeholders, there can be resistance to adhere to any sophisticated knowledge management system that aims at codifying what is difficult to be codified.

The establishment of a Community, influenced by Information, Communication, and Internet technologies, may result from the dynamics developed within other types of VNs, or may be the trigger for initiation of new VNs:

- Community-type networks may be created for information and knowledge management purposes, across a single Enterprise, Value Chain, Extended or Virtual Enterprise, or across segments of different VNs, and for building up clusters of professional expertise, by digitally connecting people belonging to different, and sometimes even competing, business organizations;
- Project-based networks may easily lead to the formation of a community of experts, who enjoy common professional interests independent of project and include others outside of the project, by using the Internet channels, Web technologies, digital information systems. Such a community may be the crucial for business or research activities, which can be established in the market and in society as a new enterprise or VN;
- E-marketplace platforms, conceived for transactional purposes, can indirectly generate communities of users having common information requirements for accessing and exchanging information through a portal gateway to the e-marketplace. And, as individual users they can, if they require, trade on-line;
- Clusters of companies in the same industry, gathering around a dedicated Portal, may develop sub-groups, networking across a Portal or through alternative Internet channels in a community style. The same holds for communities in public sector organizations that grow up and communicate in a peer-to-peer style using Intranets or email, and configure dedicated areas within an organization's portal, or by establishing cross-border Internet spaces spanning different organizations.

This potential connectivity offered by Internet technology deploys not only within and between organizations/business processes, but also it significantly enhances the connectivity between people, who are the ultimate producers and holders of information and knowledge. The possibility of communicating and exchanging information with low data cost (and also high speed over the Internet) introduces radical changes to the way information is managed and knowledge is created, and heightens opportunities to create or re-create VNs, which are all based on information and knowledge.

4. INPUTS TO A RESEARCH AND POLICY AGENDA

4.1 Lessons from Case Analysis

The adoption of complex information systems, TBPs, and E-commerce applications implies switching costs and often induces dynamics that lead VNs, or parts of them, to shift from one model to another.

Digital technologies are essentially flexible, ICT applications are not, because they are market products, whose investment, marketing and return cycles are constrained by tough market rules. These rules do not allow much flexibility for suppliers and customers after the selling/buying operations are over. When an ERP system or an E-commerce platform is delivered to the market and is purchased by a customer embedded in a VN, the adoption process has high change over costs (i.e. costs for training, learning, integration to the legacy systems, management's re-organization) and costs related to the VN's re-configuration. This lack of control of the processes, triggered by an investment in technology for optimizing the business process in organization but accompanied by a limited vision for the impact, tends to increase rather than decrease the entropy of the process.

Firms managing big supply chains and heavy procurement expenditures, (e.g. such as in the automotive, aerospace, chemicals, engineering, metals, mining and energy/utilities industries, but below the threshold to be global multinationals), are often reluctant to purchase complex ICT packages, or entertain business process reorganization, or to enjoy possible new market advantages. This is not because they are not aware of the opportunities that ICT brings, but because of the dynamics of returns on investment is not yet clearly modeled. So, the costs of re-setting a VN as well as to address its possible instability are often considered higher than the promised benefits.
This conservative behavior is not to be seen as an anti-innovation, or simply as a symptom of backwardness. This conduct is known to be well diffused in Europe (CEC 2003), because the enterprise system is dominated, to an extremely large extent, by small firms. They do however enjoy and use basic Internet technologies for information and communication and through their experience they would like many improvements to the market offerings becoming available. Today, there is more attention given by ICT business application developers and vendors to the specific requirements of small businesses, or small business networks. This is a move away from attempts to make TBP applications, which had been conceived for large multinational enterprises, extensible to small enterprises.

Yet, the strategy to force integration within a VN through tightly coupled architectures, to progress along the maturity curve for e-business that is proposed in the currently dominant literature, is not the only possibility. "Integration must be reversible or deconstructable at critical junctures to support a common range of inter-organizational interaction" (Lamb 2003).

The case analysis to detect ICT-driven dynamics in VNs shows that a solid basis for the development of e-networks is not represented by efficiency and cost-reduction objectives, rather it is by a reinforcement of the relational model, whose fundamental elements are interaction and communication (Geppert 2002, Micelli 2002).

In fact, the adoption of information technology applications to business organizations may increase transaction cost, instead of reducing, despite the views expressed in mainstream literature on the digital economy pointing to the effect as the key one, and derives from it a bias of the digital economy towards market coordination (Malone et al. 1987).

As soon as a digital application is used to ease interactivity (i.e. interactivity between people, within and between business processes, firms, networks, and with the market) and is made operational, the data information, and knowledge flowing through digital channels (or not flowing, or should be flowing) is made very apparent. In other words, when ICT opens digital channels (i.e. channels between different sources of data, information, and knowledge) to allow increase in interactivity, the information and data flowing across the digital channels depends on a selection process, which is not neutral. Rather, it reflects decisions, addressing different and alternative criteria, such as efficiency, standardization, specialization, consensus, reciprocity, redundancy, across a spectrum between hierarchy and participation.

Data information, and knowledge have an essentially relational nature; they are the kernel of any activity, and their mobility across VNs triggers dynamics which significantly impact on a VN's configuration, its stability, and its growth. A strict market-transactional approach tends to neglect such an impact, as well as it underestimates the costs associated with the re-organization of knowledge creation and transfer process across the VNs.

On the other hand, looking at online trading, where higher efficiency is expected from digital transactions, the prices that suppliers get in the e-procurement e-marketplaces are lower than the prices they get in the conventional markets from the same clients. For the latter, the lower price they obtain in the e-marketplace for supplies probably compensates for the costs of re-configuring their supply chains, and provides incentives for quality suppliers to ensure their continuing partnership.

From the consumer viewpoint, the evidence about the online market efficiency (price levels, elasticity and menu cost) is equivocal (Smith et al.

1999), which makes (low) price not the key driver to decisions to shop online.

From the online retailer viewpoint, information about consumer profiles, tastes, and preferences adds value to the electronic transaction, exceeding its market price, because it becomes an input to the customer-oriented production of the goods/service/advertising delivered by the supplier. In a way, consumers participate in the process both as consumers and as producers of information that is in turn incorporated into supply of goods and services (Quah 1999).

Another interesting dynamic is that user readiness for e-business increases gradually along with the usage of ICT, when they implement key content and objectives that address their competence and demand. So, the use of ICT creates a feedback for increased ICT usage.

Nevertheless the demand for information is higher and more diffused than the demand for e-business. A very interesting dynamic for VNs that influences their successful move to digital business is not driven by e-business models imported from outside, but is driven by the experience gained from using ICT by people and organizations with the purpose of controlling and managing the information and knowledge pertinent to them.

The frequently claimed shortage of ICT-skilled people is usually interpreted as a shortage of ICT technicians. It is certainly a problem, but technicians cannot solve the information and knowledge management problems within VNs. But those who have the right degree of competence, (that is the information and knowledge producers and owners), can address this. The moment that skilled people take control of digital technologies they know what to do with them, this is not so for information technology technicians. Key skills and capabilities are developed in the knowledge creation process, and ICT should only reinforce them, not diminish them. The separation between knowledge skills and information technology skills tends to increase the entropy for knowledge creation and its accumulation. The fully digitally integrated value chains, to be established within complex and bureaucratic organizations for e-government practices, also suffer from the same dynamics mentioned above. In many cases, the attempt to radically revolutionize an administration system, in order to create and deliver online exactly what is normally created and delivered offline, meets technical difficulties, which are enhanced by lack of consensus. In the mean time, the benefits for the public administration staff and for the citizens will be small compared to the change over cost for scattered digital applications throughout public services value chains. Access to an efficient and high quality information system is needed by citizens and public administration staff alike that provides a basis to roll out new services designed and delivered to citizens and public employees of higher utility and no loss to previous experiences.

4.2 Inputs to Policy and Research

The trade offs between the dynamic change induced by pervasive ICT infrastructures, services, and goods, and the dynamics of research and policy, gives expectations, case studies and short-term analyses a prevailing and sometimes misleading role. Actual and potential network effects make the measurement of any impact (resulting from an application of new ICT on enterprise and VNs) much more difficult, but at the same time more important.

At first, the global and decentralized nature of the Internet was revealing the global market as a place that consumers and suppliers (be they individuals or companies), could access directly, thereby bypassing intermediaries and bureaucratic hurdles. In a way, the nature of the interaction tool (e.g. Internet) was transferred to the object of the interaction itself (i.e. commercial transactions). By the mid 1990s the attention to electronic commerce grew up, in parallel with the rising use of the Internet. Widely accepted definitions of electronic commerce were delivered by the European Commission in 1997 and by OECD in 1998. These definitions were in line with a widely accepted view of electronic commerce as buying and selling consumer goods over the Internet, and in particular intangible goods.

The initial emphasis on the e-shopping dimension of electronic commerce has subsequently been balanced by a greater appreciation of the importance of B2B applications. Awareness of the potential impact of Internet technologies on the business processes behind the front office has also grown apace. And, in terms of turnover, B2B transactions represent by far the largest E-commerce market. E-business is an umbrella term encompassing both e-commerce (one-to-many e-commerce models and e-marketplaces), marketing/information websites, and the concept of E-business services, i.e., the range of ICT support services that are associated with the customization, implementation, operational activities, hosting and maintenance of one-to-many e-commerce solutions and e-marketplaces, plus those associated with marketing/information websites (EITO 2001).

A number of business models have been developed along the path towards implementing ICT-driven business processes and supply chain integration (Timmers 1999). The Internet-driven evolution of e-commerce from one-to-one applications (e-tailing) to one-to-many applications, such as enterprise and supply chain integration, e-procurement, e-auction, e-marketplace, collaborative commerce, virtual community, is transforming into both the value-added activities of an enterprise and the enterprise itself, reinforcing the shift from the enterprise-centric vision to the multi-enterprise network or virtual enterprise.

The productivity gains through application of the Internet technologies to business, resulting from increased efficiency, reduction of transaction costs and facilitated access to the global marketplace. Those gains, experienced by the large and multinational enterprises both in services and in heavy manufacturing, were also expected to take place in small businesses through the transfer of the e-business models and connection to the Internet. Yet, the connectivity has increased, and according to Eurostat, "over 90% of enterprises with more than 10 employees are using computers, and almost 80% of them are connected to the Internet. ... no major gaps between larger enterprises and SMEs exist anymore" (CEC 2003), but the uptake of e-commerce and e-business has not increased correspondingly, although a number of interesting cases across Europe have been taken up.

Current literature[8] on the digital economy often refers to models of enterprise and VNs that are inspired by mainstream theories, such as the contractual theory of the firm, and by VN engineering methodologies, which are basically oriented to maximize exchange value and minimize costs (particularly transaction costs). These theories and techniques entrust both digital and Internet technologies with the major task of smoothing out or absorbing the friction generated across business processes and networks, and between production and the market, supporting the objective of moving up along the maturity (integration) curve.

Yet, alternative analytical approaches are shifting the attention toward access to information, knowledge mobility and management, and organizational innovation, raising new methodological questions about the impact of digital technologies on the value creation process. Evidence provided by in-field research on digital behavior of VNs finds that demand and usage of ICT reveals social interaction models, and knowledge creation and transfer processes, which are difficult to fit within digital procedural architectures.

The gap between social behavior and procedural parameters is not an indication of digital divide; the divide, in such a case, is between control over content and technology; between demand for access and supply of information; and also between standards. The more the user is able to handle and control digital applications that enable them to carry out online activities that contribute to the achievement of their objectives, the more adoption is accelerated to higher take up levels.

The adoption and use of Internet applications in small businesses is selective, being essentially driven by the true interests of the companies rather than driven by a road map to advance along the e-business maturity ladder. Many analyses across Europe confirm this observation. With regard to the Internet, small businesses are very interested in getting flexible tools for communication (email), for offering and access information (via Websites) from customers (or from partners, and the market) in general, but they are less interested in e-commerce. These basic digital applications, which are deemed the less mature steps, are actually extensively and advantageously used within firms, networks, and communities of practice. And they also represent additional effective tools for the acquisition of information, engaging in partnerships, as well as knowledge sharing. An insight into these relational phenomena and non-procedural knowledge management practices supports a broader vision of the enterprise and of the VNs (Montresor and Romagnoli 2002). This leads to two major conclusions that can be drawn:

- VNs are a dynamic reality, and ICT can leverage a shift from one type to another;
- VNs encompass multiple dimensions that can be associated with different models of interaction, which reflect different models of knowledge creation.

Therefore, there cannot be just one e-business model or maturity curve. But the possibility for transforming the adoption of technology into innovation gets lost if an ICT application or platform does not fit with the characteristics of relevant dimension and configuration for a VN.

Alongside ICT and Internet deployment over society and business, the research and policy paths have been proceeding from an Information Society to E-commerce and from E-business to the Digital Economy. In the near future the next milestone is the Knowledge Economy and Society, which is marked by the notion of access.

The access to and exchange of information and knowledge, which are central to the dynamic change occurring in VNs when they go digital, are the crucial socio-economic and policy issues in the Knowledge Economy and Society. Yet, access to information depends either on an efficient ICT infrastructure, or on the quality of the information digitally accessed over the Internet, or both. The improvement of the quality of the digital content made available over the Internet depends on:

- Technology (for data mining, content aggregation and management; search engines, interfaces, portals, etc);
- Content (building, organization, supply, and evaluation).

An inadequate supply of Information Technology solutions and products, for accessing and for providing structured information, drives demand for information towards intermediaries. For intermediaries cost of services sometimes reflects a chaotic supply of free information than it does a higher quality and added value for their service. In this respect, it is very important that e-Government services can supply quality online to Public Sector, which represents an invaluable mine of resources to citizens, enterprises, VNs, and also to the Public Administration.

From the viewpoint of research, Information Technology needs a new impetus to achieve greater momentum to carry forward the original aspirations and purpose of the Internet.

5. CONCLUSIONS

In this chapter we traced the dynamics stimulated by the ICT in VNs to encompass all those cases where value is created both in and outside the market, and in the quasi-market. We looked: either at the value, which can be measured by means of variables (such as cost, price, turnover, efficiency, productivity, profit); and looked at the value which is created in a non-transactional; or pre-transactional dimension, based upon information and knowledge exchange and sharing which can be subsequently used by partners to create measurable exchange value.

Indeed, there is a tight interplay between the two kinds of value, which is fuelled by ICT, because the introduction of ICT systems and applications into a VN does activate systemic effects and feedback that lead to the reconfiguration of a VN (or parts of it), and lead to the reconfiguration of the ICT system or application. The dynamics generated by ICT deploys horizontally (i.e. implementation of the objectives targeted by the ICT application in the VN characterized by a certain coordination model) and vertically (i.e. change of VN typology, move to a different interaction and coordination model, modification of the ICT application, integration to other ICT applications). The main reason to this dynamics depends on the radical changes to the way information is managed and the way knowledge is created in a digital environment (i.e. Environment, where information and communication technologies allow information organization, delivery, retrieval, processing at a low cost and high speed across the VNs, which are all based on information and knowledge).

As soon as a digital application is used to ease interactivity (i.e. interactivity between people, within and between business processes, firms, networks, and with the market), the data information, and knowledge flowing through digital channels, (or not flowing, or should be flowing), is made very apparent. When ICT opens digital channels between different sources of data, information and knowledge, the information and data flowing across the digital channels depends on a non-neutral selection process. Rather, it reflects decisions, addressing different and alternative criteria (i.e. such as efficiency, standardization, specialization, consensus, reciprocity, redundancy, across a spectrum between hierarchy and participation).

The case analysis to detect ICT-driven dynamics in VNs shows that a solid basis for the development of e-networks is not represented by efficiency and cost-reduction objectives, rather it is represented by a reinforcement of the relational model, whose fundamental elements are interaction and communication. Data, information, and knowledge have an essentially relational nature; they are the kernel of any activity, and their mobility across VNs triggers dynamics which significantly impact on a VN's configuration, its stability, and its growth.

The adoption of information technology applications by business organizations may increase - instead of reducing - transaction cost. Efficiency-driven and market coordination approaches tend to underestimate the costs associated with the re-organization of knowledge creation and transfer process across the VNs in ICT symbiosis.

Nevertheless the demand for information is higher and more diffused than the demand for e-business. A very interesting dynamic for VNs that influences their successful move to digital business is not driven by e-business models imported from outside, but it is driven by the experience gained from using ICT. Evidence provided by in-field research on digital behavior of VNs finds that demand and usage of ICT reveal social interaction models, and knowledge creation and transfer processes, that are difficult to fit within digital procedural architectures.

Alongside ICT and Internet deployment over society and business, the research and policy paths have been proceeding: from an Information Society to E-commerce; from E-business to the Digital Economy; and now to the Knowledge Economy and Society. This is marked by the notion of access to and exchange of information and knowledge, which are central to the dynamic change occurring in VNs when they go digital. Yet, access to information depends either on an efficient ICT infrastructure, or on the quality of the information digitally accessed over the Internet, or both. In the move towards a developed Knowledge Economy and Society, information technologies have to regain momentum in research, in order to provide an adequate basis to digital services addressing the demand of information and knowledge sharing of citizens and VNs.

ACKNOWLEDGEMENTS

The work presented in this chapter is based on research carried out by the author from mid 1990's on the economic and policy impact of ICTs on network organizations, particularly in the small business area. The research work has been mostly developed within projects sponsored by the European Commission Esprit and IST programmes, in particular DEEDS (Digital Economy: Policies Exchange and Development for SMEs; www.deeds-ist.org) and BEEP (Best eEurope Practices; www.beep-eu.org). My acknowledgements go to the projects' teams, to the experts, and to the European Commission officers I've been lucky to work with. A special thanks to Olana Bojic and Julia Culver Hopper, my colleagues at Nomisma-NET, for their most valuable background support.

NOTES

1. That is relevant when accounting the value -added of the Public Administration.
2. For a survey of definitions of Supply Chain, see Vanharanta H. and Breite R., 2002
3. https://secure.hbs.edu/isc/cmp_data_glossary.jsp
4. The sources are mainly European Commission funded research projects that have been carried out by the author from 1993 onwards in the field of innovation policies, e-commerce, e-business, e-government. To mention some of them, in chronological order: MINT (1993-95) - The Transnational Evaluation of MINT -Managing the Integration of New Technologies- Programme (European Commission, DG XIII -SPRINT-EIMS); COMPETE (1996-98) - Improving competitiveness of SMEs through business engineering and targeted research technologies (European Commission, DG III - Esprit); G7 10 WGs (1996-98) - Contributing to the definition of G7 Pilot Project n.10 A Global Market-place for SMEs (European Commission, DG III – Esprit); DEEDS (2000-03) - Digital Economy: Policy Exchange and Development for SMEs (European Commission, DG Information Society, IST); BEEP (2001-03) - Best eEurope practices (European Commission, DG Information Society, IST). In particular, we remind to the BEEP knowledge system (www.beep-eu.org), to access a wide data base of cases in the four domains of eEurope 2002: work and skills, digital SMEs, social inclusion, regional development
5. For a survey of selected contributions, see DEEDS 1st and 2nd Annual Report, 2002-2003, in www.deeds-ist.org
6. For example, see the NTD International case in www.beep-eu.org
7. For example, see the Koncraft case in www.beep-eu.org
8. We refer to the literature spread off since 1995 with the special issue on Electronic Commerce of the "*Journal of Computer Mediated Communication*" (JCMC), edited by C. Steinfield (Michigan State University), including the article of R.T. Wigand and L. Benjamin [Wigand and Benjamin, 1995], followed in 1997 by the special issue on E-commerce of "*The Information Society*", organized by R.T. Wigand [Wigand, 1997], and by the production of R. Kalakota and A.B. Whinston [Kalakota and Whinston, 1997]. The approach to e-commerce inspired by the transaction cost theory, pioneered by T.W.

Malone, J. Yates, and R.I. Benjamin [Malone, Yates, Benjamin, 1987], and addressing the (re)configuration of the Industry Value Chain, has been further developed by many authors, leading to the flourishing of the e-business models literature and applications [Timmers, 1999]. OECD research and policy documentation develops this approach, focusing on the productivity gains associated to ICT [OECD, 2003].

REFERENCES

1. Antonelli C., *The Microdynamics of Technological Change*, Routledge, 1999.
2. Antonelli, G., "Macroeconomic and employment policies for structural change in the European Union", in *European Economic Integration and Italian Labour Policies*, Antonelli G and De Liso N. ed. Aldershot, Ashgate, 2003 (forthcoming)
3. Bakos J.Y., Brynjolfsson E. Information technology, incentives, and the optimal number of suppliers. Journal of Management Information Systems 1993: Fall, Retrieved July 2003 from: http://ccs.mit.edu/papers
4. Boughzala I., Zacklad M., Cooperation Engineering for the Extended Enterprise, HCP'99 Human Centered Processes 10th Mini EURO Conference, Brest (France), 22-24 September 1999, Retrieved July 2003 from: http://www-hcp.enst-bretagne.fr/bibliography/articles/hcp99/Boughzala_Zacklad_hcp99.pdf
5. Brynjolfsson E., Hitt L.M., Computing Productivity: Firm-Level Evidence, Retrieved July 2003 from: http://ecommerce.mit.edu/papers/ude
6. CEC, 1997, A European Initiative in Electronic Commerce, COM (97):157:
7. CEC, 2003, Adapting e-business policies in a changing environment: the lessons of the GoDigital initiative and the challenge ahead, COM(2003): 353final.
8. e-business W@tch, 2003, e-business indicators, Retrieved June 2003 from: http://www.ebusiness.watch.org
9. EITO- European Information Technology Observatory, 2001
10. Fariselli P., 2003, Pointing the Policy Sextant to Hold the Course from the Digital to the Knowledge Economy, DEEDS 2nd Annual Report, Available in http:www.deeds-ist.org
11. Fariselli P., 2002, "E-Commerce for Development", in *E-Commerce for Development, OECD - Development Centre Studies,* Goldstein A. and O'Connor D. ed.
12. Fariselli P., Oughton C., Picory C., Sugden R., "Electronic Commerce and the Future for SMEs in a Global MarketPlace: Networking Opportunities and Public Policy Requirements",. In *Small Business Economics*, vol.12, Issue 3, May 1999, Kluwer Academic Publishers
13. Fariselli P. (ed.), *Innovating SME Business Practices. The Compete Methodology and Tools*, Pendragon, 1998
14. Geppert L., 2002, ICT and SMEs: Role, Needs, Requirements, and Purchasing Process. Retrieved June 2003 from: http://www.deeds-ist.org
15. Haake J.M., 2001, Applying Collaborative Open Hypermedia Concepts to Extended Enterprise Engineering and Operation, OHS7 - The Seventh International Workshop on Open Hypermedia Systems: August 15, 2001, Also available in http://ww.salzburgresearch.at/suntrec/ohs7/papers/haake_ohs7.pdf
16. Kalakota R. and Whinston A. B., *Electronic Commerce*, Addison-Wesley, 1997.
17. Kraemer K.L., Gibbs J. and Dedrick J., 2002, *Impacts of Globalization on e-Commerce Adoption and Fir Performance: a Cross-country Investigation*, CRITO - Center for Research on Information Technology and Organizations, University of California - Irvine, Retrieved June 2003 from: http://www.crito.uci.edu.

18. Lamb R. Intranet boundaries as guidelines for system integration. International Journal of Electronic Commerce 2003; Summer:7: 4
19. Laubacher R. and Malone T.W., 2002, Retreat of the Firm and the Rise of Guilds: the Employment Relationship in an Age of Virtual Business, MIT Initiative on Inventing the Organizations of the 21st Century, Working Paper #033, Available in http://ccs.mit.edu
20. Malone T.W, Yates, J., Benjamin, R.I. Electronic markets and electronic hierarchies. Communications of the ACM 1987; 30; 6(June).
21. Masini A., 2003, Knowledge Integration and the Development of IT Capabilities: Configurations of ERP Adopters in the European and US Manufacturing Sector, London Business School, Operations and Technology Management Working Paper Series, Available in http://www.london.edu/otm/Working_Papers/working_papers.html
22. Micelli S., 2002, Network Technologies and Local Networks: Evidence from a three Years Survey in North-East Italy, Retrieved July 2003 from: http://www.deeds-ist.org
23. Montresor S. and Romagnoli A., 2002, Re-configuring the Firm from a System Perspective; Retrieved July 2003 from: http://www.deeds-ist.org
24. Mowshovitz A., 1999, The Switching Principle in Virtual Organisation, in Organisational Virtualness and Electronic Commerce, Proceedings of the 2nd International VoNet Workshop, September 23-24, Institute of Information Systems, Department of Information Management, University of Bern, Available also in http://www.virtual-organization.net.
25. Nichols F., Communities of Practice. Definition, Indicators and Identifying Characteristics, Retrieved June 2003 from: http://home.att.net
26. OECD, Seizing the Benefits of ICT in a Digital Economy. Meeting of the OECD Council at Ministerial Level, 2003
27. OECD Ministerial Conference A Borderless World: Realising the Potential of Global Electronic Commerce, Ottawa, 7-9 October 1998
28. Perogianni, M., 2003, B2B Internet trading platforms: Opportunities and barriers for SMEs: a first assessment, European Commission, Enterprise Papers No. 13, Enterprise Directorate General
29. Polanyi, K., "The economy as instituted process", in *Trade and Markets in Archaic Societies,* Polanyi, K., Arensberg, C., and Pearson, H. ed. Glencoe: 1957
30. Quah, D.T., The Weightless Economy in Economic Development, LSE Economics Department, Centre for Economic Performance, Discussion Paper no.417, London School of Economics, 1999.
31. Smith M.D., Bailey J., Brynjolfsson E. (1999), Understanding Digital Markets: Review and Assessment, Retrieved June 2003 from http://ecommerce.mit.edu/papers/ude.
32. Soete I. and Weehuizen R., 2003, The Economics of e-Government. A bird's eye view, European e-Government Conference, Como, 7-8 July 2003, Available also in http://europa/eu.int/information_society/eeurope/egovconf/index_en.htm
33. Timmers P., *Electronic Commerce: Strategies and Models for Business-To-Business Trading*, John Wiley and Sons, 1999
34. Ark V. B., 2002, Understanding productivity and Income gaps in the OECD Area: are ICT and Intangibles the Missing Link?, Retrieved June 2003 from: http://www.niesr.ac.uk/epke/bart_van_ark.html
35. Vanharanta H., Breite R., 2002, A Supply and Value Chain Management Methodology for the Internet Environment, in DEEDS 2nd Annual Report, Also available in http://www.deeds-ist.org
36. Wigand R.T., Electronic Commerce: Definition, Theory, and Context, in "The Information Society", 1997; 13:1

37. Wigand R.T., benjamin L. Electronic commerce: Effects on electronic markets. Journal of Computer Mediated Communication, 1995; I:3.
38. Wood A., Milosevic Z., 1998, Describing Virtual Enterprises: the Object of Roles and the Role of Objects, Objects, Components and the Virtual Enterprise An Interdisciplinary Workshop at OOPSLA'98, Vancouver, Canada, Also available in http:/www.cs.ted.ie/Virtues/ocve98/proceedings/003.html

Chapter 13

MULTI-CHANNEL MARKETING

Optimizing Performance in an ICT-Enabled World

Hugh Wilson[1], Matt Hobbs[2], Chris Dolder[2] , Malcolm McDonald[1]
[1]*Cranfield School of Management,* [2]*IBM Business Consulting Services*

Abstract: This paper address the marketing side of demand management by exploring the potentials of multi channel approaches for customer interactions. We suggest various frameworks for suppliers to consider when developing a strategy for investment with the right mix of channels.

Key words: Multi-channel marketing, Integration, Customer interactions

1. INTRODUCTION

The ICT revolution in supply chain management which forms the primary focus of this book can equally be viewed as a revolution in demand chain management, and so is as much a marketing issue as a logistics, purchasing or ICT one. Viewed from the perspective of a marketer within an individual enterprise in the chain, this revolution has led to a bewildering mix of ICT-enabled channels to the customer – web sites, e-hubs, call centers and so on – to complement traditional offline channels. Where once the channels to market were a straightforward matter of standard industry practice – "in our industry we sell via distributors" or "We serve our customers via a direct sales force" – managers now find themselves competing as much on innovative routes to market as on innovative products or services. Tesco.com, easyJet, First Direct and electronic components distributor RS Components are just some of the more obvious examples of the effective strategic use of early-mover advantage in new channels to market.

This leads practitioners to ask unfamiliar strategic questions, such as how best to choose the right channel for the right job and how to fit them together so as to add value to the customer as well as optimize margins. In this quest, one might expect there to be a ready supply of off-the-shelf marketing theory for them to turn to. But while the typical marketing textbook may contain chapters on managing individual channels such as the service center or the retail chain, practitioners look in vain for a chapter explaining in simple, practical terms how to design the right channel mix in the first place.

This is the problem we aim to address here. We focus on the relationship between the organization and its immediate customer, leaving to other papers the equally vital issues of examining the whole supply chain and how it might evolve in the future. We draw on traditional channel theory such as the standard work by Stern et al (1996), complemented by some relevant thinking, which has emerged as a result of the e-commerce revolution, to provide a new synthesis that aims to be both rigorous and practical. In doing so, we have found it necessary to invent some new frameworks to fill gaps in this previous thinking, which are grounded in our consulting experience at IBM and Cranfield School of Management.

We will start in the next section with a key point that adds some unavoidable complexity to the topic: the key strategic question is not which channel to use, but which channel combination, as often the customer is best served through a judicious mixture of communication mechanisms.

2. THE NEED FOR MULTI-CHANNEL INTEGRATION

Once upon a time, banking meant going to a bank. Then ATMs, cashback and call centers appeared, but the branches didn't go away. Then the Internet, mobile commerce and interactive TV were added to the several ways you could access your account, but the branches didn't go away. And while text messages telling you that you're about to go overdrawn might be useful, no one has yet suggested the pure-play text-messaging bank.

We live in a multi-channel world, at work as at home. The business-to-business sales force is still struggling to work out its relationship with the call center and the Internet channel. And although Amazon's pure-play model flourishes for some product-market segments, many others like Dell have found that a remote marketing model of mail order, the telephone and the Internet still has to be supplemented by a sales force to build relationships with major accounts.

This channel profusion would not be possible, of course, without the technological developments, which underpin it, from database marketing and computer-telephone integration to the Internet and personalization software.

But change is also being driven by the changing customer. Longer working hours and travel times leave less time for shopping, making travel arrangements or insuring the car. In the decade to 1996, the number of hours spent in shopping malls by the average American dropped from 7 hours a month to two and a half (Sheth and Sisodia 1997). And when purchasing at work, flatter organizational structures and consolidated buying functions put a similar emphasis on purchasing convenience. So we have a strong motivation for using remote channels to save time.

Also as the seller also stands to gain through the lower transaction costs of the call center and, particularly, the Internet, predictions abounded of the death of the sales force and the decline of the retail store. But these predictions have shown few signs of being fulfilled in the near future. While for some interactions such as configuring a computer, impersonal but precise information on a web-site may be just what we're looking for, in other aspects of the relationship with the supplier we want the human touch. While as rational consumers we have explicit needs relating to product functionality, price or service, as emotional humans we cannot avoid our complementary needs for security, justice and self-esteem (Schneider and Bowen 1999). So when we have a complaint, or we seek reassurance that the price we've been quoted is a fair one, a human has more chance of satisfying us than a computer.

So our customers use different channels for different purposes – but all in the same relationship, and indeed in the same purchase. But as organizations, we don't seem to have noticed. Behaving as if channels are mutually exclusive alternatives to each other, we run them as silos, supported by their own management and IT infrastructure, and judged in isolation by the level of orders placed through them or their total cost. While we may know the conversion rate of a web site, we are unlikely to be able to measure its contribution to offline sales, or the proportion of its leads that originate from the sales force.

To be fair, many if not most companies have instituted projects to reduce this silo mentality, from the complex IT integration projects that provide a single customer view, to metrics and reward systems based on customers rather than channels. The potential benefits they seek, and some report, include:

– Transaction costs: Given a pure-play choice between, say, an Internet channel and a call center, the customer may choose the latter. But if multiple channels can be used in the same relationship, the customer may be very happy to use the Internet to look up product specification

information or a database of common service problems – particularly if they know that personal assistance can be gained while they're online if it's required, a facility as yet provided by only 2% of call centers (IBM 2002). As well as such transaction cost savings, there can also be other cost advantages of integration, such as removal of duplicated tasks;
- Service enhancement: As customers, we expect that an organization's left hand will know what the right hand is doing: otherwise, how can it be trusted with our business? So if a salesperson's first topic in a meeting is dealing with a complaint we made to the call center yesterday, we may not pledge our undying loyalty, but at least it gives us one less reason to leave;
- Personalization: If the supplier goes the extra step of using integrated customer data to tailor products, services or the sales process, though, we may perceive genuine added value;
- Data analysis: We have little chance of understanding the customer in the absence of an integrated view. A vital input into most decisions on how to treat the customer is their value to the organization. An integrated view across channels and products can enable estimates of the customer's current profitability – sometimes with the conclusion that as few as 10% of customers generate over 100% of profits (Stone et al 1996). This analysis can inform decisions on how hard to work to keep the customer. Other analyses enabled by an integrated view include segment membership; propensity to defect and propensity to purchase a product not currently held (Wilson et al 2002).

There is, then, no shortage of potential benefits from optimizing multi-channel performance. But moving beyond plausible argument to a sound business case is not easy. The difficulties divide broadly into three areas:
- The case for change: Anticipated benefits generally depend on customers perceiving additional value and therefore giving us more business. And even projects based primarily around reducing cost need to check that customers won't see the changes as giving them a less good deal. But as the history of the dot-com boom shows, it's easy to gloss over this in the business case with market share or retention estimates that prove wildly wrong. While customer behavior can never be predicted with certainty, some kind of rational basis is needed for judging what benefits will be perceived by the customer base;
- The case for priority: There are a number of options, including:- adding personalization to the web channel; rationalizing the call centers; segmenting the customer base through CRM analytics. The range of potential projects at the customer interface is generally wide, and a strong case can be made for many of them. But with limited resources and

equally pressurized management time, projects need to be carefully prioritized. And this prioritization needs to take into account not just the benefits, but also risk factors such as the technological difficulty and the extent of organizational change required;

- The case for investment: If a project is selected as high priority in principle, there is still a need to build a sound financial case. Even assuming customers like a new channel initiative, we can end up with poor financial impact due to the cost of service, cannibalization, or a host of other issues. There is no short cut to building a financial model of the current channels and using this to predict the impact of a new project.

In this complex area, in which details differ considerably from one organization to another, we proceed by reporting some frameworks for addressing these three areas, which the organizations we work with have found most useful, though they may need adapting to individual circumstances. The frameworks are summarized in figure 13-1.

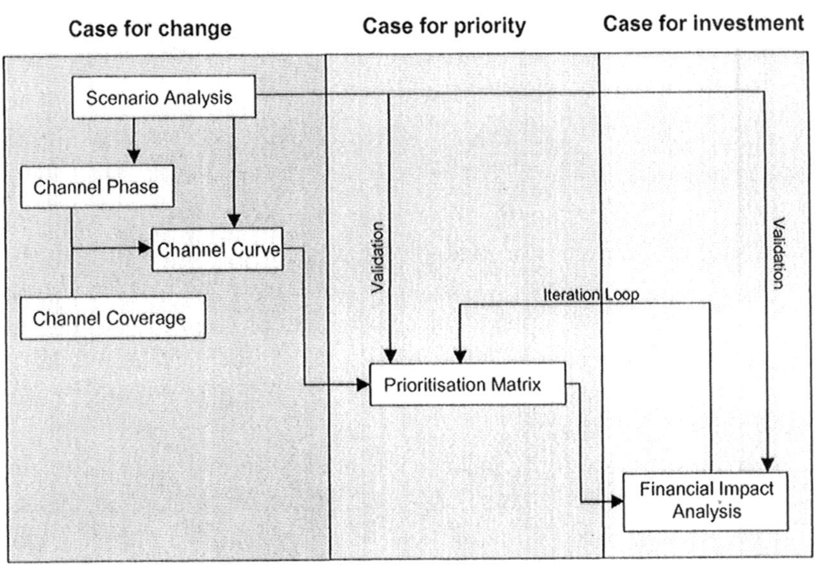

Figure 13-1. A Process for Optimizing Multi-Channel Performance

3. THE CASE FOR CHANGE

We will start with the case for change. First, we need to understand the present: how our customers currently use the various channels on offer. The channel phase table and channel coverage diagram provide simple ways of summarizing this.

We will then look behind this 'how' question at the 'why': why customers are making their current choices of one channel rather than another. This is an essential step if we are to offer them a better proposition and thereby modify customer behavior in the future, and is covered by the channel curve. The channel curve allows a future proposition – say, a new channel or channel combination – to be compared against the customers' underlying buying criteria, so we can check in advance that it is likely to be welcomed.

Finally, we will examine scenario testing, which helps to flesh out a future proposition and ensure that the customer experience is well thought through.

3.1 How Customers Are Using Channels: Channel Phase Table

A table of channels against phases of the purchase process, as illustrated in figure 13-2, provides a simple starting-point. The proportion of each phase – such as the purchase itself – that is conducted using each channel is estimated as a percentage. For example, this business school (loosely based on Cranfield School of Management, with illustrative figures) has estimated that 85% of the times, customers first get the idea of going on a short course for managers from a mailing. Although exact figures may be available on which channels are used to place the order itself, filling in the other columns will generally require making some estimates.

Next, some proposed figures can be added (in red italics, i.e. lower figures in each box in the figure 13-2). While not set in stone at this stage, these can be used to explore initial ideas on optimizing the mix of channels, in order to save costs, give more value, or both. Where a lower-cost channel can be used to perform a task at no disadvantage to the customer, for example, its percentage can be increased. In the case of the business school, some of the proposed changes are:

- Initiate dialogue: Mailing course brochures is an expensive and imprecise way of telling the business school's existing customers about courses they might be interested in. Its limited experience so far suggests that many customers would be just as happy to be sent carefully tailored emails, and just as responsive. Recommended facilities on the website might prove an even better way to steer the customer in the direction of the courses they need;

- Negotiate/tailor*:* For tailored courses run for particular companies, an element of face-to-face discussion is likely to continue to be desirable. But it may help the customer to feel in control of the negotiation if they can experimentally assemble a possible course from standardized modules using a configuration facility on the web-site;
- Exchange value*:* The means by which the course is delivered can also be reviewed. There is scope for efficient distance learning of standard components of theory such as accounting basics, but the business school envisages that most of the course delivery will continue to be face-to-face, where the scope for teamwork, networking and a rich discussion of practical issues is greatest.

Purchase phase

Channel/Medium	Initiate Dialogue	Exchange Information	Negotiate & Tailor	Commit	Exchange value
Face to Face		5% *5%*	10% *10%*		95% *90%*
Mail	85% *60%*	60% *40%*	20% *10%*	20% *10%*	
Tele	5% *5%*	10% *10%*	40% *20%*	80% *70%*	
Web	5% *15%*	20% *40%*	0% *25%*	0% *20%*	5% *10%*
Email	5% *20%*	5% *10%*	30% *35%*		

Illustrative figures only

Text Colour Key

- %age of current business using the medium to perform that task in the sales cycle
- Future target

Figure 13-2. Channel Phase Table – a Business School

3.2 Another View of Channel Use: Channel Coverage Analysis

To complement the channel phase analysis, we have often found it useful to draw a channel coverage diagram as shown in figure 13-3, which illustrates a major UK telecom company. This summarizes how different channels are used by different customer groups and for different parts of the product range. The model provides a valuable internal perspective on channel use.

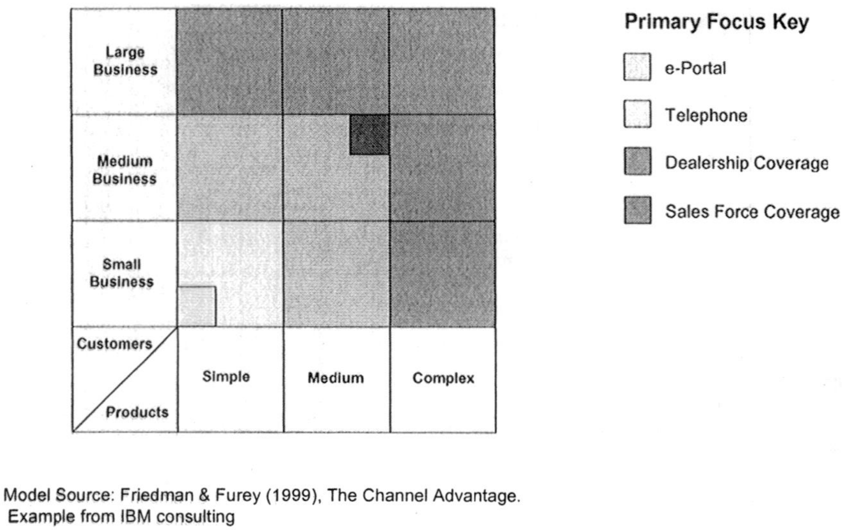

Figure13- 3. Channel Coverage Diagram – UK Telecom Company

This diagram shows that there is an extremely heavy use of the sales force channel in this company. Sales forces are almost always the most expensive channel to operate and so broad use across all customer segments and product lines, regardless of whether this is necessary, is a major cost concern. Moreover, although the Web and dealership channels are sensibly positioned on the diagram, they are in an immature state. The size of their boxes reflects the proportion of business coming through this channel, so for example only about 10% of simple products purchased by small businesses are bought over the Web. The telephone channel is also underused: its success with simple products for small businesses could be extended upwards on the diagram to medium businesses, as well as rightwards to medium-complexity products. It is clear that further work is required to understand why the expensive sales force is being so heavily used across all customer segments and products. In this example, the products are split according to product complexity, but in other cases the division might be by major product/service lines. Similarly, the definition of customer groups will vary in different cases: here it is based on company size, but in other cases it might be based on a needs-based customer segmentation. The aim is to choose the main ways in which channel use varies.

Once a 'present state' analysis is completed a desired 'future state' can then be mapped to quantify the likely magnitude of change required in the new channel mix. Crucially, supported by the correct numbers, this model serves to form a key input into the investment case. We will return to this later.

3.3 Understanding and Modifying Customer Behavior: the Channel Curve

Suppose we decide, on the basis of these analyses, that we would like more customers to use a call center or web channel in order to lower transaction costs. Clearly we have to look at this issue from the customers' point of view: what is driving their current behavior, and what would entice them to change? The channel curve analyses this, comparing how different channels rate at meeting the customers' buying criteria. It is illustrated in figure 13-4 for the books market. First, the main buying factors are listed along the bottom of the chart, with weights indicating their relative importance to the customer's buying decision out of 100. These represent the key factors taken into account by the customer in deciding where to place their business – here, the cost of purchase (including mailing costs where appropriate); convenience, or the total time taken for the purchase; the ability to browse for the book you want; and so on. The ability of each channel to deliver against each factor is then assessed judgmatically on an 1-10 basis: the higher the score, the better this channel meets this buying factor. It can be seen that taking all the factors together, the Internet and physical stores have the best matches to the needs of this particular hypothetical segment of the book market. In reality, different segments will be best matched to different channels, which would show up clearly if the channel curve were drawn for each segment's buying criteria, which by definition are different from each other. Hence, this chart may need drawing up for each major segment.

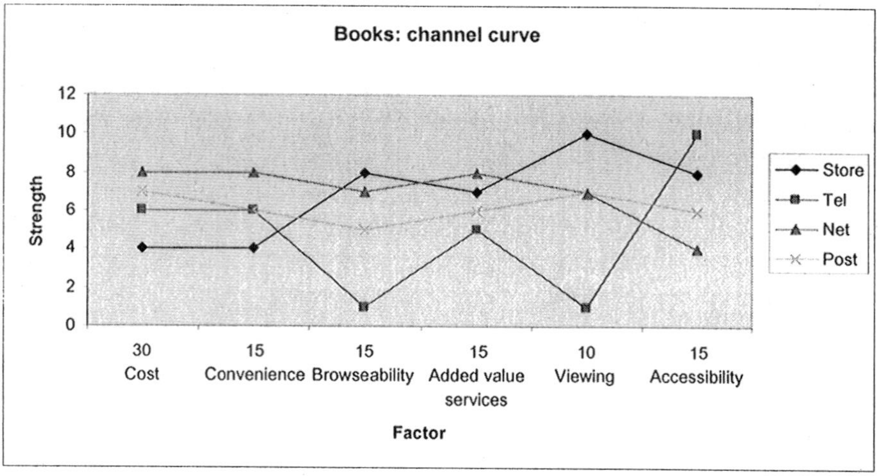

Note: A separate analysis may need to be done for each major segment.
Source: McDonald and Wilson (2002), The New Marketing

Figure 13-4. Channel Curve

If the channel phase table suggested that a particular phase of the purchasing process needs particular attention, the analysis can be focused on that phase. In this case, the factors listed are those specific to that phase, which determine which channel a customer would prefer to use for that phase of the relationship.

This example compares the inherent qualities of different channels, rather than specific competitors. But if a particular channel is emerging as a candidate for improvement, it can be useful to benchmark the organization's current offering through the channel against one or two major competitors, by adding a line for each competitor to the channel curve.

All this can be used to define a future proposition on the channel curve. This may simply be a question of adding a channel to the mix, or extending its capabilities to handle a wider part of the customer interaction. But more often, it will involve improvements to an existing channel offer. So two lines will be needed in this case for the channel in question: one for the current situation and one for the envisaged future one.

Where will the data come from to fill in the channel curve? Data warehouses, however sophisticated, rarely have this kind of information on customer needs. Broadly, there are two approaches:

1. Market research. This does not necessarily have to be formal: in B2B contexts, it can be just as effective to assemble a small team of customers, or customer representatives such as distributors or recently retired purchasing managers, and run a workshop focused around filling in the channel curve from their perception;

2. Failing that, an internal workshop with a small group of those close to customers, drawn from multiple business functions, can typically make a fairly good estimate of customer needs. In this case, comparing the resulting channel curve with the percentages in the channel phase table can make a reality check of the results. If the channel curve suggests, say, that the call center is ideal for placing orders, but very few customers are using it, then we have clearly got our data wrong.

3.4 Bringing the Channel Proposition to Life: Scenario Testing

The channel curve provides a basis for working out a proposition which will broadly have appeal to the target market. In order to flesh this proposition out, though, we need to do more than just define the right combination of product, service and channel. We also need to consider how the proposition will manifest itself to our customers as they make their various contacts with us.

One way to do this is to illustrate the future proposition with 'scenarios' (Sonderegger et al 2001) – descriptions of how a typical customer from each major customer segment might go about performing some task in their relationship with us.

In the example shown in figure 13-5, a bank customer is considering the purchase of an additional product. On the left of the figure, the segment is brought to life by naming and describing a typical member of it. Then a table is drawn up, as shown on the right, describing a walkthrough of the interactions between this named customer and the bank as he conducts the purchase. The walkthrough teases out ways in which the channels need to interact and support each other in order to deliver the ideal experience.

Name: Peter Muller

Job Role: MD of an IT company in Frankfurt.

Background: Herr Muller has been a standard current account customer for the last seven years and in that time his interest in investments and his financial sophistication have grown.

Current Portfolio: Peter now has both a standard current account and flexible fund account. He holds cash, funds and securities in his accounts.

Situation: He has a good relationship with his broker, but also uses direct channels, especially the Web, for convenience.

Scenario: Buying Alternative Investment Products (AIP)

Task	Channel	Experience	Implications
Initiate dialogue	Mail	Concise, signed personal note from broker, two weeks before annual review meeting, suggesting that Markus look into AIPs, with leaflet & Web URL	Integration of mail & face-to-face channels
Exchange information	Web	Peter looks up his flexible fund account briefing on AIPs. He models impact on his investment portfolio using what-if facility	Extensions to Web-based portfolio management tool
Negotiate/tailor	Face-to-face	Peter discusses options with his broker. Together they decide on appropriate level of risk	Broker access to personal portfolios in meetings
Etc....	Etc....	Etc....	Etc....

Figure 13-5. Scenario Testing

Scenario testing, then, helps to round the picture of what changes is desirable in the channels we use and the way they interact. And taken with the channel curve, it can help to ensure that we know what we're in for when embarking on a new channel proposition. Indeed, by this stage there are probably many more potentially beneficial changes to our channel mix on the wish list than the organization can hope to implement at once. This raises the issue of prioritization, which we turn to next.

4. THE CASE FOR PRIORITISATION

All sorts of exciting possible strategies may have emerged by this point. As a result of the previous steps, the organization will have a list of possible channel innovations, and for each, a summary of its attractiveness to customers via the channel curve. But as one always has a limited budget and limited management time, it is now necessary to prioritize between the options available. Although the channel curve summarizes which approaches are likely to succeed in the market as a whole, an organization will typically not have the resources to pursue all viable options – some of which may, in any case, deliver little value to shareholders, however welcomed by the market. It is this dimension of attractiveness to the organization, which we have not yet considered fully. The answer is to use a portfolio matrix, which looks both at the attractiveness of an offering to the customer, and its attractiveness to us. For this purpose, we suggest using the matrix of figure 13-6.

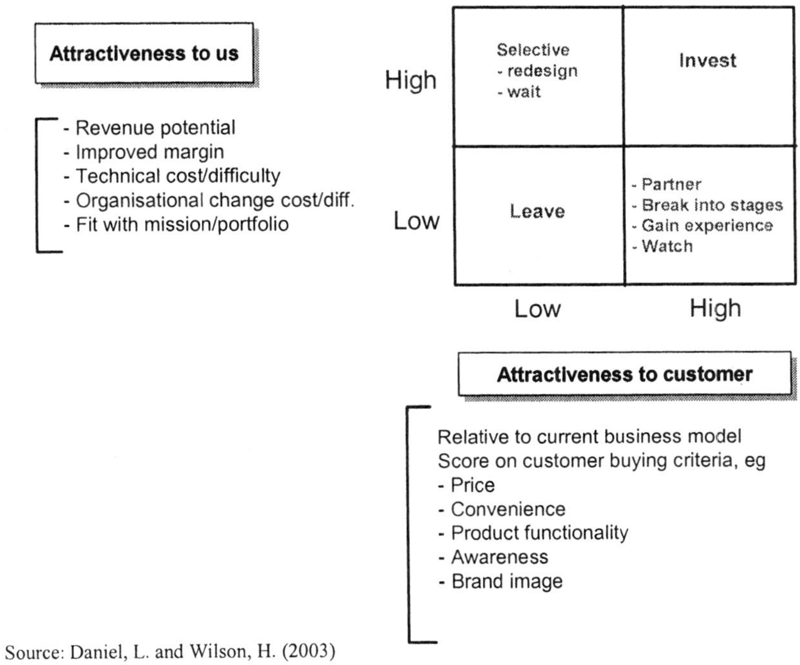

Figure 13-6. Prioritizing Channel Projects

Developed by Dr Liz Daniel at Cranfield School of Management and the authors, the matrix compares channel projects against two dimensions: attractiveness to us and attractiveness to the customer. In many ways, it is just like the well-known directional policy matrix, which compares market attractiveness against business strengths. But in assessing the horizontal axis - attractiveness to the customer - we compare not rival competitors but rival channel strategies. So if say we are considering selling via a Web/call center combination rather than a direct sales force, how would this channel strategy compare in the customer's eyes with their current way of doing business?

You may have spotted that we have already done much of the work to fill in this matrix. We have compared various channel strategies against customer needs using the channel curve. So the horizontal axis simply corresponds to the position of the proposed route to market on the channel curve compared with the best alternative. A better curve than the alternatives will correspond to a position well to the right of the matrix. To be precise, the position on the horizontal axis can be determined mathematically using the channel curve data by comparing the weighted average score of the proposed channel with the best of the alternative channels (If we are improving our channel offering through an existing channel, the comparison is with the curve for our current offering).The vertical axis uses a similar multi-factor calculation. The figure shows some typical factors, which reflect both the potential return and risk factors such as the degree of technical and organizational difficulty, but these will need to be reviewed carefully for each organization to reflect its priorities. The scores can be made, as with the channel curve, on a judgematic 1-10 basis. Clearly this is not an exact process, but it provides a practical way of reducing the list of potential projects to a few, which require the drawing up of a full investment case, the topic of the next section.

An example of the application of the matrix is shown in figure 13-7, for an insurance company examining its distribution policy in a particular European country.

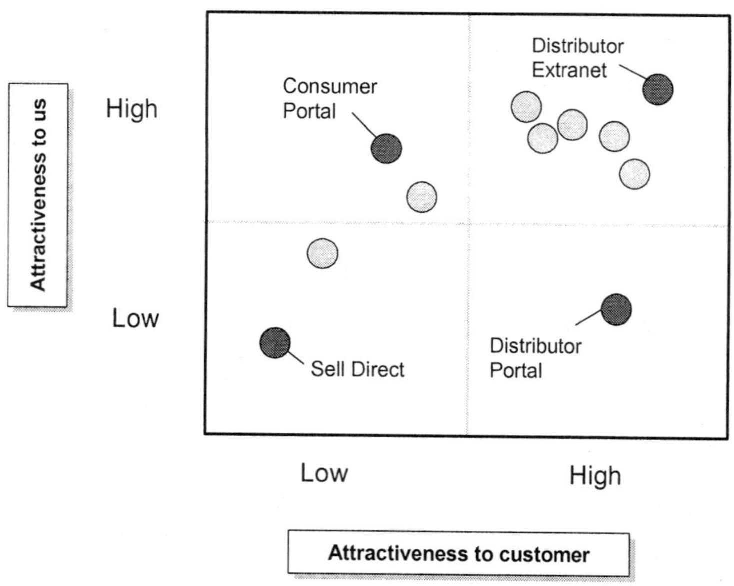

Figure 13- 7. Prioritization Matrix – Insurance Company

The company drew the following conclusions about the four sample projects that we have labeled in the figure:

- *A distributor extranet* was a potential project to provide Web access to the financial advisers who distributed its products, as an alternative to the call center and account managers currently available to them. A clear win-win for the highly technical advisers, this was implemented as soon as possible, along with some of the other projects in the top right, starting with those closest to the top right corner;

- Another easy decision was on '*Sell direct*' in the bottom left quadrant, which the company decided not to proceed with. This was actually scored so low on customer attractiveness that it was off the left-hand edge of the scale. The internal team doing the scoring believed that the few customers who were yet prepared to buy long-term savings products without any form of advice would probably prefer to use a fund supermarket to compare different offerings. Its low margins and channel conflict issues gave it a low score on the 'Attractiveness to us' axis too;
- *Distributor portal:* A new web-based intermediary between the insurance company and its financial advisers was likely to reduce prices, and forming such a portal was also outside the organization's current skill-set, so this scored somewhat low on 'attractiveness to us'. But because it was attractive to customers, the company decided it would happen anyway in due course. So it decided to partner with another company offering this service;
- *Consumer portal:* The company felt it possible that one-stop shops to buy savings products by telephone and Internet might become more attractive to consumers, as, say, car insurance has changed. The company took a watching brief, to see whether customer views would move this to the left over time.

We have found that using this tool provides a welcome structure to the board's decision-making debate and rapidly sorts out which projects are the top priorities. It also forces the proponents of a project to ask the right questions early on, increasing the quality of proposals put to the board. But it does not remove the need for a detailed financial investment case for projects that seem, on this judgmatical basis, to be candidates for go-ahead.

5. THE CASE FOR INVESTMENT

For prioritized projects, there is a need, next, to develop a detailed investment case. As well as refining the project's position on the prioritization matrix, this enables the project's return on investment (or net present value, according to the preferences of the company) to be assessed for funding purposes.

The means of doing this will vary according to the nature of the project and the company's conventions, but generally, some kind of cost/revenue model will need to be built for the company's marketing and distribution channels, incorporating revenues and costs for each channel as well as costs which are not channel-specific.

By modeling future years as well as the current year, the impact of the project on costs as well as revenues can then be estimated. Anticipated revenues, particularly, will still need management judgement, but at least the channel curve means we have some kind of rational basis for making this estimate. If a large increase in market share is anticipated, for example, this should be backed up by a significant improvement on the channel curve.

One thorough approach to building such a cost/revenue model is to use customer lifetime value (CLV) (Blattberg et al 2001), assessing the impact of a project on the sum of the customer lifetime value across all customers. A CLV approach is worth working towards. But for most organizations today, this is simply not possible, as they do not have appropriate data: most struggle even to understand current profitability by customer or by channel.

So a simpler model of costs and revenues will generally need to be constructed and used to assess the impact of a project. The model can concentrate on those parts of the overall picture, which are most likely to be impacted by a proposed project. Is it tackling fixed costs, acquisition costs, ongoing transaction costs or retention costs? Are its revenue objectives based on improving post-sales service and thereby retention rates, or improving conversion rates and thereby market share?

To start, a baseline needs to be created of the current costs and revenues across the chosen channels. While measures will vary according to the company and channels under review, some typical measures are listed in figure 13-8.

Once a baseline has been captured, it is then time to consider the likely program costs. Here it is important to consider all aspects of the proposed project(s) identified in the prioritization matrix. Typically these should include any external costs to outside vendors and internal costs. It is important to document assumptions made in making these estimates of both the set-up and on-going costs.

Finally, total program benefits should be measured in terms of cost reduction and/or revenue enhancement. Here the specific measures will depend on the company, business unit and channel(s). However, it is often advisable to design multiple scenarios for the program benefits, thus helping to mitigate risk and assess the likelihood of an attractive break-even and return on investment (ROI).

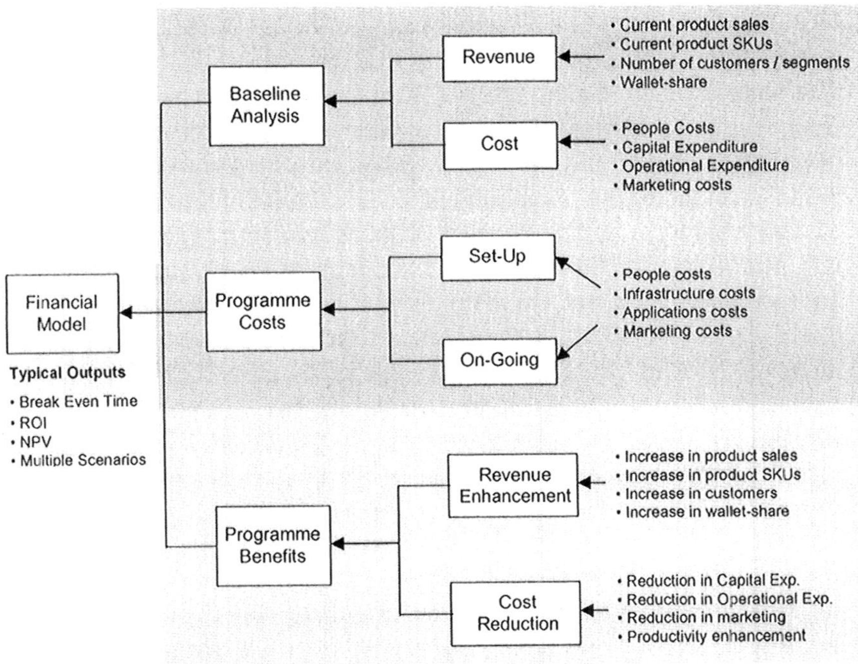

Figure 13-8. Financial Impact of Channel Strategies

An example of a successful multi-channel program investment case is provided by a large fast moving consumer goods (FMCG) company. It was suffering from a downturn in profitability despite owning an excellent range of household brands. Closer inspection of its existing channel operations showed that an expensive sales force was being employed across all customers, from the corner shop to the national supermarket chain. This cost drain was compounded by a reliance on lengthy paper order forms that needed to be re-typed, resulting in errors, delays and process inefficiencies.

The sales force was supported by a call center operation. However, the call center staff's main functions were to take 'ad hoc' customer orders and deal with the wholesalers; fielding after-sales service enquiries originating from sales force relationships was secondary. Moreover, the call center operation was not using a system integrated to back-end systems. This made it impossible to deal with real-time customer enquiries about their orders.

For this company, therefore, the cost of maintaining the channel infrastructure was extremely high in relation to best practice in the industry. Furthermore, it seemed a fair guess that the poor non-integrated customer experience across multiple channels was losing the company revenue.

A total multi-channel program was therefore initiated, split into three work-streams:

- Deploy a field force effectiveness solution: firstly the sales force was cut back and re-focused on the most profitable accounts. Their effectiveness was improved by e-enabling the sales force with wireless field sales devices that would allow them to enter orders, easily prioritize their accounts and give real-time 'added-value' information to customers;
- Build and deploy an extranet: all customers were given access to an extranet making available the full range of products, product information and order history. This enabled customers to order when they wanted and get direct access to valuable point of sale information and of course product details. This also applied to the wholesalers. In turn the company was able to dramatically cut costs from the sales force and back-end administration support;
- Streamline and re-focus the call center: In the old model, the call center was generally reactive and essentially supported the administration of order entry. The call center applications were replaced so that they could integrate fully with the other channels and back-end systems. Moreover, the development of the extranet helped the call center not only to support the two primary channels to market, the direct sales force and extranet, but also to proactively contribute to the 'identification', 'qualification' and 'sell' stages of the sales cycle.

The investment case was very attractive. Restructuring and process redesign meant that operational costs were slashed by 40% across the channel mix, while revenue increased by 11%. The revenue increase could be directly attributed to better segmentation and targeting, and more effective use of the direct sales force and call center customers.

It is instructive that the original sign-off for the investment case, though, was based on a simple break-even analysis. Often there is a temptation to 'boil the ocean' with immense detail and pages of assumptions. However, there are dangers in making the business case too complicated with too much detailed analysis founded on numbers that inevitably are mere estimates. The appearance of precision can lead to these numbers being treated with excessive reverence, and can obscure the main argument. In our experience, it is better to keep a relatively simple modular approach to multi-channel projects and ensure that large projects are split into separate phases or work streams, each with their own manageable investment cases.

6. CONCLUSIONS

Much was made in the dot-com boom of the lower transaction costs, which could be gained online. We were told that airline tickets cost $10 to sell through travel agents, versus $5 via call centers and $1 via the Web, with similar figures being widely quoted for the cost of bank transactions or the cost of share brokering. While all of this is true and potentially invaluable, the transaction costs are only one of the issues to be considered. Often, the cost-effectiveness of acquisition and retention runs in the opposite direction, with higher costs online. Many a start-up during the dot-com boom ignored at its peril statistics that the average online retailer was spending £66 to acquire a customer, against £13 offline (Hayward 2000). They also discovered that the scant attention to customer service that left 25% of email queries entirely unanswered, and only 21% answered within two days, was having a high cost in low retention rates (Resource Marketing 2000).

This is one reason why we are increasingly seeing multi-channel strategies, which use different channels for different aspects of the relationship, such as acquiring the customer, for example, through the sales force and providing service through the Web and call centers, but with personal visits to handle complaints and to ensure that the relationship is not damaged.

But there remains much scope for fine-tuning the emerging multi-channel strategies. While product innovation continues apace, we believe that the dominant theme of the next ten years will be innovation in the route to market - the channels by which the customer is communicated with and the product delivered. While the channel revolution in financial services, retailing and IT is far from over, in many other industries it has hardly begun.

As always, though, the devil is in the detail. Developing a successful channel strategy is far from easy, and the frameworks we have discussed are offered as ideas to consider rather than universal prescriptions. But equally, project implementation on complex initiatives at the customer interface presents numerous difficulties, amongst them the challenge of large CRM system developments (Wilson et al 2002), the hazards of changing corporate culture (Clark 2002) and the puzzle of determining the right organizational structure for a multi-channel proposition (Gulati and Garino 2000).

But two things are clear. Firstly, you can't hold back the tide of change. If a new channel proposition – whether revolutionary or evolutionary - is in the customers' interests, then it will happen sooner or later. And when it does, we need to be ready: there is mixed evidence on whether first-mover advantage beats being a very early follower, but it certainly doesn't pay to be much later (Robinson and Min 2002). Secondly, when defining and implementing our channel strategy, we need to keep an unremitting focus on satisfying customers, or we risk creating economic terrorists who will advertise our failings to all who will listen. Genuine value networks can only be built if each participant in the network orients their strategy around the value delivered to their downstream customers, as well as the value received from them. Such techniques as the channel curve do not guarantee that we will achieve this, but at least they will help us to ask the right questions.

ACKNOWLEDGEMENTS

This chapter is based on an IBM/Cranfield School of Management White Paper entitled 'Optimising multi-channel performance' by the same authors (Hobbs et al 2003). Permission to reuse this material from IBM Corporation and Cranfield University is gratefully acknowledged. We would also like to acknowledge the substantial role played by numerous colleagues in creating and synthesizing the ideas contained in this chapter and in providing a supportive environment for this collaborative work, including Dr Liz Daniel and Professor John Ward at Cranfield School of Management, and Rod Street at IBM.

REFERENCES

1. Blattberg, R.C., Getz, G., and Thomas, J.S., *Customer Equity: Building and Managing Relationships as Valuable Assets*. Harvard Business School Publishing, 2001.
2. Bucklin, C.B., Thomas-Graham,P.A., Webster, E.A. Channel conflict: when is it dangerous? McKinsey Quarterly 1997; 3: 36-43.
3. Clark, M., The relationship between employees' perceptions of organisational climate and customer retention rates in a major UK retail bank. In press for Journal of Strategic Marketing, 2002.
3. Daniel, E.M., Wilson, H.N., Action research in turbulent environments: an example in e-commerce prioritisation. In press for *European Journal of Marketing* special issue on Action Research, 2003.
4. Friedman L.G., and Furey T.R., *The Channel Advantage*. Oxford: Butterworth Heinemann, 1999.
5. Gulati, R., Garino, J. Get the right mix of bricks and clicks. Harvard Business Review 2000; May-June: 107-114.
6. Hayward, D., Leading article, Computing, 30 March, 24, 2000
7. Hobbs, M., Dolder, C., Wilson, H., McDonald, M., Optimising Multi-Channel Performance, White Paper, IBM United Kingdom Ltd, Portsmouth, 2003.
8. IBM, Call Centre benchmarking report, 2002.
9. Kim, W.C., Mauborgne, R. Creating new market space. Harvard Business Review 1999; January-February: 83-93.
10. Resource Marketing, Quoted in IT-analysis.com newsletter, 4 April, 2000.

11. Robinson, W.T., Min, S., Is the first to market the first to fail? Empirical evidence for industrial goods businesses. Journal of Marketing Research 2002; 39(1): 120-128.
12. Schneider, B., Bowen, D.E., Understanding customer delight and outrage. Sloan Management Review 1999; 41:1: 35-46.
13. Sheth, J.N., Sisodia, R.S. "Consumer behaviour in the future". In *Electronic Marketing and the Consumer*, Peterson, R.A. ed., 17-38. Sage, Thousand Oaks, 1997.
14. Sonderegger, P., Delhagen, K., and Dorsey, M., *Cross-channel scenario design.* Cambridge, MA. Forrester,, 2001
15. Stern, L., El-Ansary, A., and Coughlan, A., *Marketing Channels*, 5th Ed. New Jersey: Prentice Hall, 1996.
16. Stone, M., Woodcock, N., Wilson, M. Managing the change from marketing planning to customer relationship management. Long Range Planning 1996; 29(5): 675-683.
17. Wilson, H.N, Daniel, E.M., McDonald, M.H.B, Ward, J.M. and Sutherland, F. *Profiting from eCRM: Making the new marketing work.* London: Financial Times Prentice Hall, 2002.
18. Wilson, H.N. and McDonald, M.H.B., *The New Marketing: Transforming the Corporate Future.* Oxford, Butterworth Heinemann, 2002.

Chapter 14

A ROADMAP OF MANUFACTURING SYSTEM EVOLUTION

from Product Competitive Advantage towards Collaborative Value Creation

Yongjiang Shi

Centre for International Manufacturing, Institute for Manufacturing, University of Cambridge, UK

Abstract: Manufacturing has been experiencing dramatic evolutions in recent decades. The boundary of a manufacturing system is extended from a factory towards various types of network relationships. The mission for a manufacturing system is also transformed and re-defined. This chapter, based upon recent case studies and research in industry, foresees an evolutionary manufacturing system from which four main types of manufacturing systems are identified and analyzed in order to understand the motivations behind them, their intention and characteristics. This chapter suggests that classical manufacturing strategy theory should be adapted into a contingency strategy to respond to the manufacturing evolutions and treat different systems interdependently. This chapter also suggests a new conceptual framework for further research into new manufacturing evolution and contingency strategy.

Key words: Manufacturing system, Global manufacturing virtual network

1. INTRODUCTION

During the last decade, the boundaries of manufacturing systems have been extended from a factory to a pan international manufacturing network (Shi and Gregory 1998), as well as to the supply network (Lamming et al 2000), the value network (Bovel and Martha , 2000), and the global manufacturing virtual network (Li et al. 2000). This has been driven by intensified competition, the fragmented market, globalized collaboration, and acceleration in technological innovation. What then are the implications of the manufacturing evolution to practitioners and academicians? Has the mission of manufacturing system been changed, or the value creation mechanisms and business principles changed? Also what are the new relationships between effectiveness and efficiency within manufacturing industries in the new information, communication technology and knowledge era?

This chapter seeks to answer these questions, by identifying the evolution of the manufacturing system and some key challenges to classical manufacturing strategy theory, especially from the perspectives of networked manufacturing systems and value-based manufacturing strategies. The analysis and discussion draws upon recent eight-year case studies conducted by the Centre for International Manufacturing of the University of Cambridge that involves a wide range of industrial sectors. The original purpose of the case studies was to focus on networked manufacturing systems from either geographic expansion or vertical externalization or both perspectives so that the new manufacturing systems' behavior and design process could be understood. The frameworks introduced in this chapter are preliminary yet explore a general picture of the evolutionary development of the corporate international and the collaborative manufacturing systems. The chapter provides a comprehensive understanding of current theories and practices from international and inter-firm viewpoints, as well to seek the meaning of the driving forces and developing trends of system evolution.

2. MANUFACTURING SYSTEM EVOLUTION MATRIX

Manufacturing systems have evolved from the traditional input-output transformation model into various kinds of network-based relationships. During the last twenty years multinational corporations (MNCs) have attempted to globalize their geographically dispersed factories by co-coordinating them into a synergetic network (Flaherty 1986, Ferdows 1997, Shi and Gregory 1998). This transformation has changed basic manufacturing functions and effectiveness from the orientation of product-based competitive advantages towards the orientation of network strategic capability development, which drives the manufacturing system beyond the factory wall and the strategy beyond product focus.

Besides MNCs' international expansions, it has become more popular for all types of companies to downsize and outsource their non-core business tasks and to set-up inter-firm collaborations (Lambert et al. 1998, Lamming et al. 2000, Brewer et al. 2001). This development has pushed manufacturing systems further into new relationships beyond the traditional concept of the firm that owns and internally operates its factories. It is commonplace that a company may own only a very small portion of a supply chain yet they are still strategically able to coordinate, or integrate, the whole supply chain to deliver competitive products to targeted markets. It is equally interesting to note that there are increasing observations about geographic clustering emerging worldwide (Piroe and Sabel 1984, Porter 1998). These clusters actually form different supply networks – some of them are internally self-sufficient in a region and others are virtually integrated with other clusters. The two types of supply networks demonstrate that inter-firm collaborations have materialized as a new type of manufacturing system.

Combining both developments, as the figure 14-1 illustrates, a new type of manufacturing network can be derived with both the characteristics of international and inter-firm relationships. The new combination provides a new operational environment for a manufacturing system by means of which strategic resources may bet accessed, optimized and operated. The global manufacturing virtual network (GMVN) was suggested as a means to explore the new generation of manufacturing architecture (Li et al. 2000, Shi and Gregory 2002). Many other researches into global outsourcing and partnership also seek to develop a system with similar architecture and strategic capabilities pursuing higher value and innovation (Normann and Ranfrez 1993, Parolini 1999, Bovel and Martha 2000).

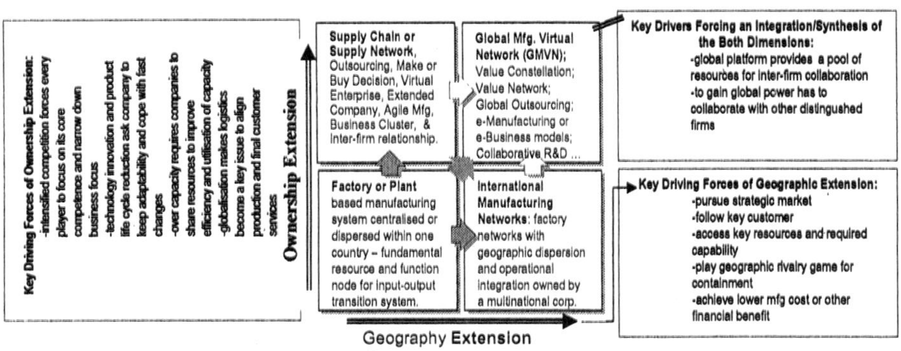

Figure 14-1. Manufacturing System Evolution Matrix and Key Drivers

Why does a manufacturing system have to evolve into a complex relationship? A company has no other choice: In many circumstances even when there are traditional product-based competitive advantages, which achieve the order winning criteria, this still cannot satisfy the necessities of new corporate demands. For example, the following case study will clearly demonstrate this. A very successful order-winning UK aerospace engine company was deeply dumbfounded when it compared its profitability with its American competitors and shocked by the resultant huge pressures from its shareholders. The reasons for such surprising behavior were not only their financial failure with a successful product but also the huge threat to their traditional ideas relating to the excellence of their product design, technology, engineering, and production. The company had indeed been very successful and gained many more orders and a greater market share than any other competitor in the world because of its modularized engine platform, advanced manufacturing technologies, and the best engine performances. However, at the same time, while the company enjoyed the advantage of more orders and production activities, its competitors quietly changed the rules of the game. In the new state-of-play the UK Company's princely engines were recognized as a commodity and integrated into its competitors' solution packages. Importantly the original competitor companies were as a result fully engaged in a new service business that provided a total solution for power to the airline companies. Additionally they had also outsourced a large portion of their manufacturing needs to suppliers in the Far East, and as a consequence redefined the manufacturing system as a value creation system rather than a product-oriented production or transition system. This was the time when the UK Company realized that the former 'best practice' manufacturing capabilities were no longer good enough to face the challenge of future competition and value creation. They realized that the manufacturing system and its strategy would have to be extended to match the changed rules of market competition. This means that the manufacturing system must be capable of not only providing competitive products but also finding a proper position in an innovative solution to final customers.

Using this case study together with an understanding of the trends of the manufacturing system towards a network relationship (Figure 14-1), the following lessons can be learned: (1) The conventional manufacturing strategy that focused on a product and its effective factory is probably no longer good enough, especially for creating higher business value. (2) The manufacturing system has been extended into a new operational space – international and inter-firm relationships – mainly because of the new competition game and strategy. (3) The manufacturing system's boundary changes imply that the manufacturing strategy also needs to be changed in terms of its contents and process.

Based on the three basic directions that the manufacturing system is heading towards in the matrix (Figure 14-1), the following four sections of the chapter will review each evolutionary direction, and analyze the main impacts on the missions of the manufacturing system.

3. FACTORY – THE CLASSICAL SYSTEM

R.OWEN, a British utopian socialist, might have introduced the term "manufacturing system" in 1815, to mean "factory system". Since then, the term m anufacturing system has been typically used to represent following two related concepts. The first one is a system approach to manufacturing that emphasizes the dynamics and optimized action - "Manufacturing systems approaches seek to optimize the initial design to commercial product time, the design lead time and factory door-to-door time, the manufacturing lead time by considering the whole factory as a system and simplifying and optimizing the performance of this complete system". (Williams 1988)

The second meaning is focused on the boundary of manufacturing systems, stressing that the manufacturing system is a unified assemblage of hardware including workers, production facilities, material-handling equipment, and other supplementary devices. Focusing more on its dynamic aspect, the manufacturing system can be defined as the conversion process of the factors of production, particularly converting raw materials into finished goods with the aim to maximize productivity.

It had been long time for many people to concede that the mission of a manufacturing system is to maximize productivity or efficiency, mainly because of the shortages of goods and capacity in society. Although Skinner's paper (1969) alerted people to pay more attention to manufacturing effectiveness or manufacturing strategic aim/mission, the time when manufacturing became a real strategic concern for product or business was the mid-1980s (Hays and Wheelwright 1984, Hill 1985). In the aftermath the factory-based manufacturing system, its effectiveness, and the strategy for a product's competitive advantages have attracted much more attention in both the practitioner and academia worlds.

Figure 14-2 demonstrates a process of manufacturing strategy that ensures that a factory will provide its products' at competitive advantages in marketplace. The basic assumption of the strategy is that a company that can create and capture value if its products can win orders in a competitive environment. Based on this rule, manufacturing managers design an effective system that will win orders to beat competitors, which transforms manufacturing missions from productivity or efficiency towards the competitive advantages in terms of lower cost, faster delivery, better quality and be more flexible for customer preferences. This fundamentally transforms manufacturing management, especially the manufacturing system design.

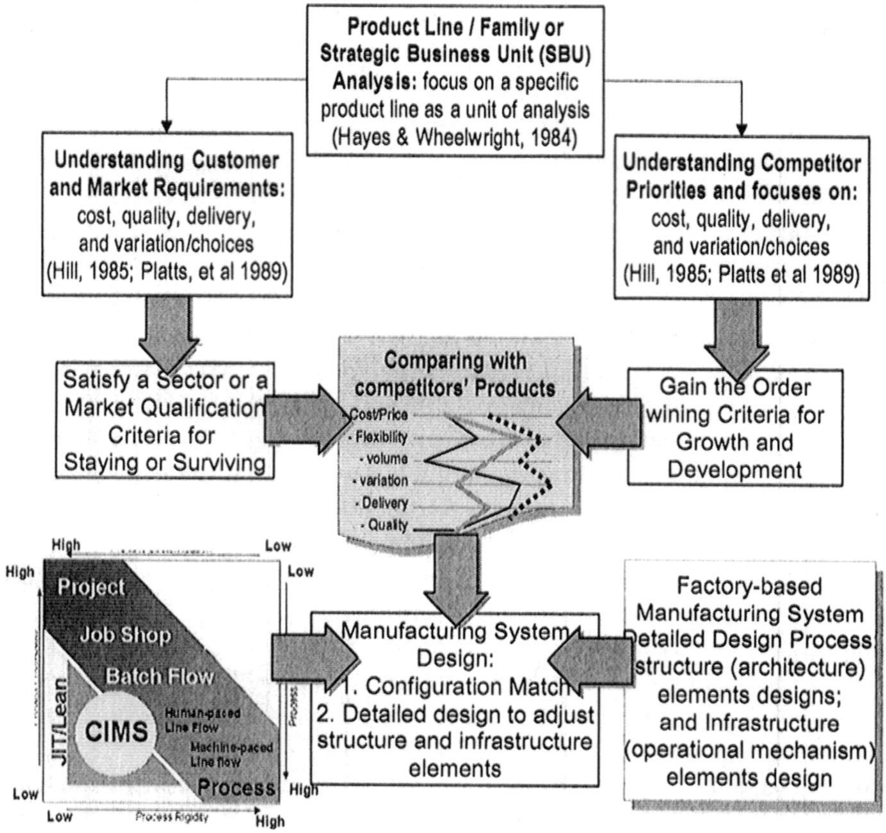

Figure 14- 2. Manufacturing Strategy Process for a Product Family

4. INTERNATIONALIZATION

Although international operations can be very different from domestic ones, it is not clear yet whether an internationalized manufacturing system will have any impact on manufacturing strategy. Different researchers have quite different solutions for dealing with manufacturing globalization. Sprague (1990) suggests a global manufacturing decision grid by globally implementing Skinner's ideas. The consequence might be that each dispersed factory should follow its own local market demands and therefore there is no commonality between factories if the targeted market or customer is different. On the other hand, MacCormack et al. (1994) developed another framework for global manufacturing site locations. As high technology and regional trade blocs are moving international manufacturing towards small scale solutions that are decentralized to serve local markets with a higher flexibility, not only does site location become a major issue for manufacturing internationalization but also the manufacturing system converges towards a globalized identical system with computerized flexibility. Both solutions hold their ground but lead to different directions. They indicate that there is not yet a comprehensive picture of manufacturing globalization, and perspicacity is required.

Since the early 1980s, many academics have begun to analyze in detail a globalized manufacturing network. Hout et al. (1983) studied several global competitors' strategies, and found that the key factor in their success was to rationalize their product lines and their manufacturing and distribution system in order to gain economies of scope from the global scale. So that global players can access different markets and achieve global efficiency it is very critical to position their manufacturing facilities strategically. Kogut (1985) and Porter (1986) built this idea into the value-adding chain concept, and suggested that global-based companies needed to disperse the value-adding chain geographically in order to access the most appropriate resources and, at the same time, to achieve economies of scale

Other interesting research findings include the global competition game played by two rivals (Hamel and Prahalad 1985, Yip 1992). The authors suggested that a better way to protect one company's home market is not only by defending yourself domestically but also through an attack on your rival's home market. This will not only release the pressure on your own home market but also enable you to penetrate into new foreign markets. The global platform therefore provides a much wider operations space to compete, harvest new markets, organize resource, and enhance capability.

Flaherty's work (1986) was very valuable with the development of another type of global coordination mechanism that goes beyond the geographically dispersed value chain. Although this coordination mechanism has been in use for a long time by the service industry, establishing McDonard's, Kentucky Fried Chicken, Pizza's Hut and so on, her detailed studies on the micro-electronic industry analyzed the globally coordinated mechanism in manufacturing operations. The analysis provided a sound basis for the observation and exploration of international manufacturing systems. Based on her empirical research, she suggested that a network with particular patterns of geographic dispersed facilities and a shared common infrastructure and mechanism could lead to a synergy advantage in the network.

Ferdows (1997) observed international manufacturing networks from another perspective and found that factories had different strategic roles, e.g. off-shore, source, server, contributor, outpost and lead, in the network. Ferdows attempted to link strategic motivations to the role and capabilities of each factory. The problem with this model is perhaps that it over emphasizes the strategic role of separate factories compared with the holistic characteristics and strategic capabilities of the network, as well as complexity of the roles from the strategy point of view.

Cambridge Manufacturing Group's work focuses on the identification of different network configurations on the platform of manufacturing dispersion and coordination (Shi and Gregory 1994, 1998). Eight manufacturing configurations were identified through extensive case studies and the relationship between network configurations and strategic capabilities were explored in order to explain the current transformation towards a more globally integrated or coordinated configuration. The research provides a wider scope for the manufacturing network system on the global platform and maps an evolutionary route of the manufacturing system toward globalized networks.

Figure 14-3 (a) illustrates two dimensions - geographic dispersion and coordinated interdependence – to map typical international manufacturing network configurations. Based on more than twenty worldwide case studies of MNCs, a new vision for a manufacturing system can be transformed from the traditional factory based single site model (Hayes and Wheelwright, 1984) towards international manufacturing networks. To understand the patterns of international manufacturing networks, a 'configuration map' and seven typical network configurations are identified. These configurations provide a structured view of international manufacturing networks and their evolutionary paths. The configuration map is compatible with established research but provides much more depth in manufacturing.

At the same time, five strategic capabilities of networks have also been identified in order to understand the new mission of a manufacturing network:

- Strategic resource accessibility: the ability of the network to capture some required manufacturing resources;
- Thriftiness ability: the ability of the network to support higher efficiency;
- Manufacturing mobility: the ability of the network to deploy and reconfigure resources swiftly;
- Learning ability: the ability of the network to capture and disseminate internally generated knowledge;
- Supportiveness of the manufacturing network to its individual factories and products in terms of cost, quality, delivery and flexibility.

There is a strong relationship between network configurations and the strategic capabilities, as figure 14-3 (b) demonstrates. The capability profiles of each typical configuration not only clarify that different network configurations can generate or reserve different capabilities but also provide a set of practical tools for network assessment, auditing and design.

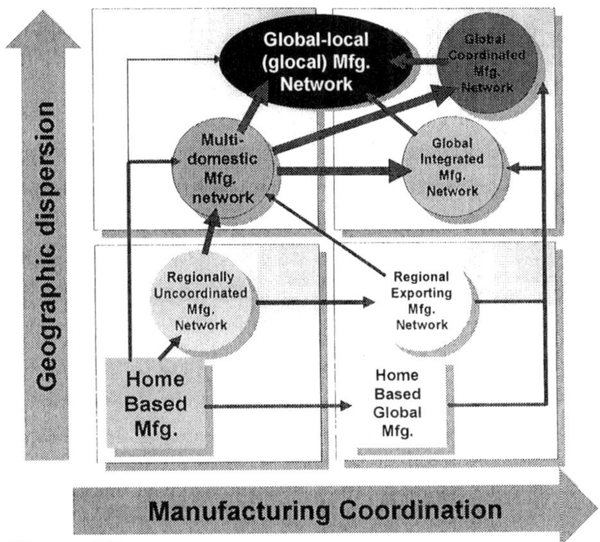

Source: Shi and Gregory, 1998

Figure 14- 3(a). International Mfg Network Configurations and Capabilities

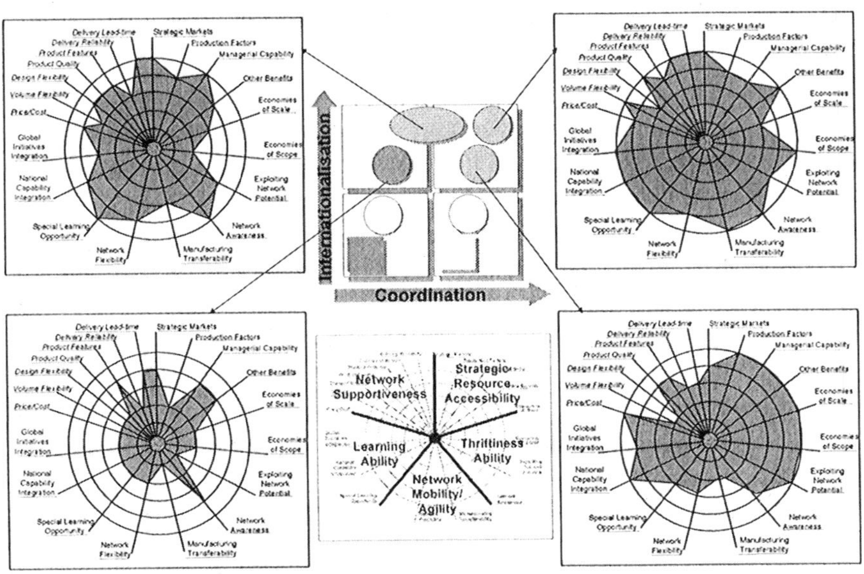

Source: Shi and Gregory, 1998

Figure 14.3(b). International Manufacturing Capability Profile

In summary, it is obvious not only that internationalized manufacturing networks are different from a factory based manufacturing system but also their missions are changed when the boundary of the system extends. The aims of an international manufacturing network can include the traditional manufacturing system's effective and efficient missions and yet can also reach well beyond them. Market and resource proximity, capability allocation and mobility, pre-emption and quick response to competition, and international diversity integration and learning drive its aims and compatible capabilities. From these perspectives, the international manufacturing system has its own missions, context, and content, and therefore its own design process, which implies a new compatible manufacturing strategy process, is necessary.

5. EXTERNALIZATION

Manufacturing internationalization was recognized as a process of the internalization of business activities (Rugman 1980), but recently in almost every industrial sector there is a clear trend toward externalization from the traditional vertical integrated firms. The externalization, or outsourcing, makes supply chain/network subject even more important. There are currently three streams of supply network studies that roughly represent the major characteristics in the subject area. All of them raise the same serious challenge to the traditional concept of the firm, especially in the field of ownership and control of resources. At the same time, the new conceptualization of the supply network (Harland et al. 1999) also has a strong impact on the manufacturing strategy theory.

The main contents of supply network management are not new, at least to production/ operations management. The first stream can be traced back to the traditional inventory models and control mechanisms, which used to streamline production flows between production stations or workshops. Forrester (1961) explored the bull whip effect between firms that widened the scope of studies in industrial dynamics. After developments following Porter's value chain (1985) and Japanese JIT system (Ohno 1988), especially the concept of Lean Production (Womack et al. 1990), inter-firm integration along the supply chain, or value chain, gradually emerged. Although there are many researches still focusing on restructuring and streamlining the supply network, fundamentally they all can be traced back to the classical manufacturing strategy and operations management root, that is seeking a product's competitive advantage, or even just efficient smoothness. Related logistics and supplier development and even whole supply network architecture design can still belong to the classical framework of manufacturing strategy (Slack and Lewis 2002).

But the second stream of supply network research is quite different. Its origins are not only in strategic management about collaboration and value creation but also associate with recent business practices like outsourcing. It focuses on core competence. And, is especially the case with respect to the recent developments in the electronics industry which include the separation between original equipment manufacturers (OEMs) and their contractual electronics manufacturers (CEMs) or electronics manufacturing service (EMS) providers. This new development goes beyond the traditional make or buy decision; it is creating a new player and even new industries that have specialized competences and innovative collaboration potentials. But the trends of specialization and collaboration between firms is not limited to the electronics industry: Ford, that used to run the most comprehensively internal integrated supply chain (Ford and Crowther 1922), also separates a large portion of manufacturing that forms Visteon to provide a professional manufacturing service to all OEMs, even including GM. This type of industry dynamics – outsourcing, specialization and collaboration – creates another type of supply network which is no longer just for the old product family but is also for a new value proposition and a new strategic position in the supply or value network (Bovet and Martha 2000).

The third stream comes from clustering studies (Piroe and Sabel 1984, Keeble 1998, Porter 1998, Teece et al. 2001). This does not directly link with supply network management, but it does expose many critical characteristics of the networked system, for example the cluster's contribution to innovation and competition for vertical integrated companies. The clustering phenomenon provides a demonstration to the firm of a complementary model that is another type of organization bridging market demands and firm's resources. It also validates the supply network with its own life power, especially in the dramatically changed and innovative environments.

There are many overlaps between the three streams. Nevertheless, it is clear that the boundary of a supply network is ill-defined from that of a focal-firm based rigid network, with integrated up and down stream tier suppliers to a virtual enterprise and industrial cluster, which has increasing dynamics, as well as an equally important position for each player with specialized core competence, and also a loose collaboration relationship in network.

Like the international manufacturing network, the externalized inter-firm supply network has extended its boundaries from the factory based manufacturing system along the ownership dimension (Figure 14-1). This extension makes the supply network a new unit for study that has more features than the classical factory. While the focal-firm based supply network can still be designed by following the classic manufacturing strategy process, this is not the case with the virtual enterprise, value network, or clustering. This is because it is beyond its comprehension, as well as being beyond the design and planning schools of strategy (Mintzberg, et al. 1997). The new supply network, in contrast to the international manufacturing network, also has new missions, architectures, mechanisms, and a strategy process. Apart from providing support to competitive advantages, the supply network strategy (Lamming 2000) seeks more opportunities in external markets, a time share of other companies' resources and capabilities that lead to internal innovation, growth and higher value.

6. GLOBAL MANUFACTURING VIRTUAL NETWORK

During the internationalization and, at the same time, externalization illustrated in the up-right grid of the figure 14-1, there are two typical trends. One is the multinational corporation (MNC) dominated by international strategic alliances; another is the high-tech small and medium sized enterprise (SME) dominated virtual enterprise.

MNC's internationalization or globalization has paralleled with externalization featured by focusing and specialization. This limitation of the manufacturing process span helps its global expansion through a rapid worldwide dispersion. Consequently it opens collaborative linkages to both global and local players. This is the reason why Intel "copy exact same" model can be seen worldwide. The cloned manufacturing factories are more efficient to reduce investment risk, reach target markets, and to replicate innovation/ knowledge. At the same time, strategic alliance relationships between MNCs are also developed to replace the traditional internally vertical integration. Tier structure, global strategic account management, matching of their global network configurations, and co-evolution have become main streams of management concern and academic research interests in the arena of the global supply chain management.

High-tech SMEs, on the other hand, have passed from the stage of intra-cluster integration and exporting towards a more inter-cluster global based virtual network. In electronics, telecommunication and, especially the software sector, firm-based cross-cluster collaborations develop very fast. Unlike MNCs' strategic alliances, high-tech SMEs evolve towards a higher innovation oriented network, which does not intend to have a very long-term and stable business relationship but nevertheless mutually understand and trust each other. Once a project (or an opportunity) is identified, a network is set up operationally as an effective supply chain to efficiently deliver an integrated product. As the result of information communication technology, this global reach through the virtual network has become a complementary business model for globalization sharing with MNC's globalization and alliances.

More recently, it is more interesting to note a new convergence of the two models in the electronics industry. In this case, large scale de-mergers of the supply chain, OEMs and EMS providers, for example, all pursue a global scale opportunity development and are widely open for collaboration. The matured industry infrastructure provides further incubators for inspired innovators, system integrators, and customer solution providers, which fosters the global manufacturing virtual network (GMVN) to grow from a more innovative idea through exploitation towards a more dynamic and diversified business reality.

Many research cases from different industries (Shi et al. 2003) highlight the following interesting implications:

- manufacturing collaboration – especially when based on the professional contract manufacturing service providers – is a cross-sector activity. It has fundamentally changed manufacturing industries, inter-firm relationships, and the ways in which companies can compete;
- collaborative manufacturing, or virtual manufacturing, offers some unique advantages, especially in innovation (by encouraging new-comers with radical ideas) and efficiency (by sharing existing fixed costs);
- many questions remain concerning GMVN including: its nature, the new managerial skills required to control it and its implications for manufacturing practice and theory.

Is GMVN a new manufacturing system, a new manufacturing environment, something in between them or a combination of both? This might be one of the key characteristics of GMVN having two equally important aspects of successful collaborative manufacturing. One is the collaboration between two or more companies to form a new, virtually-integrated network to deliver products/services to customers and value to companies. If successful, these virtual organizations may evolve into international strategic alliances (ISAs) featuring stable, inter-firm relationships and long term commitment between partners (Doz and Hamel 1998, Child and Faulkner 1998). The second aspect is the existence of a latent business network or, environment, in which every company has its unique competence and is eager to collaborate with other companies in order to form a practical supply chain. The new environment changes a company's behavior and makes it more open to collaboration. This second aspect is usually ignored because it is difficult to analyze and rapidly changes, but it is at least as important as the first.

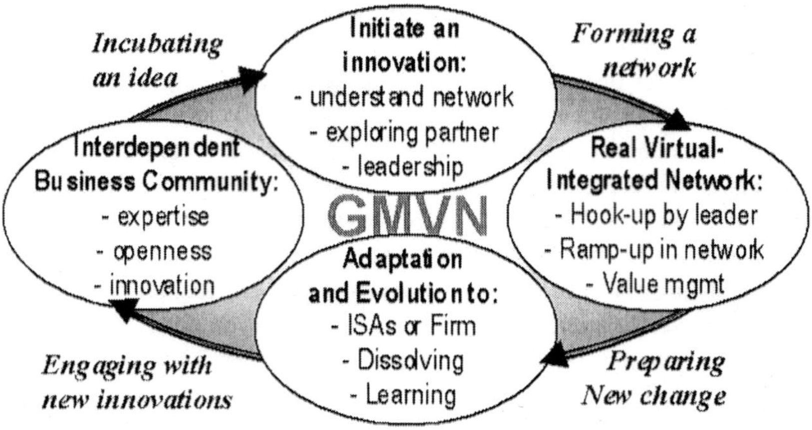

Figure 14-4. GMVN Formation, Development, and Adaptation Life Cycle

Figure 14-4 links these two aspects to present a possible process or life cycle for GMVN - from the existence of a collaborative environment; through an innovative initiative by one of the companies to the virtual integration of required resources from different specialized companies. This virtual organization then adapts and evolves in response to competitive pressures and, in doing so, influences the collaborative environment. The GMVN concept embraces the whole process rather than just some parts of it and gives manufacturing managers a new perspective to their resources and management space.

In summary, it is clear that the GMVN or international collaborated manufacturing system also has its different contexts, aims/missions, architecture, mechanism, attributes, capabilities and related strategy process, which raises the same question if the classical manufacturing strategy should be adapted to match the new challenges.

7. IMPLICATIONS TO MANUFACTURING STRATEGY

Each of the above four types of manufacturing systems has different characteristics. They have very deep impacts on the further understanding about the manufacturing system and the further development of classical manufacturing strategy. They imply that manufacturing strategy should be changed to a more contingency-based model rather than a universal recipe covering all scenarios. The above discussion might also be able to draw the following general conclusions:

- When a manufacturing system context changes, the system boundaries have to be adapted to reorganize new functions in order to satisfy external requirements;
- When manufacturing boundaries are changed, the system contents or building blocks must also be changed, which intimates new architectures, dynamic mechanisms and attributes or behaviors of the system;
- When an external requirement and an internal attribute are changed, the missions or aims of the system should also be re-defined or re-explored, and if a system has such radical changes, its design process or strategy formulation process must also be evolved.

Facing up to the emerging challenges of manufacturing strategy, there appears to be two different types of policy in operations management research work. One is to develop a more generic strategy seeking to cover a wider range of content (Slack and Lewis 2002). Another is to develop more specific strategies that seek to adjust to different types of systems under a contingency plan, which is supported by this chapter. This section discusses the differences between the policies and suggests that the boundaries of the manufacturing system do need to be extended and furthermore new manufacturing strategies for different systems do need to be researched and developed.

7.1 Theoretical Limitation of the Manufacturing Strategy in Corporate Strategy

It is interesting to combine Skinner's thoughts on manufacturing effectiveness (1969) with Ansoff's corporate strategy model (1965). Figure 14-5 introduces Ansoff's corporate growth model, which considers strategic development through the relationship between products and markets. The classical manufacturing strategy theory mainly deals with business growth through an existing product penetration into existing markets – one fourth or half of the corporate strategy in Ansoff's matrix.

Figure 14-5. Identification of the Relationship between the Manufacturing Strategy and Ansoff's Corporate Strategy

This fitness is appropriate for the traditional manufacturing context in which a long product life cycle, stable technology innovation, and difficult adjustment of manufacturing capabilities are predicted. Either product or customer can be clearly identified with their characteristics and boundaries in strategic decisions. But nowadays, the higher value or fast growing business is not based on this type context; many manufacturing firms, because of global competition, have to capture much more adventurous innovative opportunities by coping with or creating unpredictable demands.

This indicates that would be too small a business space for manufacturing strategy to focus only on existing products or an existing market or even both together. From Ansoff model (Figure 14-5), manufacturing strategy actually can play very critical roles in every part of the grid. For new market developments a greater potential is provided by the manufacturing system. This may be from perspectives of international dispersion, especially by globalizing existing products, and function similarity, but mainly by exploiting the products' core technology or functions to penetrate into other related yet new market segments. For any new product development manufacturing can have even wider contributions from global product development (platform or modularity), opening up product ranges, and strengthening the economies of scope. The latter is directed to the whole supply/value chain operations and the total solution care appropriate to existing customers. Even in the diversification of business development, manufacturing strategy also can have a strong influence on corporate strategy and business growth. The power can come from not only the manufacturing capability, internal business mechanisms and their replication in other business units, which are the most critical implications of resource-based view in strategic management, but also identification of new business opportunities through building up, scanning and positioning in a wider value network that is based on demands of final customer. This might mean a completely new territory for the firm but will still have a strong relationship with current business (Bovet and Martha 2000).

Therefore, classical manufacturing strategy will continuously play its rightful role in many manufacturing companies and help them sharpen products' competitive edge and gain more market share. But for more and more companies seeking higher value creation and an innovative product/service or solution to new customers, the classical manufacturing strategy is not aggressive enough to generate new ideas, capture potential opportunities, move to a better position in a supply/value chain, and cultivate the manufacturing system itself with new strategic and evolutionary capabilities. From this perspective, the paper suggests that different scenario based manufacturing strategies will be more effective for extending the manufacturing function and designing a more effective manufacturing system than the classical or a universal model.

7.2 A Contingency Manufacturing Strategy for Different Manufacturing Systems

Based on Ansoff business development model (Figure 14-5) and the four types of manufacturing systems analysis in figure 14-1, it is significant to make a link between both and develop a contingency manufacturing strategy.

The contingency manufacturing strategy can be structured from both corporate strategy and manufacturing system perspectives. As corporate strategy has covered much wider issues than manufacturing system themselves can independently solve, it might be more effective to build a contingency manufacturing strategy based on characteristics of the manufacturing systems:

- International manufacturing networks – can mainly contribute to new market development and partially support new product development in the corporate strategy development matrix(Figure 14-5);
- Supply networks (inter-firm alliances) – can significantly contribute to new product development, especially to define company's position in the supply network, and support diversification;

- Global manufacturing virtual networks – can mainly contribute to the value network or constellation in order to develop hidden business opportunities and run globalized collaborations.

This contingency manufacturing strategy pushes the manufacturing system towards not only the network relationships but also to a business level directly linked with corporate strategy. There could be four types of manufacturing strategies dealing with different levels of manufacturing systems from factory/ plant level, to international dispersed factory network, to inter-firm factory network, and further to GMVN. It is worthy to note that as the network relationship becomes more complex and is a dominant feature the focal company's role in a network is reduced, as well as the strategy becomes more emerged than planned (Mintzberg et al. 1997).

7.3 How Far the Boundary Should Be Pushed: Effectiveness of Manufacturing System

It is arguable that the extension of a manufacturing boundary leads to confusions in decision hierarchy between the manufacturing function, business, and corporate levels. Many gaps are emerging between the various levels when the manufacturing environment dramatically changes. Even so if there is no serious debate, from both discipline and inter-discipline aspects, academic opportunism cannot establish a fundamental transformation between paradigms. However, on the other hand, any ignorance of the emerging challenges will cause further fragmentation of the body of knowledge, like a buzz-word brand replacing a deep understanding of a phenomenon, and damage theoretical integration that will lead to eventual failed delivery of a synthesized knowledge that can be implemented.

It is important to realize that there is no intention in this paper to change manufacturing strategy to a whole business strategy that covers every business function. Other functions like marketing and even new product introduction (NPI) should still play their parts in the overall business strategy. But the manufacturing system and strategy, as they coordinate resources to satisfy external requirements, have to or should open a wider window upon engagement with new opportunities for higher value and growth.

When Skinner first raised the question about manufacturing effectiveness and argued the differentiation between efficiency and effectiveness, it is very similar to the challenges posed by the question of how far a manufacturing system boundary should been extended. Skinner pushed it to product - the outcome of the manufacturing system, which fundamentally changed people's mindset about manufacturing and operations management. It is very logical to define a system's mission first and then design it and later operate it. The effectiveness should lead the efficiency, or doing the right thing first and then doing it correctly (Skinner 1985).

The current debate about the boundaries of the manufacturing system and strategy is still about the system effectiveness. The effectiveness represented by product is no longer acceptable, mainly because even a very successful product, with many orders and larger market share, nowadays may still not be able to generate high value, as a previous case demonstrates. More fundamentally, the core challenge may be not only the profitability issues but more crucially a life-warning signal that indicates a business course or a rule-of-the-game has been changed. If the company fails to sense the change and still focuses on its product order-winning criteria, its innovation power will be damaged and sustainability will be jeopardized. Especially in the new knowledge economy the manufacturing effectiveness and its determinants have been fundamentally evolved from physical product-based competition towards the new competition of potential opportunity identification and flexible organization of global resources. One of the key performance criteria is the value to the stakeholders. The manufacturing system, as the real power engine of manufacturing companies, has to upgrade its mission or effectiveness from order-winning towards value-creation as well as appropriation.

The new evolution of the effectiveness of manufacturing system implies a new definition of manufacturing which represents the full cycle of business process from understanding markets through product and process design to operations and distribution, taking into account economic, financial and people issues. Manufacturing also defines the cycle boundary starting from a market and also ending at it. Today it is rare for a company to own its full cycle. Manufacturing strategy therefore should decide its position, collaboration, and more importantly the dynamics adapting itself to its environment.

7.4 Relationship between Positioning and Capability in Strategy Development

One of the new contributions of Slack and Lewis's (2002) most recent book is their "strategic reconciliation" – a balanced view between market requirement (or strategic positioning) and operations resource (capability development). Their process-oriented strategy demonstrates the strategy formation mechanism. The reconciliation actually not only solves a long-term debate between Porter's positioning school (1985) and a resource-based view (Panrose 1959, Teece et al. 1997) in the strategic management area but it also provides a procedure that may be implemented. This is an important contribution from operations management to strategic management, which illustrates the richness of manufacturing and operations management from both practical and theoretical points of view.

However, based on this paper, the reconciliation does not yet hit the core of the problem in the contingency manufacturing strategy. In the new context, a manufacturing company not only needs reconciliation mechanisms but must adapt and have new special capabilities to allow for dynamically positioning in a better place in a value network so as to create as well as capture higher value. This asks for new types of capability to do two different things – one is to develop a new capability enabling an effective positioning; another is to develop a new strategy process, which is beyond the reconciliation but syntheses both. The synthesis mechanism is the core of the contingency manufacturing strategy in order to cope with both the dynamic context and system complexity.

7.5 New Frameworks for Research and Manufacturing Strategy

New challenges to manufacturing system and strategy create a new context for manufacturing strategy research. This can be demonstrated by a three-dimension space as figure 14-6 (A), which provides the vision highlighting key issues and their relationship between supply/value chain, internationalization, and externalization or collaboration.

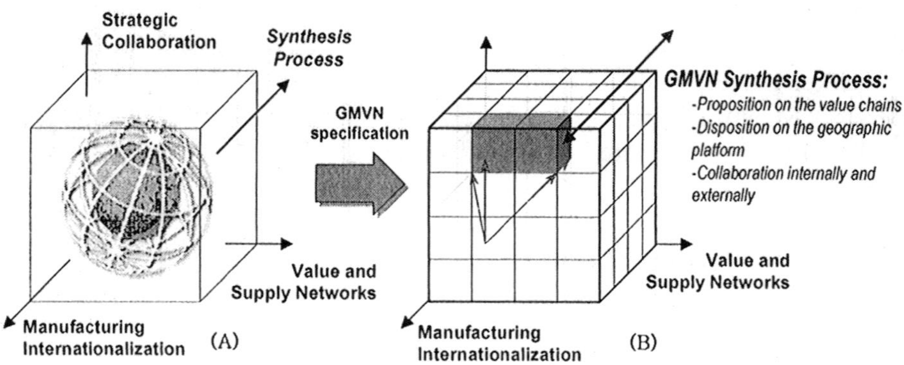

Figure 14-6. The Building Blocks of New Manufacturing Network Systems

The generic three-dimension strategic environment indicates the key decision zones, and the future manufacturing system should be built with these four basic building blocks:

- Supply and/or Value Creation Network – Manufacturing system and its tasks must be defined along the supply/demand or value-creation networks by proposition, configuration and optimization for higher value and competitive advantages;
- Strategic Collaboration (Externalization) – A spectrum of collaboration including intra-firm coordination and inter-firm co-operation has to be evaluated and decided. In the inter-firm collaborations, it also has a wide span of choice from strategic alliances, for longer term commitment, towards virtual community, or arm length trading relationship development for more flexibility;
- Manufacturing Internationalization – Manufacturing system is no longer a single-site factory. It has to decide not only geographic expansion or dispersion but also the internationalization evolution process and cross-cultural integration;
- Strategic Synthesis Process – The above three dimensions cannot work independently in the current global competition environment. Therefore, it is essential to synthesize them into an integrated manufacturing system with a systematic strategy process and most appropriated technology including a cyber platform.

Figure 14-6 illustrates a challenging framework for the future research work. An understanding of each dimension and its interaction relationship will be the key to secure knowledge of manufacturing networks.

The strategic synthesis process will be the core of the strategy formulation process of the contingency manufacturing strategy. As a contingency strategy, the scenarios can be divided into four types of manufacturing systems from factory/plant, towards international manufacturing network, inter-firm supply network, and global manufacturing virtual network (GMVN). Figure 14-6 (B) presents the GMVN holding a very strong position in three dimensional space. It implies GMVN should have a very dynamic mobility on the value network, a very strong collaborative relationship, and a responsive wide dispersion or reach in the world. Therefore, the process of contingency manufacturing strategy will depend upon each type of manufacturing system, and have its own context and content in decision zones. There is no doubt that more research work needs to be done in order to understand, develop and validate these speculations.

8. SUMMARY

This paper is based on a recent eight-year case study undertaken in many multinational corporations in the aerospace, automotive, engineering, garment and electronics industries. The original objectives were directed to the evolutionary process of the manufacturing system, beginning with a factory-based system and then moving toward a globally collaborative inter-firm network. It attempts to explain why manufacturing companies evolve into this m ore s elf-uncontrollable relationship. T he questions r aised from the research beyond any answer that can be provided by classical or the existing theory of manufacturing strategy. Some preliminary findings highlight the following points:

As the manufacturing system context changes, the manufacturing systems with new boundaries, mission, architecture, mechanism and capability have to be adapted to reorganize new functions in order to satisfy external requirements.

As a manufacturing system evolves into quite different relationship oriented networks, a universal process of manufacturing strategy process is not specific enough to fully satisfy the strategic planning and system design; a contingency manufacturing strategy therefore is demanded.

In a manufacturing strategy process and effectiveness analysis, higher value creation has been suggested to be the new aims of manufacturing system; and the strategic positioning and dynamic mobility in a value network have become the major tasks for manufacturing strategy.

In a manufacturing system, taking advantage of and leveraging other companies' existing resources seems to be more important than owning these resources – this implies that various types of inter-firm collaborations are therefore the preferred architecture.

In collaborative situations, understanding the dynamics and keeping a balance are more and more critical in the spectrum of collaboration from one end of the corporate hierarchy to international strategic alliances, and further towards networked virtual enterprises, geographic clustering, and a free market end; which suggests that a contingency picture is required

From the perspective of evolution, a greater understanding is required about the choices between an internal vertical integration, and strategic outsourcing, sub-contracting, and manufacturing hollowing-out. These know-why areas are now becoming more and more important than the single competence of knowing how to do them.

These new challenges pose new solutions for industry. They require not only a new architecture of the manufacturing system based more on an inter-firm relationship but also require a new strategy process with a context-based contingency framework. To achieve the new tasks, the new conceptual framework with three main dimensions for the networked manufacturing system and the new contingency manufacturing strategy process are needed in order to guide further research, integrate different knowledge disciplines, and eventually manufacturing business practices.

REFERENCES

1. Ansoff, H.I., *Corporate Strategy*, New York: McGraw-Hill, 1965.
2. Bovet D., Martha, J. From supply chain to value net. Journal of Business Strategy 2000; July/August
3. Bovet D., and Martha, J., *Value Nets: Breaking the Supply Chain to Unlock Hidden Profits*, John Wiley and Sons, 2000.
4. Brewer, A.M et al (eds.), *Handbook of Logistics and Supply-Chain Management*, Amsterdam: Pergamon, 2001.
5. Child, J., and Faulkner, D., *Strategies of co-operation*, Oxford University Press, 1998
6. Doz, Y., and Hamel, G., *Alliance Advantage*, Harvard Business Press, 1998.
7. Ferdows, K. Making the most of foreign factories. Harvard Business Review 1997; March-April: 73 - 88
8. Flaherty, M.T. "Coordinating International Manufacturing and Technology", in *Competition in Global Industries* Porter M. E. ed., Harvard Business School Press, 1986
9. Ford, H., Crowther, S., My life and work, Messina, 1922.
10. Forrester, J. W., *Industrial Dynamics*, The MIT Press, 1961
11. Goldman, S.L., Nagel, R.N., and Preiss, K., *Agile Competitors and Virtual Organisations: Strategies for Enriching the Customer*, Van Nostrand Reinhold, 1995.
12. Hamel, G., Prahalad, C.K. Do you really have a global strategy? Harvard Business Review 1985; July-August 63(4): 139-148
13. Harland, C.M., Lamming, R.C., Cousins, P.D. Developing the concept of supply strategy. International Journal of Operations and Production Management 1999; 19(7): 650-673
14. Hayes, R.H. and Wheelwright., S.C., *Restoring Our Competitive Edge: Competing Through Manufacturing*, John Wiley and Sons, 1984
15. Hill, T. *Manufacturing Strategy: the Strategic Management of the Manufacturing Function*, London: Macmillan, 1st ed. 1985; 2nd ed. 1993.
16. Hout, T. M., Porter, M.E., Rudden, E., "How Global Companies Win Out", in *Strategy Management*, Hanermesh, R.G., ed., John Willey and Sons, Inc., 1983
17. Keeble, D., Wilkinson, F. Collective learning and knowledge development in the evolution of regional clusters of high technology SMEs in Europe. Regional Studies 1998; 33:4:295-303
18. Kogut, B. Designing global strategies: comparative and competitive value-adding chains. Sloan Management Review 1985; Summer: 15-28
19. Labowitz and White, OIT Times, 2001; 4:2 Retrieved July 2003 from: http://www.oit.doe.gov/news/oittimes/spr01/spr01pg2.shtml
20. Lambert, D.M., Cooper, M.C., Pagh, J.D. Supply chain management: Implementation issues and research opportunities. International Journal of Logistcs Management 1998;9:2:1-19
21. Lamming, R.C, Johnsen T.E., Zheng, J, Harland C.M. An initial classification of supply networks. International Journal of Operations and Production Management 2000; 20:6: 675 - 691.

22. Li, X., Y. Shi and M. J. Gregory, "Global Manufacturing Virtual Network (GMVN) and Its Position in the Spectrum of Strategic Alliance", in *Operations Management: crossing borders and boundaries: the changing role of operations*, Dierdonck, R.V., and Vereecke, A., ed. EurOMA 7th International Annual Conference, Ghent, Belgium, June 4-7 2000, pp330-337
23. Miscioscia, L, EMS evolution has given rise to true virtual OEMs, EBN, April, 2001
24. MacCormack, A. D., Newman L. J., Rosenfield, D. B. The new dynamics of global manufacturing site location. Sloan Management Review 1994; Summer: 69-80
25. Mintzberg, H., Lampel, J., and Ahlstrand, B., *Strategy Safari*, Free Press, 1997
26. Normann , R., Ramirez R. From value chain to value constellation: designing interactive strategy. Harvard Business Review 1993; August-September: 65-78.
27. Ohno, T., *The Toyota Production System*, Productivity Press, Cambridge, MA, 1978; English translation 1988.
28. Parolini, C., *The Value Net: a Tool for Competitive Strategy*, John Wiley Books, 1999.
29. Piore, M.J., and Sabel, C.F., *The Second Industrial Divide: Possibilities for Prosperity*, New York: Basic Books, 1984.
30. Penrose, E., *The Theory of the Growth of the Firm*, Oxford: Oxford University Press, 1959.
31. Porter, M., *Competition in Global Industries*, Harvard Business School Press, 1986
32. Porter, M. Clusters and the new economics of competition. Harvard Business Review 1998; December: 77-90.
33. Rugman, A. A new theory of the multinational enterprise: Internationalization versus internalization. Columbia Journal of World Business 1980; 15: 1 (Spring):23-29
33. Shi, Y., Gregory, M., International Manufacturing Strategy: structuring world-wide manufacturing networks, The University of Cambridge, MSRG Working Paper 11/94, 1994
34. Shi, Y., Gregory, M. International manufacturing networks – to develop global competitive capabilities. Journal of Operations Management 1998; 16: 195-214.
35. Shi, Y., Gregory, M., "Global Manufacturing Virtual Network (GMVN): its dynamic position in the spectrum of manufacturing collaborations", In *Managing Virtual Web Organizations in the 21st Century: issues and challenges*, Franke U.J., ed., Idea Group Publishing, 2002
36. Shi, Y., Fleet, D. Gregory, M., Global Manufacturing Virtual Network (GMVN): revisiting of the concept after three years fieldwork, in the Proceedings of the EurOMA and POMS Joint International Conference in Como Lake, Italy, June 2003
37. Skinner, W. Manufacturing: Missing link in Corporate Strategy. Harvard Business Review 1969; May-June: 136-145.
38. Skinner, W., *Manufacturing: the Formidable Competitive Weapon*, John Wiley and Sons, 1985
39. Slack N., Lewis, M., *Operations Strategy*, Pearson Education Limited, 2002.
40. Sprague, L. G. "Strategic Analysis for Global Manufacturing", In *Strategic Manufacturing: Dynamic New Direction For the 1990s*, Moody P. E., ed., Business One Irwin, 1990.
41. Teece, D, Pisano, G., Shuen, A. Dynamic capabilities and strategic management. Strategic Management Journal, 1997; 18(7): 509-533
42. Teece, D., Dosi, G., Carroll, G., Malerba, F., Tunzelmann, N., The geography of innovation and economic clusters. Industrial and Corporate Change 2001; 10 (4): 815-816, Also available in http://icc.oupjournals.org/

43. Williams D. J., *Manufacturing Systems: An introduction to the technology*, Chapman and Hall, 1988
44. Womack, J. R., Jones D. T., and Roos. D., *The Machine that Changed the World*, Rawson Associates, New York. 1990
45. Yip, G., *Total Global Strategy: Managing for World-wide Competitive Advantage,* Prentice Hall, 1992.

Chapter 15

SUPPLY CHAIN MANAGEMENT USING AUTO-ID SYSTEMS

Preparing for Real Time, Item Level Supply Chain Management

Yoon Chang[1] , Duncan McFarlane[2]
[1]*KAIST,* [2]*University of Cambridge*

Abstract: Recently RFID technology has been raised in profile as a potential alternative to bar code technology. So called Automated Identification (Auto-ID), Auto-ID systems which incorporate RFID technology, can provide frequent, and unique item level information at all points in the supply chain. In order to do so however, these systems must be tightly linked to existing manufacturing related business information (BIS) systems. This chapter addresses the issues and opportunities for introduction of Auto-ID technology into various BIS systems in order to improve supply chain efficiencies.

Key words: Auto-ID, Supply chains

1. INTRODUCTION

1.1 Auto-ID Technology

An intelligent infrastructure, which automatically and seamlessly links physical objects to each other, people and other information systems through the global internet has been introduced by the Auto-ID Center (Sarma et al. 2001). This infrastructure combines low cost, wireless communications, with modern networked data management services. Four major components of this intelligent infrastructure are described below:

- **Electronic Tag and Reader**: refers to a family of technologies that transfer data wirelessly between tagged objects and electronic readers. When coupled to a reader network, it allows continuous tracking and identification of physical objects. Examples of this technology are Radio frequency identification (RFID) and Electro Magnetic Identification (EMID) Technologies;

- **Electronic Product Code** (EPC™): is a numbering scheme which can provide unique identification for physical objects, assemblies and systems. The EPC™ code was intended to be universally and globally accepted as a means to link physical objects to the computer network (Brock 2001a);
- **Object Naming Server** (ONS): is the "glue" which links the EPC™ with its associated data stored in a Product Markup Language (PML) file. It is currently under development based on the standard domain naming service (DNS). The ONS is an automated directory service, which, when given an EPC™ number, returns a host address on which the corresponding PML file is located (Sarma et al. 2001, Brock 2001a);
- **Product Markup Language** (PML): is intended to be a common language for describing physical objects, processes and environments. The primary objective of PML is to serve as a common reference for software applications, data storage and analytical tools for industry and commerce (Brock 2001b). PML uses the same format and structures as the Extensible Markup Language (XML).

Individual physical objects are identified with a 96-bit electronic product code (EPC) stored in devices known as "smart tags." The EPC can uniquely identify more than 268 million manufacturers, each with more than one million products, with enough numbers left over to tag all the individual consumer products manufactured for the foreseeable future. The smart tags – which are attached to, or embedded in each object – have antennae that allow them to communicate wirelessly to other devices.

Strategically placed radio frequency "readers" scan the smart tags and transmit an object's embedded identity code to the Internet, where more detailed information on the object is stored.

On the Internet, the EPC works together with an ONS and a PML. The ONS tells computer systems where to find information about any object that carries a particular EPC code. The ONS routes information to appropriate web sites, but it will be more complex than today's internet data management procedures, requiring the locating of data for every one of the trillions of objects carrying an EPC code. The PML is a new standard "language" for describing physical objects in the same way that HyperText Markup Language (HTML) is the common language that tells web browsers how to display most Internet web pages. Taken together, Auto-ID technology merges bits and atoms together to form one seamless network that allows everyday objects to interact intelligently with people and organizations in real time (McFarlane et al. 2002).

1.2 Auto-ID Data

Before beginning to examine the different BIS that might need to be integrated to Auto-ID data for better supply chain management, we define exactly what we mean by the term Auto-ID Data. By Auto-ID data we refer to any data originating from, or being stored within, the Auto-ID information infrastructure and specifically to EPC™ and PML originated information. By way of example, table 15-1, illustrates the types of Auto-ID data that might be required by 3rd party information systems.

In addition to considering which of these data types are required by different Business Information System it is important to also to consider issues such as content, format and frequency of the data.

Table 15-1. Information Available from Auto-ID Sources (Chang et al. 2002)

Information Type	Source within Auto-ID environment check?
Product type	ePC code
Product ID	ePC code
Product Data	PML core
-. Parameter;	
-. History;	
-. Transaction details;	
Operation Data	PML extension
-. Bill of material (BOM)	
Product State	Other resource (Stored in PML history)
-. Location;	
-. Temperature	

1.3 Relevant Research

A considerable amount of research work and application developments have become generally available through publications. In this section we briefly review some of the interesting features that are relevant to our work.

Manthou and Vlachopoulou (2001) have developed a conceptual framework for the integration of a bar-code system in inventory and marketing. In the framework, they also considered the issues in the development and implementation of such a system. Bar codes have been widely used for a number of years in retail sectors and other industrial areas to provide product visibility.

Mandal and Gunasekran (2002) have studied order tracking issues and stock management issues in a wine producing company. They explain an implementation case study with respect to use of SAP/R3 in conjunction with bar-coding system. If the data can be collected in real time the bar-code system provides up to the minute information of stock. However, the proposed system has some weak points: the data collection process is still manual; the bar code has a line of sight problem; and in order to track a product by item level, another system (e.g. 2D bar-code system or Auto-ID system) is needed.

Karkkainen et al.(2003) introduced a basic control system which is based on distributed programming together with wireless identification technologies. The system focuses on the logistic problems of international investment projects by providing intelligence to the delivery systems. It was designed to integrate the flow of information with material flow and to build information services that are flexibly operated in an open network. The idea proposed is near to that of Auto-ID and it shows the potential research opportunities in this area.

Joshi (2000) introduced a generalized framework for examining supply chain processes using Auto-ID technologies. The EPCs™ on tags, coupled with tag readers and the controlling software application will form an automated tracking system for real time data acquisition. Along with ONS, PML and the Internet companies can convert acquired data into relevant information and store it securely at appropriate locations.

Saar and Thomas (2003) address the issues of the use of product tags for environmental management. They point out three strategies for encouraging the development of "end-of-life" applications: One, make moves to be more competitive solutions; Two, target market-based waste management systems; Three, encourage experimental systems and the coordination between manufacturers and the waste management industries. They consider two ways that bar code and Radio Frequency Identification (RFID) tags would improve End of Life (EOL) management: (1) by increasing efficiency in recycling and reuse industries and (2) by making it easy to provide incentives for good waste management (The issues have also been raised in Parlikad et al. 2003).

In the retail supply chain, the ePC is structured to provide a migration path from the 12-digit U.P.C. and 13-digit EAN bar code systems, or Global Trade Item Number (GTIN). By Jan.1, 2005, all North American companies must expand their databases to accommodate the 14-digit GTIN structure, and such changes may require extensive system retooling. Also, different code sizes are an even larger issue for European manufacturers. In such a situation, a single global code scheme (i.e. ePC) would simplify the situation dramatically (Kay 2003).

1.4 Overview to the Work Presented in the Next Sections

This study begins with an introduction to key manufacturing BIS, identifying functions and input/output data flows. We review the issues/problems of existing manufacturing BIS systems with respect to planning and execution in supply chains. We explore the opportunities of Auto-ID based supply chain management and provide examples. We introduce specific data to achieve supply chain efficiencies, real time and item level tracking. As a conclusion, we address further issues and future requirements. It must be emphasized that this study is not focused on the different systems associated with capturing Auto-ID data but rather on the Manufacturing BIS that are likely to receive it.

2. MANUFACTURING BIS SYSTEMS FOR SUPPLY CHAIN MANAGEMENT

In this section we will briefly introduce the different classes for relevant manufacturing BIS. To prevent misinterpretation we restrict the definition of manufacturing BIS to mean only Manufacturing Execution Systems (MES), Supply Chain Management (SCM) systems and Enterprise Resource Planning (ERP) systems that are supporting the supply chain processes[1] in manufacturing enterprises. Figure 15-1 shows an example of how ERP, SCM and MES are connected together and typical data that is transferred between systems in manufacturing enterprises.

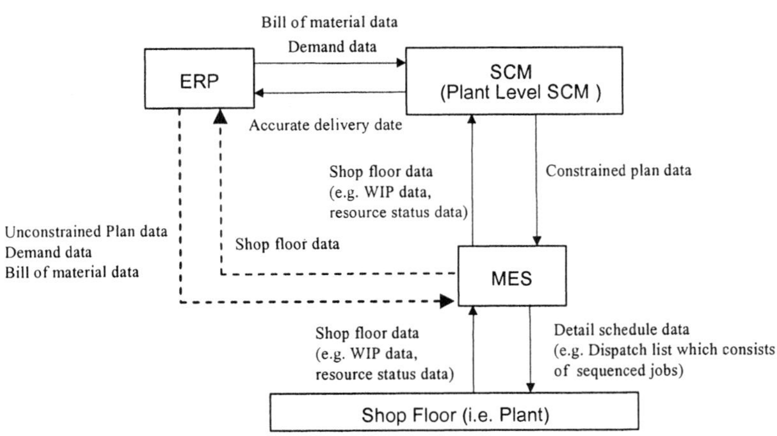

Figure 15-1. ERP, SCM and MES and Typical Data Transferred Among Systems (Dotted lines are data transferred when there is no plant level SCM)

2.1 MES

MES systems were first deployed in the 1980s and primarily focused on tracing lots of material in the production process. In the late 1980s, sophisticated MES users began to create additional customized functionality and to link it into their existing MES. We note the following MES description:

"*MES Systems collect and deliver information enabling the optimization of production activities from order launch to finished goods. Using current and accurate data, MES guide, initiate, respond to, and report on plant activities as they occur. The resulting rapid response to changing conditions, coupled with a focus on reducing non-value-added activities, drives effective plant operations and processes. MES improve the return on operational assets as well as on-time delivery, inventory turns, gross margin, and cash flow performance. MES provide mission-critical information about production activities across the enterprise and supply chain via bi-directional communications (MESA 2002)*".

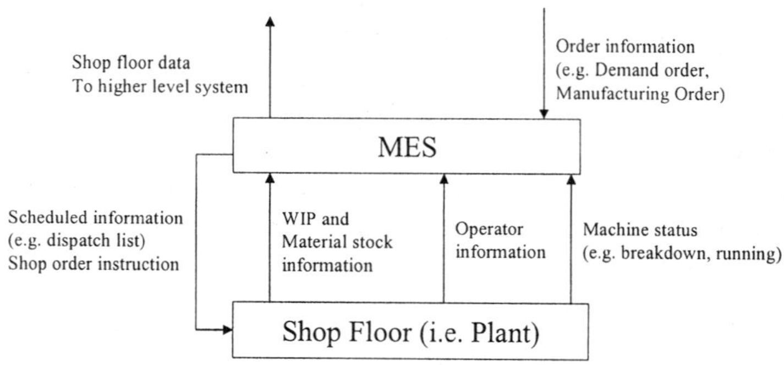

Figure 15-2. Example MES Input/Output Information

By way of illustration, figure 15-2 indicates typical input and output information flows from a MES environment. Typically, MES systems include the following modules: resource allocation and status, dispatching production units, data collection/acquisition, quality management, maintenance management, performance analysis, operations/detail scheduling, document control, labor management, process management and product tracking. We note however, that there is enormous variation between implementations.

2.2 ERP

In the same way that MES systems orchestrate the physical transactions, ERP systems orchestrate the business transactions in a manufacturing enterprise. Orlicky developed the concepts of Material Requirements Planning (MRP) in the 1970s. He realized that the computer technology could be used to enable effective manufacturing inventory management. Using demand requirement information, together with information on the product structure from the bill of materials (BOM) file, current inventory status from the inventory file, and component lead time data from the mater part file, MRP produces a time based schedule of planned order release on lower level items for purchasing and manufacturing.

A major limitation of MRP systems however, was the "assumption of an infinite capacity of resources." Manufacturing resource planning (MRP II) evolved from MRP by a gradual extension to MRP system features and in particular a consideration of finite capacity. MRP II is an extended MRP system with the functions of: Material requirements planning; master production schedule; capacity requirement; production activity control; purchasing and finance features.

In the early 1990's, MRP II was further extended to cover computer systems support for areas such as Engineering, Finance, Human Resources, Project Management etc. Hence, the term Enterprise Requirements Planning was coined. ERP can be defined, as a software solution that addresses the needs to meet the organizational goals (including manufacturing goals) of an enterprise by tightly integrating all functions of the organization.

Generally, ERP systems include the following modules:

Inventory management, plant maintenance/service management, warehousing, production planning, human resources and finance, and recently, some of ERP companies are initiating marketing of a new product called "Logistics Execution System (LES)[2]", which begins to integrate execution level functionality within the ERP environment.

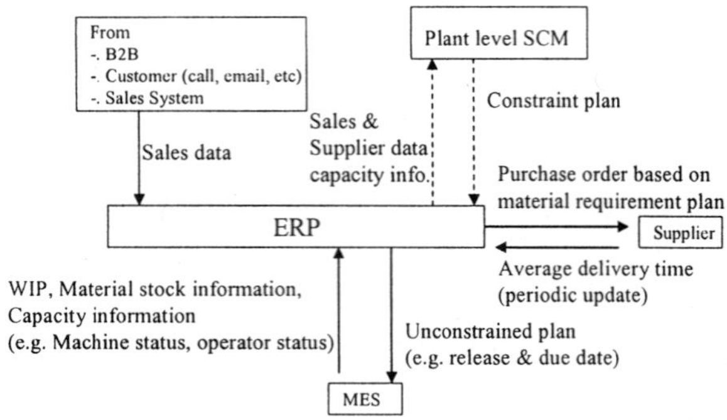

Figure 15- 3. Example ERP Input/Output Information in Relation to Planning

Figure 15-3 indicates input and output information flows from an ERP environment with particular focus on the planning function (BOM data is usually defined in ERP for easy material requirement planning). In this figure, we only include manufacturing related transactional data (i.e. planning) and do not consider other data configured in the ERP system (i.e. financing, product information). Also dotted lines show a case when there is a SCM.

2.3 SCM

A large number of software packages emerged in the 1990's that are collectively called SCM systems. The goal of the SCM system is to integrate the data between suppliers, manufacturers, warehouses, and retailers, so that goods are produced and delivered at the right quantities, and at the right time, while minimizing costs as well as satisfying customer requirements (e.g. due date). The SCM system applies optimization technologies to achieve improved plans by considering most of the factors and constraints that limit the ability to deliver on time. SCM systems simultaneously consider capacity and material requirements to generate plans (ERP considers those sequentially[3]), and create a detailed production schedule that sequences production 'optimally', to achieve organization goals (e.g. minimizing inventory, maximizing throughput, etc.) as well as achieving customers' due dates. Generally, supply chain management system consists of the following functions:

- Demand planning: aims to reduce forecast error and to suggest buffers considering demand variability;
- Master planning: provides multi-site planning. Master plan is generated based on the material, capacity, transportation and other constraints.

- Manufacturing Planning: generates plans considering material, capacity and other constraints which impact on manufacturing;
- Transportation Planning: considers dynamic transportation requirements and generate 'optimizing' transportation plan;
- Procurement Planning: constraints such as supplier capacities, costs and lead times can be modeled as part of a supply chain resulting in superior plans.

Figure 15-4 is a diagram, which indicates input and output information flows from a SCM environment (In this figure we ignored ERP. ERP can be placed next to plant level SCM).

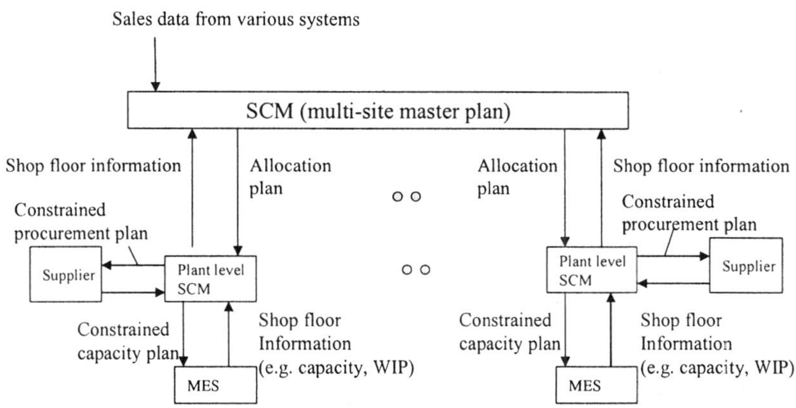

Figure 15-4. Example of SCM Input/Output Related to Planning/Execution

3. SHORTCOMINGS OF EXISTING MANUFACTURING BIS

In this section, we address issues or problems of the existing BIS with respect to planning and execution in supply chains.

Over the last ten years manufacturing industry has undergone a dramatic change. State of the art systems (i.e. ERP, SCM, MES) have been widely introduced to improve supply chain processes. However, due to some systematic problems and the gaps existing between an "idealized" stable environment and a real, dynamically changing implementation environment, disappointing results are encountered industry wide on a frequent basis. There are various reasons for such failures, but in this section we only address a few examples:

1. *The gap between planning and executions in the supply chains*: As Carpenter (2003) pointed out that most companies today can only react after disruptions, because of "the disconnect" between planning and execution IT systems in manufacturing;
2. *Manual read, and inability to perform batch read and write*: With existing bar code systems, all the data is generally read manually. So at inventory receiving points in a plant, the more materials arrive together, the more operators are needed to read the data (Or the more time it will take). This kind of inefficiency creates mismatches in delivery information (e.g. in the delivery party's system, the system will show those materials delivered) and receiving information (e.g. in receiving party's system, even if the materials have physically arrived, the system will not consider those materials until operators have input the data). This is one of the many causes of the information delay in planning and execution in the supply chains;
3. *Non real time or shift based inventory data input*: In most of manufacturing companies, inventory data are not updated in real time. There are two reasons for this: either a manual read is used or the importance of real time data is overlooked. In many manufacturing companies information updates are only done on a shift by shift basis even though they have enough available resources to update the information either in real time or pseudo real time (Many companies still view information updates as less important than production);
4. *Back flush (or data input policy)*: This is a special case for data input that is not required in real time. In a warehouse or a stocking point, when materials are taken for manufacturing, many companies follow a policy called "back flush": in which "material information (i.e. information about material quantities and materials used for certain product assembly) is updated only when all the assembly processes are ended or certain manufacturing processes are ended". This kind of procedure creates a mismatch between system information and physical status thereby impacting on planning efficiency;
5. *Line of sight issues*: In current bar code systems there are line of sight issues;
6. *Lot sizing*: In most assembly plants, there are three kinds of size related information, 'size in BOM'; 'manufacturing lot size'; and 'procurement lot size'. When an MRP engine runs, it calculates using BOM information the raw materials quantity needed for a certain planning horizon. But in procurement, there is a 'procurement lot size' (i.e. the 'customer-supplier' agreed delivery size based on both economics and convenience). The problem here is the difference that occurrs between the 'procurement lot size' and 'BOM based quantities'. To make it worse,

in manufacturing, a third quantity, 'manufacturing lot size' is used due to equipment and production process specifications as well as the plant's policy. So, it is very hard to get the correct information about stock, (i.e. material quantities required, used, left) due to, "size" differences;

7. *Hard to trace the movement of orders/materials across supply chains*: With the current system architectures within individual organizations it is challenging to trace movements of orders/materials across supply chains and thus hard to make effective decisions, which consider efficiencies in the supply chain or values in the networks;
8. *Impossible to peg orders to physical product*: In current practice, when the same raw materials are required by different orders (manufacturing orders), it is impossible to physically unite selected materials to specific orders (See example below).

(Example) A computer OEM supplier assembles a PC for many computer companies (HP, Dell, Compaq, Samsung etc.) and in assembling PCs, they use the same type of hard disk for each of those companies. With current practice at the planning stage it is impossible to allocate a specific 'hard disk' instance to a specific order item of a computer company. So when there are orders from different companies in the same planning cycle, there is no way to prevent production operatives from selecting hard disks for less important customers or for less important orders (or less urgent orders).

4. MANUFACTURING BIS DATA FOR INTEGRATION WITH AUTO-ID SYSTEMS

Motivated by the issues in the previous section, real time, item-level Auto-ID data needs to be integrated into existing manufacturing BIS in order to improve the effectiveness of companies' supply chain processes. But in order to understand and prepare for this integration, we must understand the kind of data now available to manufacturing BIS. The data needed by a Manufacturing BIS varies in character, through its source and the frequency at which it is updated. From a data processing point of view, it is divided into dynamic and static data[4].

Table 15-2 shows sample manufacturing related data (Note that due to enhancements of each BIS, BIS products have many overlapping features and data requirements). It is intended that Auto-ID data will be integrated with existing data, as defined in table 15-2[5] to improve both planning and scheduling efficiencies as well as product/order tracking performance in supply chains. Some of the existing solutions in table 15-2 already have a certain level of capability to source and to use real time or pseudo real time data (e.g. work in progress, inventory data). However, in current practice in most industries, real time data collection is not fully achieved and item level information is not used or even supported. Even if existing systems could currently handle item level, real time data, there is still a scalability issue in terms of handling the volumes of new data likely to be available.

Table 15-2. Information Available from Auto-ID Sources (Chang et al. 2002)

Data file	Description	Dynamic/Static	Data defined
Product data	Product structure and BOM data	Static	ERP, SCM, MES
Routing/Operation data	Describes the operations which must be performed in order to manufacture items	Static	ERP, SCM, MES
Resource data	Information about each resource (i.e. operator, work center, cell machine, group of machines, etc) identified for capacity planning and product scheduling purpose	Static	ERP, SCM, MES
Stock/inventory data	Details of the stores, bin locations and quantity	Dynamic	ERP. SCM, MES (Currently not many companies are collecting inventory status data in real time fashion except those who use MES)
Work-in-progress data	Details of all outstanding operations scheduled to go through work data	Dynamic	ERP, SCM, MES in many high tech companies, MES is widely used in collecting real time data
Order/Demand data	Work orders, purchase orders	Dynamic	ERP, SCM. MES
Transportation data	Information on transportation time, availability and cost etc.	Dynamic	ERP, SCM

The so called PML Core (see, Floerkemeier and Koh, 2001) addresses the product location and telemetry related information that is related to specific inventory data of the BIS system (e.g. In most SCM and ERP, inventory file contains location information, quantity and date information).

Some examples of BIS data (SCM or ERP) related to inventory and work in progress (WIP) are as follows:

- *Inventory related data* (either finished goods or raw materials): inventory_location_id, part_id, part_quantity, part_arrival_time, purchase_order_id;
- *WIP related data*:Shop_order_id (or manufacturing_order_id), operation, resource, current_location, finished_part_id, finished_quantity (or unfinished_part_id, unfinished_quantity), finished_location, completion_time, completion_flag.

PML Extensions address product-related information and process-related information (Floerkemeier and Koh, 2001). This kind of information is related to the BOM, routing/operation related data, resource data, part/ product data, etc. Some examples of BIS product-related data (SCM or ERP) are as follows:

- *BOM data*: produced_part_id, produced_quantity, routing, operation, consumed_part_id, consumed_quantity;
- *Routing related data*: routing_id, operation_id, run_time;
- *Product/Part related data*: part_id, description, unit_price.

Even if the data types presented above provide real time information of material movement (or order transactions), to provide truly real time synchronized decision-making (i.e. material and capacity synchronized decision making), real time resource status data is essential[6]:

- *Resource tracking data*: resource_id, resource_type, status, location, description, total_down_time, in_production_time.

In practice, even if ERP and SCM use real time data for WIP (or inventory status data) and resource status data, this data is mostly collected by the MES and passed on to these systems. This is because of the inherent dynamic nature of the data (i.e. Many MES systems have capability to directly interface with machines to monitor/collect real time status of machine) and the objective of each system (For example, when companies introduce an ERP system, they want to standardize their business processes while when companies introduce MES, they want to more accurately trace shop floor activities).

It is very desirable to have cost related information with other data given above so that decisions may be made that consider the efficiencies or transaction values for supply chains:

- *Cost data*: part_id, item_cost, quantities and etc. (other cost data inlcudes transportation cost, depreciation cost, direct/indirect cost of machine, opportunities cost)

Figure 15-5 briefly shows a feasible mapping between BIS and PML data sources. The figure also shows what kinds of data (either dynamic or static data) the BIS system generally uses for its own decision-making. The data files shown in the Figure only indicate the basic files required to represent manufacturing BIS needs (but more data may be required for complex modeling). Here, basic files means the minimum data required to run a planning engine (In the figure the dotted lines show possible relationships between PML and BIS). Straight lines show the relation among different data within BIS. Arrow lines show data used for planning decisions).

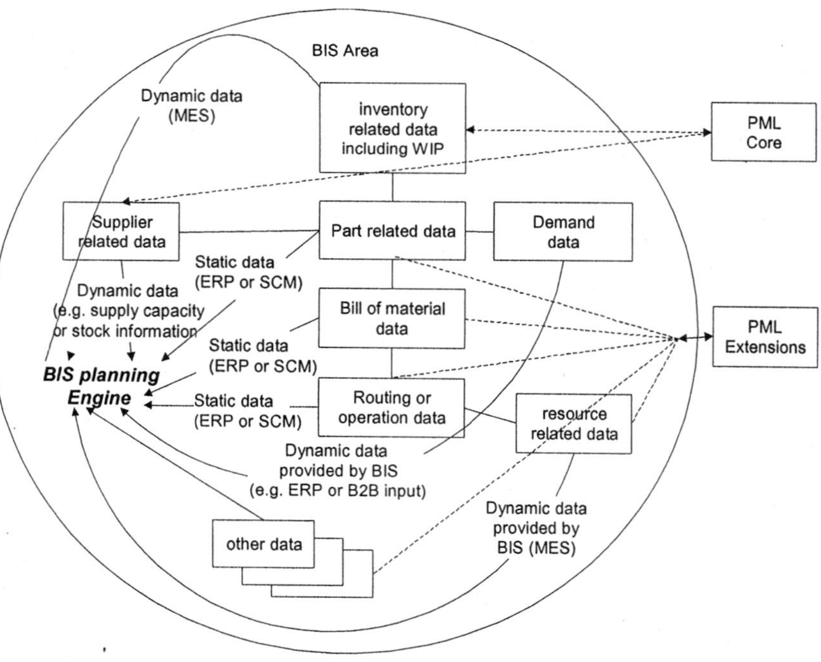

Figure 15-5. PML to BIS Mapping

A number of problems are encountered when integrating data among heterogeneous systems. To minimize problems during integration it is important to understand the existing data structures for the systems under consideration then to map data between the various data sources (e.g. PML and Manufacturing BIS data) and workflow (as in the previous section). These processes are so important because, for example, generally different systems use different names for the same item; and different systems may also use the same name for different items of data.

Table 15-3 shows an example for data mapping between PML and manufacturing BIS[7]. For mapping, data characteristics (e.g. data type and length) and its source table name (if it is a relational database), field name must be known (As in the table, different BIS systems may have different definition for the same data).

Table 15-3. Example Data Mapping

Table (BIS)	Field (BIS)	Format (BIS)	System (BIS)	PML (Field name)
Part	Part type	Char(10)	SCM	Item_ID
Item_master	Item type	Char(20)	ERP	Item_ID
Location	Location name	Char (20)	MES	Location
Location_id	Location_id	Char (10)	SCM	Location

The performance of any BIS (SCM, MES)[8] planning function is heavily dependent on the "cleanness" of data, proper frequency of data collection and other factors. So, it is better to have a data cleansing process before transferring data to different systems since it is very hard to detect 'dirty data' in production environment. It is also desirable to have meaningful aggregations of data by meaningful categories or frequently used categories to improve the performance of BIS, since the aggregation of the data makes for faster retrieval.

5. SUPPLY CHAIN PROCESS USING AUTO-ID

In this section, we present several brief scenarios for Auto-ID applications in both a supply chain management and an existing e-business transaction environment. Each company has different Manufacturing BIS and each needs to develop an appropriate workflow in order to prepare for integration with Auto-ID technology (especially data) for supply chain management. The purpose of this section is to provide example cases as a guideline.

5.1 Workflow with Manufacturing Supply Chain Focus

Many common business processes can be modeled for business transaction management (ebXML 2001). Among them, are query price and availability, query availability, request promise of availability, create sales/order forecast, create shipping schedule for JIT delivery, manage purchase order, create order, query order status, distribute order status etc.

A scenario for Auto-ID based warehouse application was presented by Joshi (2000):

"Data acquisition begins from the point at which goods arrive at inventory stocking points. Tag readers at the entry point, which read the EPCs™ on the tags, then proceed to inventory shelves, which are also equipped with tag readers. When orders from customers are processed, and packages are ready to be shipped, they are passed out on to the loading dock through another set of tag readers again which note the EPCs™ of the goods being shipped. A software application managing data acquisition for the distributor warehouse communicates with the tag readers installed at various locations in the warehouse. It receives large amounts of data from many inputs. This data has the potential to be converted into real time information. The software application gathers the EPCs™ of products being tracked by the tag readers, queries the ONS for the web address where information about the product is stored, updates that information if necessary, and maintains localized data records relevant to operations of the warehouse. The updated product information can be stored either in a centralized database or distributed databases".

The above research work is useful, but the processes only focus on simple warehouse transactions and do not fully cover manufacturing related transactions. The solution design for Auto-ID based supply chain management (using existing manufacturing BIS) can be different in each individual case. However, in order to give an idea on the nature of Auto-ID based supply chain solution with a manufacturing focus, we present two examples with a simple workflow.

Bar Code Based SCM: the Existing Systems

We first provide a brief description of a general solution for a distributed enterprise with a Make-to- Order (MTO) production environment (In this solution we ignore the transactions from the supplier side. We also assume that the existing MES system is not ready for collecting automated Auto-ID data). Referring to figure 15-6, we outline the following 6 step scenario (Step 0~Step 5).

0. Customer orders are placed through various ordering systems/channels (e.g. B2B, Telephone, email, ERP). At this stage, there is no real item visibility in current practice (Also it is impossible to provide real time shop floor visibility with existing technology in place);
1. To fulfill orders, both material information and resource information are needed. A Company which has globally distributed manufacturing plants, collects product data based on the bar code system (e.g. WIP, Inventory and shop order transactions). Material related data read by operatives is transferred to the MES. But the processed information is frequently out of date because of the limitations of manual read data entry and batch read data;

Real time resource information (e.g. machine in a plant or an automated guided vehicle in a warehouse) is also collected either using MES or LES.

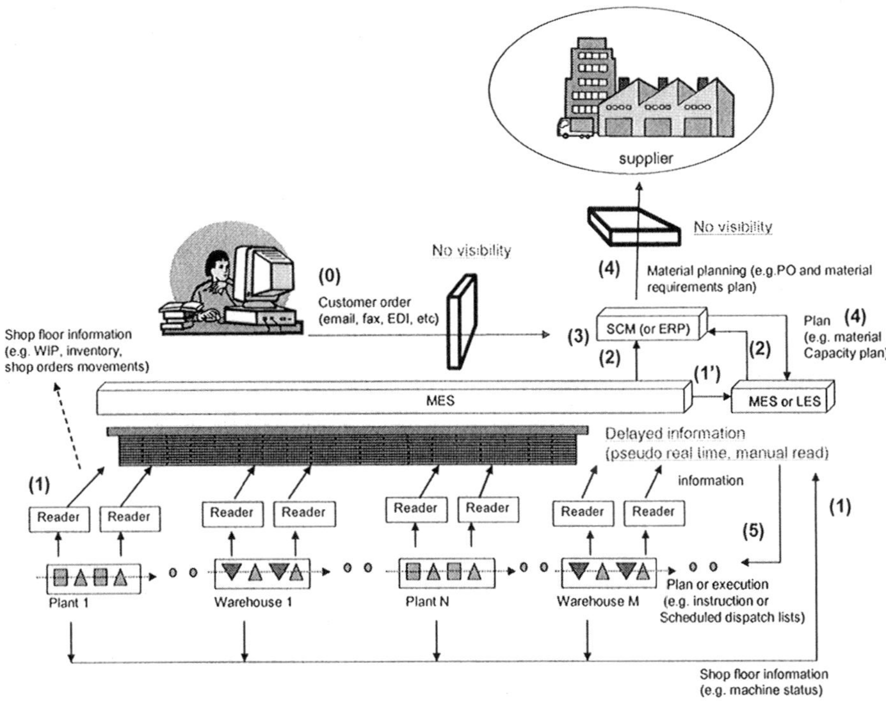

Figure 15-6. General Scenario

2. Those data collected in (1) are transferred to SCM (or ERP) as required;

3. The SCM planning engine (or ERP if it is capable to do multi-site planning) generates an allocation plan using real time information;
4. The manufacturer sends the plan to the MES level for detailed scheduling and to send material requirements to suppliers. When the manufacturer contacts suppliers, it is impossible to get real time, visibility on the supplier side of raw material inventory;
5. The MES generates a dispatch list for each work center or shop floor and supports execution of the schedule.

As shown in the figure 15-6, there are three gaps in the current bar code based system: (a) Delayed information transfer between shop floor and MES; (b) No order visibility between a customer and the manufacturer(s); and (c) No visibility between a manufacture and the supplier(s). For this reason no dynamic value-chain analysis can be made available to understand or improve supply chain efficiencies. This emphasizes the importance of visibility and real time information.

Auto-ID based SCM: the On-Going System with Simple Workflow

We now consider the introduction of Auto-ID systems. Figure 15-7 shows a potential example for Auto-ID based supply chain management. The workflow is as follows:

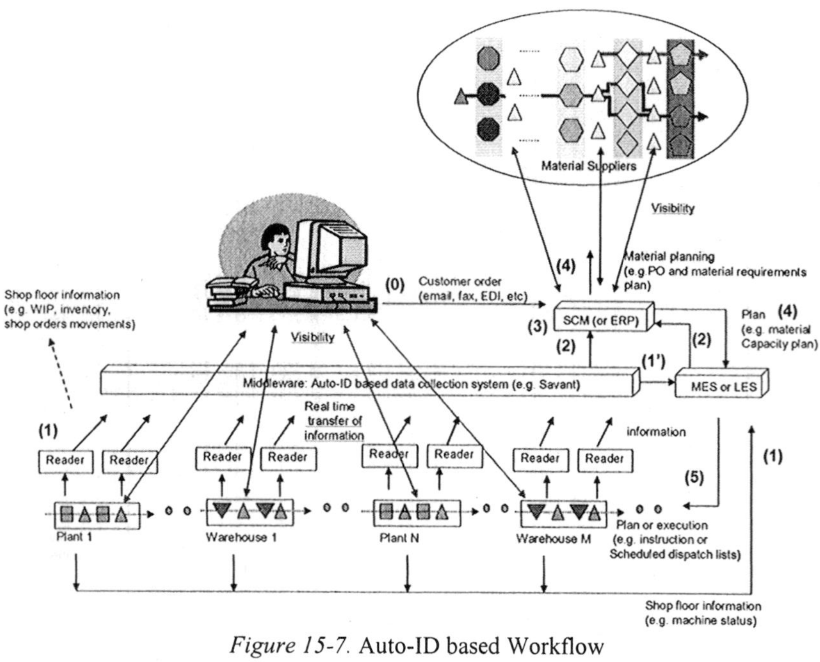

Figure 15-7. Auto-ID based Workflow

0. Customer orders are placed through various ordering systems/channels (e.g. B2B, Telephone, email, ERP);
1. To fulfill orders, both material information and resource information are needed. A company, which has globally distributed manufacturing plants collects item level, real time information (e.g. WIP, Inventory and shop order transactions) with the aid of Auto-ID technology. Material related data is read by an Auto-ID based data collection system (e.g. following the Savant™ specification[9]). Here it is assumed that existing systems (MES or ERP) are not yet ready for integration with automatic readers. In the future, it is considered that this area can be easily covered by MES or even perhaps ERP;

Real time resource information (e.g. machine in the plant or automated guided vehicle in the warehouse) is also collected either using MES or LES.

It is expected that compared to conventional bar code based data collection, Auto-ID technologies will significantly reduce the data collection time by batch read and by removing manual read.

2. Those data collected in (1) are transferred to SCM (or ERP);
3. The SCM planning engine (or ERP if it is capable to do multi-site planning) generates an allocation plan using real time information;
4. A manufacturer sends the received plan to the MES level for detailed scheduling and to send material requirements to the suppliers;
5. The MES generates a dispatch list for each work center or shop floor and support execution of the schedule.

During the above execution process, it is expected that Auto-ID technology will help to solve a number of the problems outlined in Section 3:

- Preventing 'back flush' by providing timely inventory information;
- Prevention of chaos, which could occur through different lot-sizings, by adaptability through provision of item level information with a tagged item, and adapting to the differences caused by lot sizing issues (e.g. identifying the requirement in real time, reforming a lot dynamically if needed to, etc);
- Physical tagging: By physical tagging, one can keep a strategic inventory for key customers, and if economically required, by pegging inventory to specific customers.

Auto-ID based SCM: Use of Item level Information

It is almost impossible to distinguish individual product items from other instances of the same types of products with current bar code systems. When materials are delivered from suppliers, they are placed in the same product stocking locations. There is no way of telling which item arrived first (i.e. item which arrived earlier). The following example shows a simple workflow of how to introduce item level information into planning and execution, and thus one can expect more cash flow return.

(Example) By introducing Auto-ID systems, the warehouse operator can pickup old material (item which is arrived earlier) and use it for production. By doing this the company can have more inventory turns (i.e. Earlier investment for raw materials can be re-deposited by earlier sales of finished products).

Figure 15-8 shows a simplified BOM structure for a Coupé. As shown in figure 15-9 (a), there are two demand orders for Coupés. The BOM shows that to partly manufacture a Coupé, a manufacturer needs 1 'Audio' and 1 'Body'. Currently the manufacturer use 'Sony A' type audio and 'T-body' body. With the aid of Auto-ID technologies, it is possible to get item level inventory information for each part (Figure 15-9 (b)). To satisfy demand order 'D001', the planning system checks inventory information first. The system picks an 'Audio' and 'Body', which arrived earlier than the others of the same type at the stocking point (See, 2nd and 3rd rows in the figure 15-9 (c); and 4th and 5th rows in the figure 15-9 (c)), to speed the cash flow return.

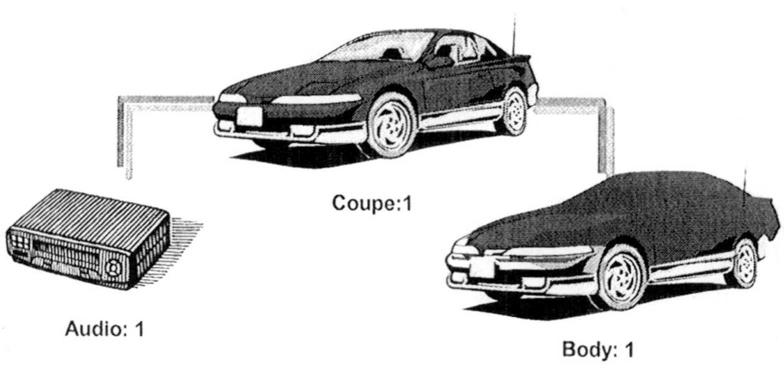

Figure 15-8. BOM(Product) Structure of a Coupe.

The solution did not include the control (dynamic) aspects of an Auto-ID based supply network. Whenever, there are disruptions (e.g. material problems, machine breakdown), which violate a manufacturing business goal (e.g. due date) within the context of the supply network, controllers will be able to detect disruptions with the aid of an Auto-ID system (i.e. whenever late orders happens or inventory inaccuracy problem occurs, Auto-ID based data collection system will detect those information) and their

MES system (i.e. Whenever machine breakdown or operator breakdown occurs it will detect those either by direct or indirect interface).

(a) Demand order for coupe

Demand order no	Customer	Due date	Quantity
D001	A	2003/09/24/00:00	1
D002	B	2003/09/25/00:00	1

(b) Inventory information for Engine and Body (assuming item level visibility)

Part type	Location	Part No	Arrived	Quantity	Part ePC (simplified)
Audio	100	Sony A	2003/09/01/00:00	1	1
Audio	100	Sony A	2003/09/15/00:00	1	2
Audio	100	Sony A	2003/09/05/00:00	1	3
Audio	100	Sony A	2003/09/20/00:00	1	4
Body	200	T-Body	2003/09/01/00:00	1	5
Body	200	T-Body	2003/09/04/00:00	1	6

(c) Example manufacturing (Mfg) order plan considering item level information

Mfg order no	Demand order no	Part ePC	Due date	Quantity	Planned completion time
001	D001	1	2003/09/24/00:00	1	2003/09/23/00:00
	D001	5	2003/09/24/00:00	1	2003/09/23/00:00
002	D002	3	2003/09/25/00:00	1	2003/09/24/00:00
	D002	6	2003/09/25/00:00	1	2003/09/24/00:00

Figure 15-9. Example of Item Level Information

After collecting information concerning disruptions, either from Auto-ID data collection system or MES or both, the MES system is able to re-generate schedules and broadcast any corrective actions to managers and operatives (e.g. reprioritized dispatch lists, change of due date, swapping of an item pegged to the important customer order with an item pegged to less important customer order based on the due date or cost).

The design described also shows disintegration between MES (i.e. for resource information) and Auto-ID based data collection system. In the future it is desirable to have a system that can handle material movement and machine capacity in the same cycle in the future.

Any solution design for integration between different levels of the management and control system it is important to appreciate control loop timings (e.g. second, shift-8 hours, days, months) for each system (e.g. Control system is usually working on a 'second; basis; MES usually works from 'seconds' to one 'shift'; ERP usually works from 'days' to 'months' and perhaps 'years').

5.2 Query Example: To Support Existing e-Business Transaction Standard

Since Auto-ID technology has a high potential to provide a immense amount of data about products in distributed locations within supply chains it is very important to know how to query only the useful information arising from the data. To query a meaningful set of data the structures of the data to be queried needs to be known, by understanding the both the query methods and the data files needed for such queries Manufacturing BIS solution providers and users can enhance the performance of existing Manufacturing BIS functionality (mainly by using real time, item level data). In this section, we present a brief example of data to support an existing business standard as well as to explain what kind of data and BIS are needed to get the whole benefit of Auto-ID.

Consider an electronics industry example using RosettaNet[10] transaction Standard. Currently RosettaNet provides so called Partner Interface Processes (PIP)[11] for globally distributed work in microelectronics manufacturing processes for silicon integrated circuit fabrication, device assembly, and device and final product testing (RossettaNet 2001). The PIPs support the process used (1) by a solution requester (e. g. customer) to query the status of a product in manufacture or test, and (2) by a service provider (e.g., manufacturer or assembler) to return the requested information. "The solution requester initiates the process by sending a work in process query through ICT and is directed to the relevant service provider(s). To facilitate communications, the query document includes both query constraint lines and incomplete results lines. After processing the query, the manufacturing service provider completes the results lines and returns the information to the requesting party".

The following shows as an example a hypothetical query to determine work in process status:

Description: Company A wants to know product AA's finished goods quantity and WIP quantity in its distributed plants at time "2:10pm, April 6th, year 2002".

Hypothetical query: **Work In Process Query**

Search all databases, and Return all results with the product type AA completion time "2:10 pm, April 6th, 2002 year",

The query result may display the following information dependent upon a further condition: Result “final product type AA”, “location”, “quantity”, “operation id or routing id”, “start time” etc.

The query will find the data throughout the “given database”, and display data with the following conditions:

- “completion time equals 2:10pm, April 6th, 2002”;
- “final product type equals AA”;
- “quantity”;
- “location” (Globally agreed location id to define specific location distributed plants in the supply chains);
- “operation id or routing id” (or manufacturing order id etc)

Hence in order to use Auto-ID data for such a WIP transaction, the following data need to be considered:

- EPC™ Code of palette/Case/Item
- Equipment id or transportation id or carrier id (e.g. Palette id)
- Lot id or shop order id (could be EPC™)
- Material id (or product type)
- Procedure id (This is related to shop order if used in manufacturing)
- Location id (i.e. site)
- Stage or operation (Current stage in the routing. This is related to routing information)
- State (e.g. shop order or lot state)
- Time (start time for the generation of alarming and calculation of elapsed time)
- Start time; End time
- Priority (e.g. customer priority); Customer id
- PO number
- Quantity
- Operator id (Manufacturing order ID)
- Instruction

Such a WIP query for distributed supply networks is far from an easy task with the existing bar code based system, because a real time capability is not available. With the Auto-ID technology (especially, real time capability by batch read and automatic read capability) information for supply networks is available in real time and will be used for improving its efficiency and agility.

5.3 Implementation Issues for AUTO-ID

Although Auto-ID applications in supply network management have high potential, there are still some issues, such as interferences with remote LAN, low read rates, and tag and reader prices etc.

Hill (2003) exposed a number of barriers for RFID applications in industry. These are:

- Tag sizes and packaging
- Tag generation and application
- Quality: no reads and misreads
- Deployment Environment: Culture/readiness for change
- Installed base of other AIDC technologies
- Required physical infrastructure
- Required information systems infrastructure and interfaces
- Technology and communications standards

Many people are convinced that RFID will not make its way into mainstream supply chains until standards are adopted. It is considered that technological standards and practical implementation methodologies are the key to the proliferation of RFID technology, particularly in the more mainstream applications, such as supply network management and logistics.

6. CONCLUSION

In this chapter, we have reviewed the basics of Auto-ID technologies, introduced manufacturing related BIS, described integrated workflow for BIS and Auto-ID technologies, and the data needed for the integration. We have also provided examples of query and manufacturing BIS functionality that could be benefited by integration with Auto-ID.

Some general further steps that should be considered by an organization preparing for Auto-ID Integration are as follows:

a) To have more of an understanding about existing BIS systems, PML (Core and Extension) and their interfaces. First of all, before any integration project begins the strategic ICT plan for supply networks must be available and understood by all partners in the network. Ignoring this prejudices stepwise evolution and future expansion;
b) Once the current ICT architecture and future projections are studied then the scope (or levels of detail) of any project may be properly defined. For example, is it critical to track all material movement? Do we need in that particular stage of processing or industrial sector item level tracking (is it

desirable or of economic benefit?) The scope should be defined dependent upon the gains expected.

c) After defining the scope for the project the key system integrations need to be defined. For example for distributed MES systems, this would be a key integration area for integration with Auto-ID data. But if there is no MES, and the supply network uses SCM or ERP systems or both, checks must be made as to whether existing SCM functions can handle real time data (the user has to consider the proper data collection interval for competitive business). If it is not necessary to collect data in real time users can resort to pseudo real time data collection.
d) Establish data mapping among systems and develop detail workflows. When designing a solution for integration between different levels of systems, one should understand the control loop feedback time frames, whether it be second or years.
e) New information provided by Auto-ID technologies (e.g. EPC™ and PML) will improve the data quality for existing BIS. It will reduce the discrepancies between planning and execution. It could prove to be very useful for improvements to agility and efficiency of the supply networks. To best achieve gains though, existing Manufacturing BIS systems need to be modified, enhanced or drastically improved in order to worthily include new Auto-ID provided information in their ICT functionality.

NOTES

1. Here, supply chain/network is meant in its fullest sense of distributed suppliers, manufacturers and service providers, involving a number of companies with distributed businesses and resources
2. LES differs from MES. LES focuses on the transactions/management of warehouse and transportation while MES focuses on the transactions/management of manufacturing plant (Some MES has capability to trace/manage transactions among distributed plants).
3. Recently many ERP vendors have introduced SCM concepts in order to improve the planning and scheduling capability of their ERP systems.
4. Static data updates only when needed (e.g. product data is updated when a company develops a new product. Routing data is updated when a new routing is added to the existing processes)
5. In this table, data defined means "data source where the data comes from"
6. These data are mainly defined in the MES.
7. The data in this table are examples. The purpose of this table is to give the reader an example of a structured approach.
8. In current practice, ERP systems are less frequently used for planning purposes than SCM and MES. Some ERP vendors claim that they have capability to deal with dynamic changes of shop floor data.

9. Savant™: Savant™ is has been developed by Auto-ID Center and OATSystems. It is a data router for Auto-ID that performs operations such as data capturing, data monitoring, and data transmission.
10. RosettaNet: is a consortium of major information technology, electronic components and semiconductor manufacturing companies working to create and implement industry wide, open e-business process standards. More information can be found at http://www.rosettaNet.org.
11. Partner Interface Processes (PIP): define business processes between trading partners. Currently PIP fit into seven clusters that represent the backbone of the trading network. More information on PIP can be found at http://www.rosettaNet.org

REFERENCES

1. Brock (a), D., The Electronic Product Code: A naming scheme for physical objects. White Paper, Auto-ID Center, MIT, 2001
2. Brock (b), D., The Physical Markup Language a universal language for physical Objects. White Paper, Auto-ID Center, MIT, 2001
3. Chang, Y., McFarlane, D., Koh, R., Floerkmeier, C., Putta, L., Methodologies for integrating Auto-ID data with existing manufacturing business information systems, White Paper, Auto-ID center, November 1, 2002.
4. Carpenter, T. Without visibility you can only react. Frontline Solutions, Jan, 2003.
5. ebXML White Paper, Business Process and Business Information Analysis Overview v1.0 2, 11 May 2001.
6. Floerkemeier, C., Koh, R., Physical Mark-up Language update, White Paper, Auto-ID center, June 2001
7. Hill, J., Evaluating RFID? Frontline Solutions, Feb 2003.
8. Joshi, Y., Information Visibility and its Effect on Supply Chain Dynamics. MSc thesis MIT, 2000
9. Kay, E., What's next step for RFID? Frontline Solutions, August, 2003.
10. Karkkainen, M., Holmstrom, J., Framling, K., Artoo, K. Intelligent products-a step towards a more effective project delivery chain. Computers in Industry 2003; 50: 141-151
11. McFarlane, D., Sarma, S., Chirn, J.L., Wong, C.Y., Ashton, K., The intelligent product in manufacturing control and management, to appear in Engineering Applications of Artificial Intelligence: special issue on Intelligent Manufacturing, 2003
12. Mandal P., Gunasekaran, A. Application of SAP R/3 in on-line inventory control. International Journal of Production Economics 2002; 75: 47-55.
13. Manthou V., Vlachopoulou, M. Bar-code technology for inventory and marketing management systems: A model for its development and implementation. International Journal of Production Economics 2001; 71:157-164.
14. Saar, S., Thomas, V. Towards trash that thinks. Journal of Ecology 2003; 6(2): 133-146.
15. MESA web pages. Retrieved May 2002 from: http:// www.mesa.org
16. Parlikad, A.K.N., McFarlane, D., Gross, S., Fleisch, E., "The Role of Product Identity in End-of-Life Decision Making", Auto-ID White paper, June 2003.
17. RossettaNet White Paper, PIP3D8: Distribute Work In Process, 17 July 2001.
18. RossettaNet White Paper, PIP3D9: Query Work In Process, 17 July 2001.
19. Sarma, S., Brock D., Ashton, K., The networked physical world: Proposal for engineering the next generation of computing, commerce and automation-identification. White Paper, Auto-ID Center, MIT, 2001.

PART 4. SYSTEMS AND TOOLS

"The challenge of human-centeredness is how to move beyond the traditional notion of human-machine symbiosis, and promote notions and concepts which deal with a variety of human-human and human-machine relationships and networks of relationships both at the local and global level. ...This challenge is how to reintegrate technological innovations ...so that technology supports new forms of work-life and living environments." Karamjit S. Gill; 'Human Machine Symbiosis'; published by Springer Verlag 1996

In the previous three parts of this book we have explored the transformation that modern businesses have undergone to become more agile and competitive in the modern marketplace. We have discussed the new requirements and how businesses are beginning to implement new ways of working. It is evident that the linear supply chains of the past are changing and now there is a trend for businesses to operate as agile nodes in adaptive value networks rather than isolated 'black boxes' in inflexible, linear supply chains.

An essential ingredient to the transformation process is symbiotic support provided by information and communications technologies. There are three primary technology enablers. The first is the communications infrastructure. It is expected that as the communications costs drop and the Internet's bandwidth continues to increase this will result in increased speed of exchange of rich content rather than infrequent exchange of aggregate data. The second enabler is hardware. Hardware not only includes faster and cheaper computers but also very small smart sensors capable of collecting large amounts of information in real time. The last enabler is advanced and adaptable software that utilize the infrastructure to support new ways of working, facilitate sense and response to real time events and provide users with the tools for fast, astute and coordinated decision making.

Unlike today's static and linear supply chains, adaptive value networks will be empowered by multi-partner processes that are intimately supported by ICT. Such processes and ICT must be event driven and also at the same time be aware of the environment through smart sensors, and protect the autonomy and self-governance of each partner in the value network. This means that every node in a value network is able to pursue individual goals and opportunities as well as dealing with local constraints whilst complying with global objectives, resulting in maximized overall benefits to both the stakeholders and end customers.

In part four of this book, we explore such enabling software technologies, which aim at facilitating adaptive value networks.

Chapter 16, written by Christian Dannegger and Klaus Dorer, presents an agent paradigm for software development and an agent server and toolset which allows developers to build agent based systems. Agent based systems are intrinsically scalable, distributed systems.

Also agents offer a higher level of abstraction which encapsulates not only traditional objects but also the strategy for how to use those objects to pursue delegated goals. For those reasons new generation agent based software systems are not only capable of more accurately modeling a supply network, but also they offer a capability to solve complex problems in dynamic domains, such as logistics execution in a value network, real time order management and planning, fulfillment execution etc. The authors discuss the shift from traditional software engineering paradigms to the agent-oriented software engineering paradigm and go on to present an agent server and development toolkit. Finally they present a real life application in the area of logistics execution.

Chapter 17, written by Grace Lin, Jun Jang Jeng and Ko-Yang Wang, presents an IT framework that enables collaboration amongst businesses participating in value networks. The lack of trust among businesses in a value network and the lack of a well defined framework and adaptable IT infrastructure are among the main reasons that prevented collaboration in decision-making across businesses to materialize. This chapter addresses this important area by presenting such a framework. This framework comprise of two main elements. The first is an agent model that provides the decision support infrastructure which aims to optimize operational business processes in a value network. The second is a flexible enterprise integration infrastructure that contains both methodologies and systems integration technology. Both elements of the framework aim at addressing the key issues of adaptable cross-enterprise integration and also enable trust among businesses participating in a value network.

Chapter 18, written by Toshiya Kaihara, presents an agent-based virtual eMarket. This eMarket has been designed and built using what is known as market-oriented programming. The author presents an approach for deriving solutions to distributed resource allocation problems by computing the competitive equilibrium of an artificial economy. The eMarket is comprised of demand agents and supply agents, which evaluate the trade-offs of acquiring different products in response to demands. The author provides an overview of the virtual eMarket approach and discusses its applicability to solving supply network management problems. The focus of the chapter is on the application areas and presents a new business model for supply chain management using the agent-based eMarket, which is supported by a mathematical model. The author then provides results of simulations utilizing the model.

Chapter 19, written by Douglas Scott, Robert Pisa and BG Lee, presents a commercial intelligent shop floor execution system. In contrast to traditional supply chain management, adaptive value networks are demand driven, and comprise of versatile manufacturing plants that are run by 'bottom-up' schedules. Factory level information should be thus visible to a large extent and leveraged by business systems. New generation manufacturing execution systems are thus a key ICT component to enable collaborative manufacturing execution in value networks. The authors present such a system in the domain of 300mm silicon wafer fabrication. They discuss the challenges for automating 300mm wafer fabrication plants and also the key requirements and challenges for achieving complete production automation.

Chapter 20 written by Charalampos Makatsoris, Yoon Chang and Howard Richards, presents a commercial agent-based platform that has been designed to enable coordinated decision making across adaptive value networks. The platform offers integrated, demand-driven planning and fulfillment functions as well as providing sense and respond capabilities. The authors discuss the challenges for coordinated planning and fulfillment in adaptive supply networks and describe the shortcomings of current ERP and MRP-II type systems in addressing those challenges. The design and functionality of the platform is presented illustrating its distributed agent architecture, which offers the functionality that facilitates both local autonomy and self-regulation. In essence a value network node can pursue opportunities by managing local constraints and by observing global objectives, when responding to real time events and external demands. The authors discuss not only how performance of a partner in a value network is increased but also the way decision coordination is achieved across a value network that leads to overall increased performance of the network itself. A case study is provided to illustrate the use of the platform in the context of order management and fulfillment planning.

Chapter 16

LIVING AGENTS

An Industry-proven Agent Server and Development Toolset

Christian Dannegger, Klaus Dorer
Living Systems GmbH, Germany

Abstract: For more than 15 years software agent has been a fascinating area of research. During the time agent systems have matured in terms of business application as well as technological support. After an introduction to the agent paradigm this paper describes the 'Living Agents' agent server and development toolset as an example of how tool support helps with setting up a business application using an agent system.

Key words: Agent system, Agent development tool, Living Agents, Agent server

1. INTRODUCTION

Value networks as they exist today are mostly handled manually using phone, fax and email. One reason is that it is hard to model the dynamics and complexity, of such value networks with 'traditional' software. The problem is not so much that information is heavily distributed but that a change in one node of a value network usually spreads across a large portion of the complete network. Moreover, changes and deviations from plans are not the exception but the rule.

A new paradigm, agent-based systems, in software development has evolved that deals with this complexity by distributing tasks and acting proactively in a network of responsibilities. In this article we give an insight of this new paradigm. We start by explaining the agent paradigm itself. Then the main part of the article explains how agent systems can be designed and what tools are needed to deal with the complexity of value networks.

1.1 Agent Definition

You can find hundreds of agent definitions. It seems to be impossible in one or two sentences only to define the power behind this paradigm. So let us first define what agents are not. Real software agents are not just renamed search buttons or a meta search engines. They are not just price finders and comparers. They are not the 3D representation of help functionality. And, they are definitely not spies. Instead agents are electronic assistants that act in the sense of a representative, broker or secretary. The following definition approximates our understanding of the important characteristics:

> "Agents are software objects that proactively operate on behalf of their human masters in pursuing delegated goals."

1.2 The Paradigm Shift

Since the Neumann-Computer and the hereby associated machine level programming the software-industry has gone through several levels of solution analysis, design and implementation:

- First there was the introduction of commands (Mnemonics) for the assembler code and the programming style was sequential;
- The idea of reusing code blocks (modules, subroutines) led to the paradigm of function orientation. The results were 3rd generation languages (3GL) like C and Pascal;
- The next step in software development was the shift from strictly functional thinking to object orientation, where functions combined with its data was hidden as black box within an object. The results were OO-languages like Java or C++;
- With the idea of software agents, researches are talking about the next paradigm shift (e.g. Jennings and Wooldridge 2001). Software assistants (agents) playing a role within an application space take over goals. Agents as a new, higher abstraction level encapsulates not only traditional objects but also the strategy how to use them to pursue delegated goals.

In each case, first there was the paradigm shift followed by technology support, meaning the introduction of new programming languages (Figure 16-1).

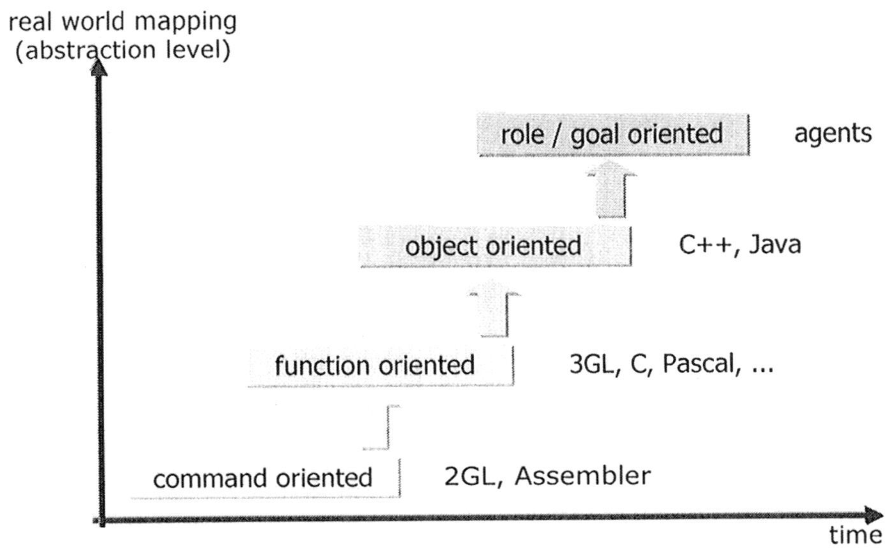

Figure 16-1. Paradigm Shift in Software Development

Agent technology is no more mysterious than Cobol, Pascal, C++ or any other 3GL or 4GL. It is the consequent result of analyzing the real world's complexity and dynamics and how we deal with it in our daily life. Each company, each organization, each social structure is a complex network in which each participant plays certain roles. Associated with each role are a number of responsibilities and rights. All of these networks do not work because of a centralized optimization and centralized decisions. It is the opposite: decentralized decisions based on decentralized optimization. Though all participants of the organization may share a common overall goal, all of them also have their own lesser goals. These goals build the decision base of a role/person. The system does work, if the right roles and the right goals are defined.

1.3 Agent Solution Space

As discussed above the agent paradigm has its origin in complex and dynamic environments. The more complex and more dynamic a problem is the more appropriate it is to be solved by agent technology. Or even more: extreme complex and dynamic challenges may only be solved with an agent approach (Russel and Norvig 2002).Less dynamic areas like strategic planning or less complex scenarios like single node in-house processes are already covered by traditional systems like ERP- and DSS-systems. The real world consists of endless dynamic challenges. Most software solutions are planning systems based on historical data for an ideal and static future. But the reality is a sequence of permanent exceptions, which makes it complex and very dynamic. Examples are production execution, where plans are made for weeks or even months and within a planning period, exceptions like supply shortages, machine breakdowns or additional orders happen abundantly. The same applies for a whole supply network, where more than ever "time is money" to react on deviations from plan. Even faster decisions are needed in the financial area, where brokers and the systems they are using have to react in seconds.

Software agents are the result of long and intensive research to build solutions for complex and dynamic environments and – at the same time – keeping these solutions manageable.

1.4 Planning vs. Execution?

A considerable number of existing software solutions are planning systems. E.g. "ERP", which stands for Enterprise Resource Planning. Planning here as in other solutions means collecting historical data, aggregating the results into general tendencies (e.g. average), combining this with guesses for the future (forecasting) and calculating out of this a plan for the future. If you agree with that and read again through this process you will recognize that at no point in time a planning system takes the actual situation into account.

All the deviations from plans, all the real time exceptions are still managed (in most cases) by humans following manual processes. When a supplier fails to deliver in time the buyer has to find alternatives very quickly. When a machine breaks down, the production manager has to deal with it. In a positive exception when a customer increases his order the whole production process has to be adjusted very quickly with lowest impact to other orders in the most effective (cheapest) way. All of these exceptions do not only happen now and then – unfortunately exceptions are the rule in real life. This process layer of exception handling and real time optimization sits on top of three other already existing layers (Figure 16-2 from bottom to top):

1. The layer of strategic planning in timeframes of 3 years and more helps to decide network strategies;
2. The layer of tactical planning looking at a period of 1 to 18 months is the basis for capacity planning and asset management;
3. The operational layer optimizes and plans with existing resources and orders 1 day, 1 week or even 1 month in advance;
4. The new layer of real time optimization within a planning period leads to the agent paradigm.

With this layer concept you begin to understand the question mark in the heading of this paragraph. It is not about planning versus execution. It is real time execution on top of the existing planning layers. Planning systems were and still are needed for the purposes mentioned in the overview above. Exception handling in the 4th layer contains a huge unused optimization potential, which can ideally be unleashed by agent technology. Agent-based automation and bottom-up optimization extends existing planning solutions by getting a calculated plan at the beginning of a planning period, dealing with the exceptions and constraints in real time and reporting back results to the planning systems at the end of the period.

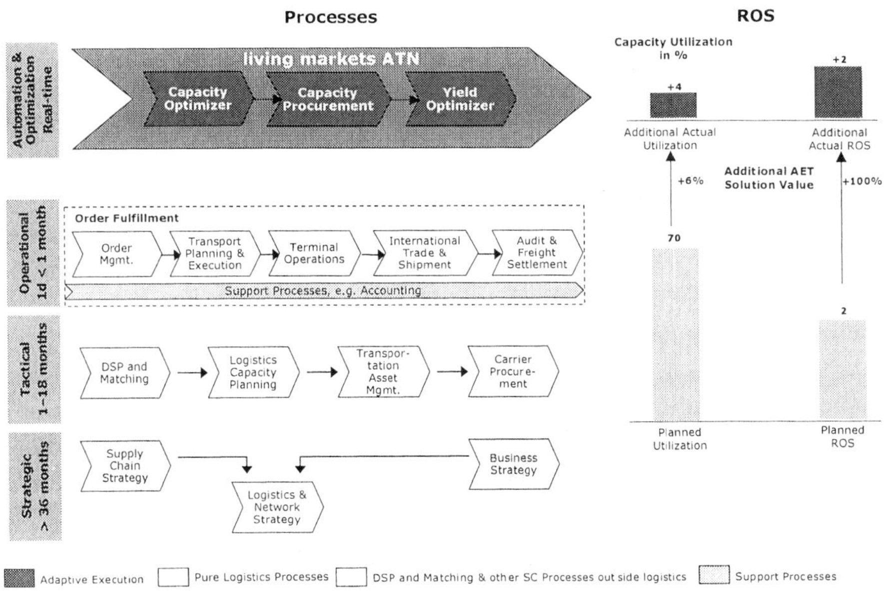

Figure 16-2. Enterprise Planning Layers

The rest of this chapter is organized as follows: section 2 describes how to design an agent application and the tool supporting the creation of the design. Section 3 explains in detail the Living Agents runtime environment needed to run agent applications. It also covers the tools that help to manage and debug running agent applications. Both sections are illustrated by an example agent application in the logistics domain. Section 3 also contains some results achieved using an agent application in the logistics domain. Section 4 draws some final conclusions.

2. DESIGNING AGENT APPLICATIONS

Agent solution architects have discussed several approaches and styles of agent analysis and design methodologies (e.g. Wooldridge et al. 2000, Odell et al. 2000, Kinny et al. 1996). All approaches are very comprehensive. Most of them, however, lack the support of well-integrated tools. Either tools are not existing at all or, as is the case for Agent UML, of the shelf tools are existing, but do not provide easy deployment integration for agent servers.

In this section we describe the design approach used at Living Systems to easily design and deploy agent applications. The tool used for this is the Living Markets Development Suite explained in section 2.2. After that section 2.3 gives an example of a design for a logistics application.

2.1 Design Approach

To be able to provide easy to use tools that are based on our practical experience we developed a very simple agent design approach having three steps (Figure 16-3):

1. First of all the roles have to be defined. Role stands for: agent, electronic assistant or any active object with a certain task. The major decision at this stage is to define the granularity of the solution – the level of detail of each agent. A higher level in a logistics scenario would be a truck or even a distribution center. A low, detailed level would be a package or a single product;
2. The second step is to assign the domain capabilities to each of the agents. Some need financial capabilities (expertise). Others need trading or logistics expertise;
3. The third step – the most important one – is to define the strategy of each of the agents in terms of goals and business knowledge. The goals and their importance are the basis to calculate the action with the maximum utility at runtime. Business knowledge is created by combining actions and perception out of the capability libraries (the domain expertise) to behavior rules. The capabilities are functional primitives created to be reusable and are the building blocks for the agent's business logic.

These three steps cover all levels of an agent implementation from the role concept to the business strategy of each agent down to the capabilities implemented in Java.

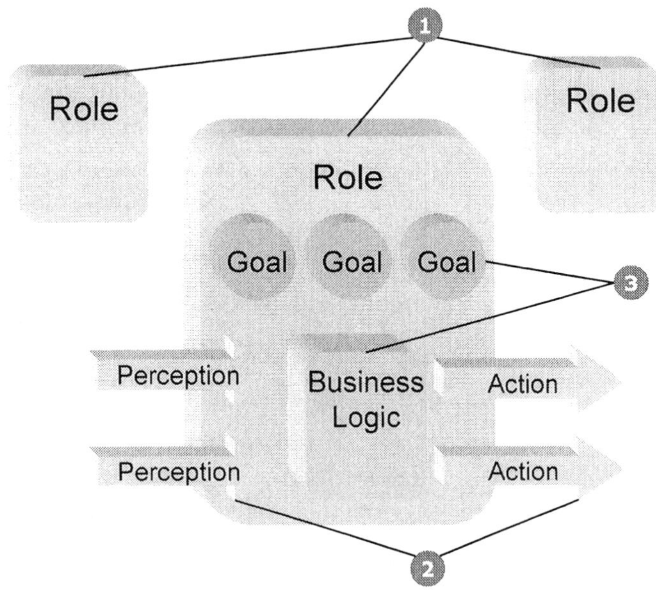

Figure 16-3. Three Steps to Design an Agent System: 1) define roles 2) assign domain capabilities 3) define goals and business knowledge using domain capabilities

2.2 Living Markets Development Suite

Designing Living Agents applications is supported by the Living Markets Development Suite (LMDS). It directly supports the steps described above.

In a first step the roles can be defined by adding new roles to a scenario. Multiple scenarios are used in large projects to allow different views on the roles. Associated with a role is a name, a description and an icon that later identifies the agents having this role. The responsibilities of a role are specified in step 3 when the goals and business logic is defined. In this first step the interactions between roles may also be defined. Usually agents interact by asynchronous communication using messages. LMDS allows specifying messaging by linking the according roles with an arrow labeled with the name of the message. This only defines that two roles exchange a message. The time at which this is done is defined in step 3.

In the second step the user can assign the domain capabilities. This will depend on the domain that the agent will be responsible for. A broker will need financial capabilities, a dispatcher will need logistics capabilities. Assigning domain capabilities either means to select from an existing library of capabilities, or to create new ones. However, it is rarely the case that capabilities already exist especially for projects in new domains. Therefore

LMDS also allows creating ad hoc capabilities that later in the implementation phase of a project have to be accomplished.

Step 3 contributes to two important properties of agents that distinguish them from conventional software objects: autonomy and proactivity. Autonomy means that an agent should be able to decide on its own what the best action is in the current situation. Proactive means that an agent not only reacts to triggers from outside but also acts on its own initiative. Agents are therefore also called active objects. The decision engine of an agent determines the degree of autonomy and proactivity. Living Agents and LMDS currently supports two decision engines: a workflow engine and an extended behavior network (EBN) engine (Dorer 1999). The first allows for defining a deterministic workflow that contains the business logic. The goals are only implicitly specified by such a workflow and the degree of autonomy and proactivity is low. Workflows are used in deterministic domains, where the flow of actions of an agent is fixed in a certain situation. EBNs on the other hand allow for specifying explicitly the goals of an agent as well as the importance and the situation dependent relevance of the goals. Behavior rules contain the effects of an agent's actions and their probabilities in order to plan proactively to satisfy the agent's goals. Multiple rules may be appropriate in a given situation. An agent has the autonomy to select whichever rule has the highest expected utility in a given situation. EBNs are used in continuous, dynamic and non-deterministic domains.

The result of this design process is an overview of all the roles involved in a solution and their communication as well as their high-level business logic. If all of the capabilities used in the business logic are already implemented, roles may already be instantiated by agents and deployed to the agent server. Connecting LMDS to an agent server supports design time deployment of agents. A simple click starts an agent on the connected platform.

2.3 Example Application

Living Systems used the design process described above in several agent-based applications (e.g. Fritschi and Dorer 2002). Here it is illustrated by an example in the logistics domain. The example is taken from a business application (Living Systems Adaptive Transportation Network - ATN) for a large logistics company. To keep the example concise we show here a highly simplified design of the real application. The general idea, however, should become clear.

The core of the application is that a dispatcher receives orders from customers to be transported from one place to another. The dispatcher's task on the one hand is to determine allocation of orders to trucks that have optimal loading of the trucks and on the other hand is to stick to all constraints. Constraints are for example pickup times, delivery times of orders and drive times for drivers. Apart from any new orders unexpected events may occur and are reported to the dispatcher. Global (re)planning is prohibited by the complexity of the domain. A local, agent-based approach is therefore necessary.

Figure 16-4 basically shows the result of step 1 of the design process. The two most important roles involved are the dispatcher and the truck. The dispatcher is responsible for allocating orders to trucks. The trucks are responsible to proactively optimize the allocation by exchanging orders between different instances (agents) of the truck role.

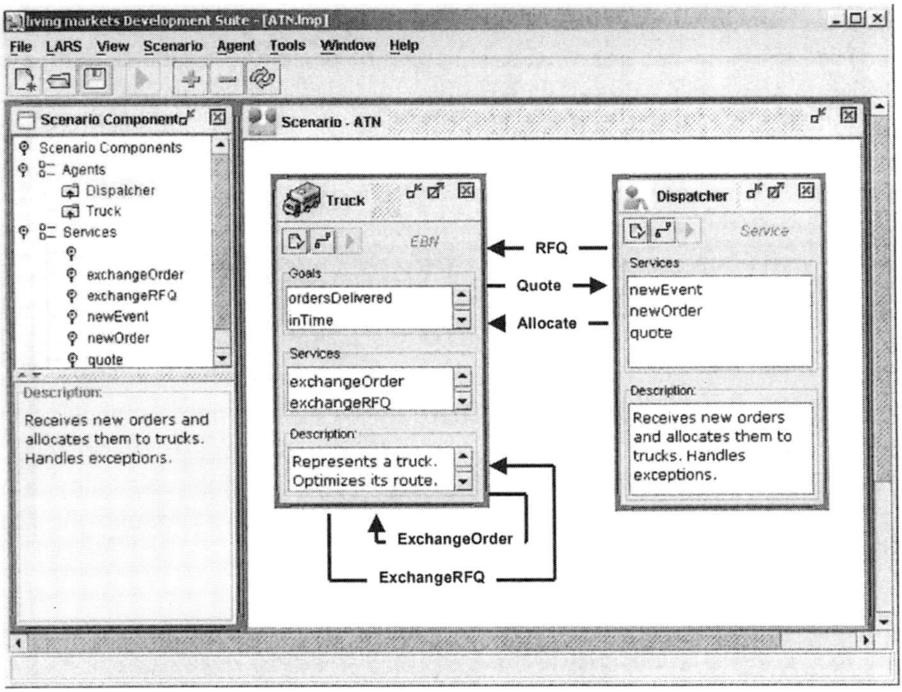

Figure 16-4. Scenario View of a Logistics Application in the LMDS

In step 2 both roles need capabilities of the logistics area, the dispatcher also needs market capabilities. The needed specific logistics capabilities differ of course from truck to dispatcher. The truck role needs the capability to get the distance between two locations or cost estimations for a route. The dispatcher needs the capability to run a request for quote (RFQ) among all the trucks he is responsible for. The capabilities, if not available, are the pieces that have to be developed later in the implementation phase. If properly implemented, the RFQ capability can be reused in other market scenarios.

The logic engine, goals and business logic of the roles are defined in step 3. The dispatcher in this example has straightforward deterministic behavior. Whenever a new order arrives it starts an RFQ and assigns the order to the truck with the lowest quote. Therefore the dispatcher's business knowledge is represented using a workflow engine. It uses a standard workflow for RFQs. Goals are not explicitly represented. The truck role uses an EBN engine. Its goals are that orders are delivered, are in time and are achieved with low costs. Behavior rules contain rules for exchanging orders with other trucks or rules for quoting on a new order. By using the EBN engine these rules are evaluated continuously and may fire proactively without triggers from the outside.

The design phase for business applications also contains a database design to store application data and the object-oriented design for complex capabilities. As mentioned above, the design phase is followed by an implementation phase in which outstanding capabilities are accomplished.

3. RUNNING AGENT APPLICATIONS

It is a long way from reading about agent technology, and being fascinated by the new concept, to be able to implement reliable solutions for industry. To run an agent application you need at least one component: an agent server or agent runtime system that provides the necessary infrastructure and services for the agents. Other helpful tools allow for remotely observing and managing the runtime system by visualizing the state of the agents (see also Dannegger et al. 2002).

3.1 Living Agents Runtime System - LARS

An agent server is needed as the basic platform and runtime environment for agents. Agents can only run on an agent server. This means that you have a well-defined environment for the agents. The agent server runs as a black box and does not need to be changed by agent developers. The Living Agents Runtime System (LARS) is an agent server that has been developed since 1997. More than 30 customer projects have been developed based on the LARS agent server. An agent server has to exhibit several properties that are described next.

3.1.1 Standards

Two different groups of standards can be distinguished, which have to be recognized by agent solution architects:

- Firstly there are web application and Internet standards like HTTP, XML, J2EE, JMS, Corba, SOAP (Web-Services), JSP, JDBC, JNDI. All of these and other standards are used, implemented or supported by Living Agents;
- Secondly there are agent standards defined by FIPA (Foundation for Intelligent Physical Agents, http://www.fipa.org), promoted by e.g. AgentLink (European Union funded organization, See http://www.agentlink.org) and practically used by the agent community. Agent standards are also promoted by projects like agentcities (http://www.agentcities.org) and agent related activities of the W3C (World Wide Web consortium, http://www.w3c.org) and the OMG (Object Management Group, http://www.objs.com/agent). The most important standard here is the ACL (Agent Communication Language) of FIPA, which standardizes the message exchange between agents.

For Living Agents XML is one of the most important standards. XML is used for configuring agents and their business logic and strategy, and XML is used as communication syntax between agents and to the outer world. Being aware of the performance issues with XML, LARS optimizes internal XML conversions by suppressing them whenever possible. It still provides transparent XML communication to external systems.

It is beyond the scope of this paper to discuss all the standards, their status and their special advantages in combination with software agents but greater detail may be found by referring to the various web pages relating to standards and the associated organizations.

3.1.2 Agent Layers

The layered architecture of Living Agents allows for optimizing, enhancing or exchanging each of the layers separately (Figure 16-5). The layers start from the bottom with the communication layer. It implements and allows for configuring one or more of the standard communication protocols like Socket, JSocket, JSecureSocket, RMI and JMS. The next layer decrypts incoming and encrypts outgoing messages. It supports a configurable key length of at least 4096 bits, which means an ability to increase the security level. The XML layer parses XML messages into internal XML objects for further processing. Outgoing XML messages in object format are converted back to XML documents.

The standard logic layer buffers the incoming messages in an agent's inbox, processes them if they are system messages and forwards them to the business logic if they are regular messages. The business logic layer is the one containing the customized part of an agent. Different decision engines can be plugged in as described above in section 2.2. The persistence layer is the database and file system interface for an agent. Some agents may need database persistence. Others are only needed at runtime without database access.

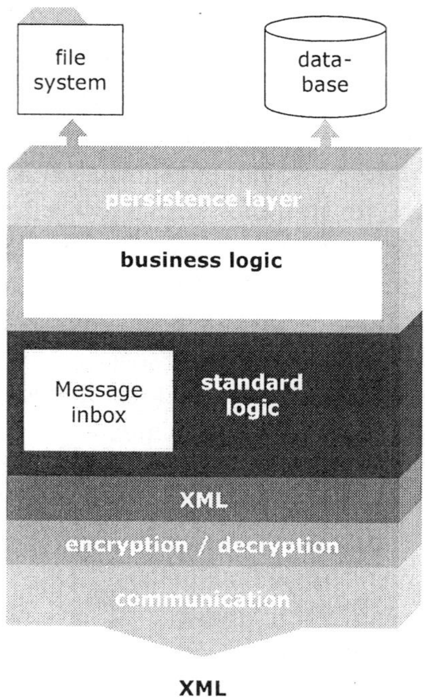

Figure 16-5. Layers of an Agent

3.1.3 Logic Engines

As mentioned in section 2.2 agents are active objects. The core principle of agents is to sense the environment and then act accordingly. Between the sense and the act there is a decision, 'what to do' – accordingly choose the next action that maximizes the utility for the agent and by that for the whole solution. Choosing the next action based on the current knowledge can be very simple in certain environments and can be very complex for others. Depending on different tasks or problem types there are different approaches to model the solution strategies as there are:

- Deterministic service-oriented, reactive workflows allowing traditional sequences, branches and loops;
- Goal-oriented proactive extended behavior networks for continuous domains;
- Decision trees which can learn from sample data;
- ...

The logic engine framework of Living Agents very easily permits extension of this set of action selection strategies. All logic engines make use of the capability libraries with perceptions and actions written in Java. The business logic of the respective logic engine always is stored in an XML-structure. This gives the flexibility to either use a tool like LMDS to design and maintain an agent's business logic or to edit the XML-representation directly. It also opens the ability for other tools to be used or developed in the future to create this XML-based business logic.

3.1.4 Footprint / Micro-Kernel-Architecture

The footprint of LARS for a system can be very small. It depends on the workload we put on the agents. A single agent can even run on a PDA, whereas the same technology is scalable to support high performance, heavy loaded internet platforms like eBay.

This is possible through the micro-kernel architecture of LARS, which means that the kernel consists only of the absolute minimum to start and run an agent server. Almost all system functionality is implemented by so called system agents. If you do not need particular system functionality you simply do not start the respective agents.

3.1.5 Scalability

Scalability is a core characteristic of agents. Each agent always holds the data it needs for decisions and communicates with other agents through Internet protocols. This even allows, if needed for its task, an agent to work exclusively on one single server. On the other side an agent can share a server with hundreds or thousands of other agents. An agent architecture allows you to make use and leverage each piece of resource.

Mobile agents can move to another server if the current one is overloaded or is scheduled for maintenance and by that balance the load across all resources of the IT infrastructure. Following an individual strategy to balance the load this is managed by load balancing agents on each agent server.

3.1.6 Flexibility

An installation of an agent-based solution can be distributed across several computer systems. Reasons for that are strategies for load balancing, fault tolerance, firewall security, and solution integration.

3.1.7 Integration

By hiding protocols within a single agent, agent technology is ideal for heterogeneous integration challenges. Data transformation ideally is done by XML-based XSLT technology. The interface itself sits within an agent. Web-Services and the underlying SOAP-protocol based on HTTP-requests are the perfect fit for agents. Agents ideally implement web-Services. Traditional integration solutions like bulk processing by exchanging data via structured files or using API-libraries can be added to agents by adding these capabilities to the respective integration agent.

Last but not least Living Agents of course fit into an existing EAI-infrastructure like e.g. TIBCO, MQ-Series, webMethods, Entire-X or other message-based platforms.

3.1.8 Reusability

The permanent goal of software developers is to increase the reusability of once developed functionality. The clear separation of business strategy (goals and behavior rules) and functional primitives (capabilities) is a major step into this direction. All capabilities are organized in libraries. Graphical agent development tools like LMDS to build specific business agents use them. All details of a Java-based capability are hidden while implementing or changing an agent's strategy.

3.1.9 Security

Each agent can be configured to accept only messages of a certain type from certain agents from certain agent servers. This security shell is outside the agent's logic and neither affects the logic nor does the logic have to take care about the "agent private firewall".

3.2 Composer and Visualizer

An important step between the design and running an agent application is the instantiation of roles with agents. Using the Living Markets Composer and Visualizer (LMCV) does this. It loads the role definitions of LMDS and allows for defining a setup of any number of agents per role. This is done by dragging and dropping the agent icon to a composer window.

The LMCV can also be used to visualize agent states during runtime. This is especially useful if the agent represents a role with position information. In the logistics example from above this can be used to display the truck agents and their position on a map (Figure 16-6).

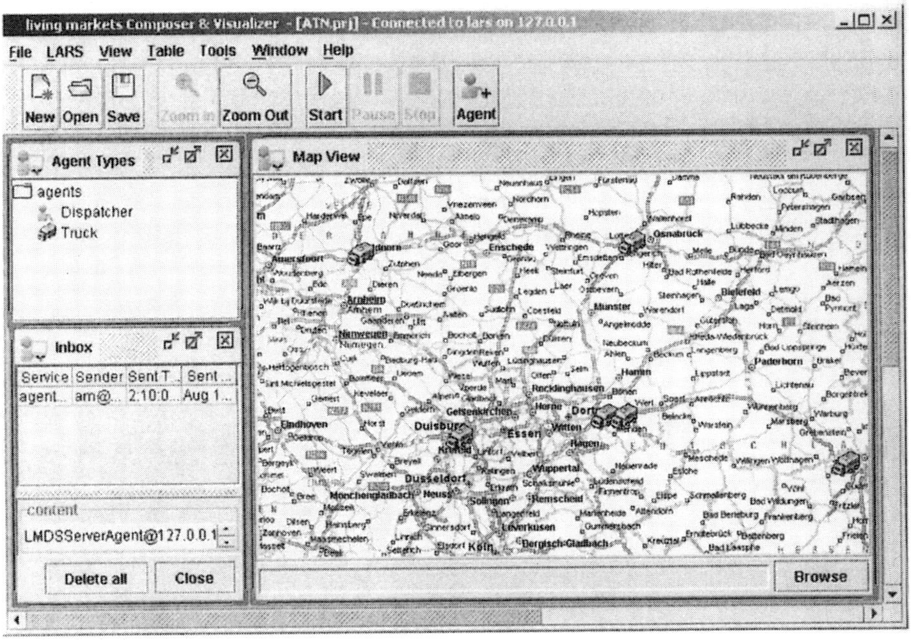

Figure 16-6. The Living Markets Composer and Visualizer

3.3 Control Center

The Living Agents Control Center (LACC) is a tool for managing the agent servers and the agents running on them (Figure 16-7). It connects to an agent server as if it was an agent. All communication is done using agent messages. LACC allows for monitoring the state of the platform and the agent running on it. It also allows for changing the state of the agents and platforms by sending messages to the agents. Finally it integrates debugging functionality by observing the messages on the platform.

The LACC displays a list of all agent servers to which the current agent server is connected. It can show a list of all agents running on a platform. Any agent can be asked to send description information about it containing version information, some messaging statistics and a list of services the agent provides.

Any administration task is performed by either manually sending the appropriate message or by pre-configured messages stored in an XML configuration file. Changes may be starting or stopping agents, migrating agents to other platforms or changing the state of agents in any other way by sending appropriate messages. For example, one can change the log level of the agent or manually trigger a service of an agent.

The LACC also integrates debugging functionality with respect to the message traffic on the server. It therefore allows for tracing specified messages on the server done by a debug agent on the server platform. The trace can be used to create charts of the message traffic to detect bottlenecks. It can also be used to create message sequence diagrams to verify protocols and the order of messages exchanged between agents.

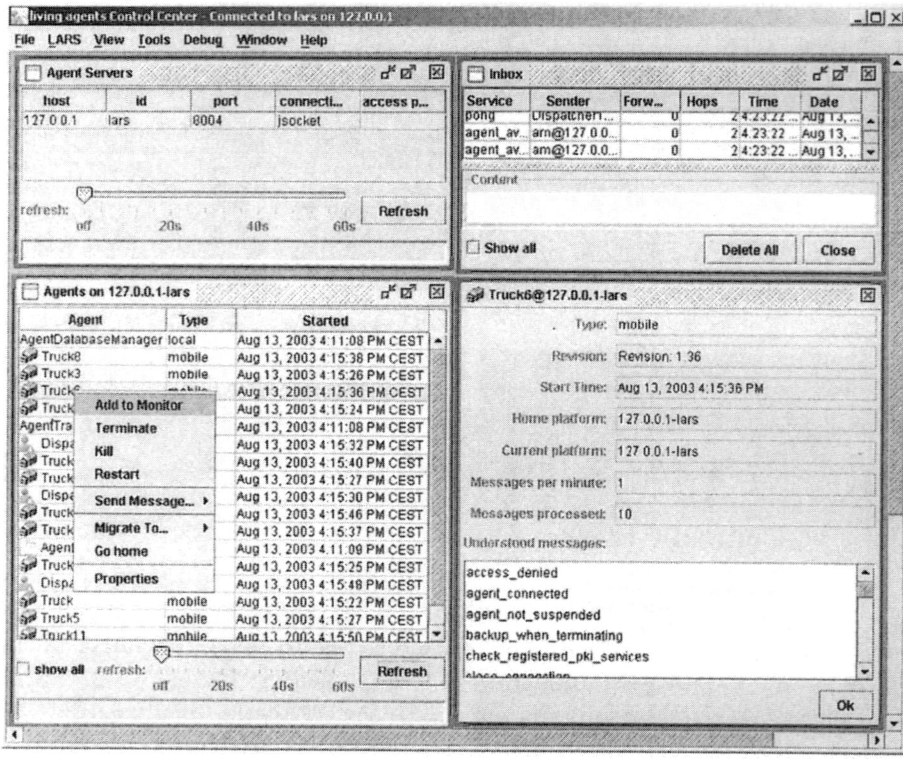

Figure 16-7. The Living Agents Control Center

3.4 Running the Example Application

As described in section 1 and 2 one advantage of agent-oriented software engineering is its very natural way to design an agent application. Does such a design result in a system that produces the desired results in the dynamic domains it is designed for? In this section we show some results achieved in the example logistics domain. The results described here are not results of the simple design used as illustration in section 2. They are the result of the more complex agent design of the Living Systems ATN software used in an industry application.

Figure 16-8 shows the results gained by running ATN on real data of a big logistics company[1]. The goal was to reduce the overall driven kilometers to transport specified orders and to keep the number of constraint violations (time delays) as low as possible.

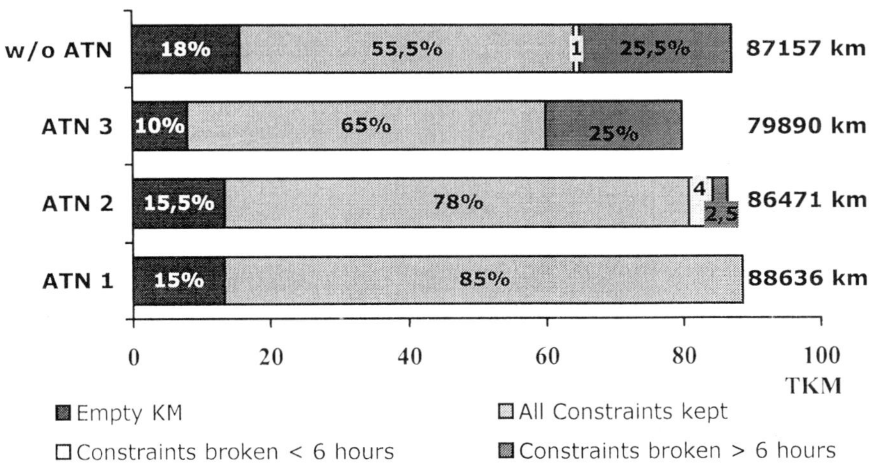

Figure 16-8. Results on Real Logistics Data of Human Dispatchers (w/o ATN), and Running Agent-Based ATN on Same Service Level (ATN 3), Higher Service Level (ATN 2) and Total Fulfillment i.e. No Constraint Violations (ATN 1).

The top bar (w/o ATN) shows the result achieved by a company's dispatcher employees. The first part of the bar represents the empty kilometers for all trucks, the second the kilometers driven with all constraints kept, the third kilometers driven with orders violating pickup or delivery times with less than 6 hours and the fourth with more than 6 hours.

The second bar (ATN 3) shows the result of ATN running on the same service level as the employee dispatcher controlled result. This means those violations of pickup or delivery times for orders are about the same amount when compared to human dispatchers. The overall driven kilometers have been reduced by about 9%. Especially the empty kilometers that is reduced considerably.

The third bar (ATN 2) shows the result when running ATN on a higher service level. It shows that when producing about the same amount of driven kilometers this can be achieved by very much lower amounts of constraint violations, this means delivering more orders in time.

Finally the bottom bar (ATN 1) shows running ATN with complete fulfillment, i.e. no constraint violations allowed picking up and delivering all orders in the specified timeframe. This only produces roughly a 2% higher amount of kilometers having a much higher quality of service.

A big difference to conventional dispatching systems is that the agent-based system is able to deal with real time changes like delays of trucks or new or changed orders. Due to its local and decentralized optimization there is no need to re-plan all orders and trucks. Changes are optimized locally by the affected agents and then spread across the network to get closer to a global optimum. With this, the reaction time of the system to such unforeseen events is between fractions of a second to few seconds.

4. CONCLUSIONS

Experience to date has shown software continues to reach greater levels of complexity. One reason is that it deals with increasingly complicated and dynamic domains. Value networks are an especially demanding example due to their distributed and dynamic nature. The agent paradigm has a number of appealing properties to deal with this. In this paper we have exemplified two of them. Firstly, the design of such agent systems is straightforward, because the agent design maps very closely to the roles and responsibilities of existing value networks in the real world. Secondly, the properties of agents like distributed collaboration, autonomy and goal-directedness exhibit a runtime behavior that is necessary to handle the abstruseness of value networks.

As with other paradigms the success is highly dependent on the availability of tools and infrastructure supporting the user in applying the paradigm. The Living Agents toolset supports an agent developer in the phases of design, deployment, debugging and maintenance of an agent system. This is a necessary condition to run applications in an industrial environment. So far the results obtained from an industrial application of agent technology are encouraging. As described in this chapter a saving of 9% of the overall kilometers of a logistics company is quite a success. Equally valuable is the ability to optimize the logistics network in real time.

This takes the daily exceptions and opportunities into account without a huge manual intervention, as it is done today.

Now agent-based systems have to make the step from successful proofs of concept to smoothly running real-sized value networks. One precondition to achieve this is to define standards for methodologies and processes to design and implement agent-based systems. Another is to establish industry wide standards for the interoperation of agents for which there has been considerable progress. This will eventually reduce a customer's dependency on one specific agent system provider.

NOTES

1. The data was a subset of the complete transport orders in the LTL (less than truckload) business of the company.

REFERENCES

1. Dannegger Ch., Kluge Ch., Katzenberger R. Agents in Real-World Applications: living markets. AgentLink Newsletter 9, 2002.
2. Dorer K. Behavior Networks for Continuous Domains using Situation-Dependent Motivations. Proceedings of the 16th International Joint Conference on Artificial Intelligence (IJCAI'99), 1233-1238, Morgan Kaufmann, Stockholm, 1999.
3. Fritschi C., Dorer K. Agent-oriented software engineering for successful TAC participation. Proceedings of the First International Joint Conference on Autonomous Agents & Multi-Agent Systems, AAMAS 2002, July 15-19, Bologna, Italy, ACM, 2002.
4. Jennings N.R., Wooldridge M. "Agent-Oriented Software Engineering." In *Handbook of Agent Technology,* J. Bradshaw, ed., AAAI/MIT Press, 2001.
5. Kinny D., Georgeff M., Rao A. A Methodology and Modelling Technique for Systems of BDI Agents. Proceedings of the 7th European Workshop on Modelling Autonomous Agents in a Multi-Agent World. Springer: Berlin, 1996.
6. Odell J, Parunak H.V.D., Bauer B. Extending UML for agents. Proceedings of the 2nd International Workshop on Agent-Oriented Information Systems, AOIS 2000, June 5-6, Stockholm, Sweden, iCue Publishing, 2000.
7. Russel S. and Norvig P. *Artificial Intelligence: A Modern Approach*. Prentice Hall, 2nd Edition, 2002.
8. Sycara K., Paolucci M., Van Velsen M., Giampapa J. The RETSINA MAS infrastructure. Autonomous Agents and Multi-Agent Systems 7(1/2), July/September 2003, K. Sycara, M. Wooldridge, eds., Kluwer Academic Publishers, 2003.
9. Wooldridge M., Jennings N.R., Kinny D. The Gaia methodology for agent-oriented analysis and design. Journal of Autonomous Agents and Multi-Agent Systems 3 (3): 285-312, 2000.

Chapter 17

ENABLING VALUE NET COLLABORATION

Grace Lin, Jun Jang Jeng, Ko-Yang Wang
IBM Business Consulting Services, IBM Research, IBM Global Services

Abstract: The concept of Value Net collaboration in which enterprises share values/risks and make business-critical decisions has attracted significant interest. However, lack of trust and support for cross-enterprise integration for collaborative decision-making has so far prevented this vision from becoming reality. This chapter examines the latter issue. Emerging technologies that support Value Net partners to connect their business processes and enable distributed decision making will be discussed. These technologies, and the trust they enable, fuse the boundaries between enterprises and between business and IT. They represent the critical first step to enable the Value Net to be adaptive and sense and respond to environmental changes.

Key words: Adaptive technologies, e-business, Adaptive e-business Integration, Integration technologies, Distributed decision, Distributed control, On demand business, On demand Infrastructure, e-Utility, Value Net, Sense and Respond, Sense and Respond Value Net, Sense and Respond Value Net Optimization, Enterprise Collaboration, Business Ecosystem, Business Activity Monitoring, Fusion of business and IT

1. INTRODUCTION

The business environment is changing at a rapid pace, fueled in part by the e-business evolution and in part by competition and globalization. In this environment, companies must be able to adapt quickly to change if they wish to compete, survive and thrive. Being adaptive requires specific capabilities, not the least of which is the ability to recognize events, the status of customers, partners, competitors, and the global environment.

Further, the capability must exist to analyze a broad variety of situations that will ultimately lead to valuable and productive strategic insights. Flexibility will be demanded, as will be intelligent information and adaptability. Agility will be a vital component of the enterprise of tomorrow if changes are to be managed, challenges to be met and a competitive edge to be retained.

This new model demands that participating enterprises in the Value Net support interconnected and inter-operable business and information systems that have the ability to "plug and play" at both the business process and IT infrastructure levels. This requirement fuses the boundary between business and IT and demands enterprises to acquire capabilities to quickly transform process changes into IT supports to maximize the value of the IT and enable new business capabilities. The virtual enterprise requires information sharing, process inter-operability, and coordinated integration and collaboration across not just its own organization, but also across all members of the Value Net. Through information integration, the partners in the Value Net selectively share critical information and knowledge, such as filtered demand and supply data, coordinated forecasting, and other key factors that enhance the enterprise and make it more productive and efficient. As a consequence, they gain better insight into critical events. The technology that enables this Value Net interaction is called *flexible enterprise integration*.

The Value Net partners can improve their supply chain performance by taking advantage of the deep collaboration that is enabled by *flexible e-business integration* to engage in a level of collective decision-making that maximizes the performance of the whole Value Net (Strozniak 02). This collaborative optimization involves the selective sharing of critical events and information, the distribution of decision rights, the redistribution of work, and the better allocation and use of all the resources available in the Value Net.

2. TECHNOLOGIES FOR FUSING BUSINESS AND IT

Technologies to fuse business and IT enable enterprises to respond to environmental changes by adjusting their policies or operational models and by quickly deploying the solutions supporting these changes. These technologies can be grouped into the following four categories:

1. **Right-time intelligent information:** Getting critical information when it is needed, while it is needed, in the form that it is needed. Maximize the use of technologies that utilize real-time predictive and proactive analysis

and formal/informal collaboration that will improve value chain performance;

2. **Value Net optimization**: The analytic and distributed decision framework that companies need in order to optimize not only their internal supply chain, but also to collaboratively optimize the performance of the Value Net. This will enable the sharing of risks and rewards between the Value Net partners;
3. **Adaptive enterprise integration**: Technologies that enable enterprises to connect and reconfigure themselves quickly, and at a lower cost, to support the integration of processes, data and systems within and between enterprises;
4. **Business process automation technologies**: technologies to quickly transform operational business specification into code or code components. This is an emerging area that will not be covered in this chapter.

2.1 Right Time Intelligent Information

The effectiveness of a Value Net depends on its ability to recognize critical business events "in time" and before competitors (Havensteirn 2003, Havensteirn 1992). This information can be gathered by sensors that are planted or configured to probe operational systems to recognize data and events of interest in business processes, inventory systems, and on the manufacturing floor. Trends and events are also generated by processing and mining both structured and non-structured data, such as text, from across the Internet or the Value Net. Today's business intelligent information technologies are capable of identifying patterns, discovering associations and uncovering implicit knowledge within a veritable sea of data.

Furthermore, cross-enterprise real-time message systems will alert all involved parties and allow them to engage in multi-party communication in real time. These interactions can be pre-programmed into the operational process and designed to ensure that collaboration happens at the right time and in the right way. This will help to ensure that the tasks associated with, or which are triggered by, the collaboration will be quickly carried out at all the involved enterprises in a manner that enables the quick resumption of workflows. This level of just-in-time collaboration, which involves all key enterprise decision-makers, will enable the Value Net to significantly improve its responsiveness.

2.2 Value Net Optimization

Value Net is a promising business model that enables enterprises to quickly acquire new or critical resources and competencies at minimal cost. Companies in the Value Net can pool their resources and capabilities to create new capabilities or offerings that would individual enterprises might otherwise find unaffordable. This model also allows Value Net members to bring the enhanced capability of the whole Value Net, rather than just a piece of the enterprise, to their clients! Furthermore, small companies can join forces to create a formidable virtual enterprise that is capable of competing with much larger companies that might otherwise dominate them.

The Value Net optimization design also enables participating businesses to leverage each other's strengths and to collaboratively achieve higher overall performance levels. It gives the enterprise partners an opportunity to increase the speed and flexibility of their response and enables a connection with their partners for collaborative execution. As a consequence, the entire Value Net can respond to environmental challenges in a much better and more effective way by utilizing shared capabilities and resources.

2.3 Flexible Enterprise Integration

2.3.1 Rapid Infrastructure and Application Integration

Today, infrastructure and application integration is a costly and time-consuming effort (Robertson 2002). Integrating a package application into any enterprise environment can cost millions dollars and account for major application implementation delays and customer dissatisfaction. This area has been the focus of significant technology development in recent years. On demand, Web Services, Grid Computing, Autonomic Computing, Legacy Revitalization, and Integration Technologies are all being vigorously pursued by technology companies. Their objective is to create capabilities that will be able to quickly integrate IT systems with other parts of the total "package" that will, in turn, enable lower integration and system integration costs, introduce a new level of flexibility, and which will enhance resource allocation and utilization. Rapid infrastructure and application integration technologies will enable enterprises to quickly integrate or utilize data, applications and infrastructures based on an open, integrated, autonomic and flexible framework. It is a critical foundation of the on demand business.

2.3.2 Flexible Process Integration

One key to the success of the adaptive enterprise is a business design that will allow operations be created and managed on a common enterprise business design model. This encourages a disciplined process specification, which makes the operational management system easier to automate. The common semantics support of the process, process interfacing points and the shared understanding provided by a disciplined approach accelerates the planning and design phases of process improvement. This also greatly simplifies the implementation of applications to support the process by enabling the generation or re-use of run-time process components, which are based on operational specifications. The rapid support of business process and integration enables the enterprise to respond to change rapidly.

We will discuss more details on the Adaptive Enterprise integration in Section 4.

2.4 Business Process Automation

Business Process Automation technologies enable enterprise to quickly implement the business design changes. This requires the enterprise to rethink how the business operations are defined, how the governance models are setup, and how the processes are executed. Most efforts divide the process models into multiple levels (for example, strategy and measure, operational model, execution, and infrastructure) and develop automatic mappings among the levels. The success of the business process automation depends on a common standard for operational business design and business artifacts. The process standards being developed include: OMG's (www.omg.org) UML based Business Process Definition model, WfMC's (www.wfmc.org) workflow standards, OASIS's (oasis-open.org) ebXML BPSS and BPEL, BPMI's (bpmi.org) BLML, W3C (w3c.org) Choreography and RosettaNet PIPs (Partner Interface Processes). These standards vary in the representation levels and coverage, and most lack the capability to address business artifacts as the first class objects in the model.

In section 3 and 4, we will focus on the Value Net Optimization and Flexible Enterprise Integration.

3. VALUE NET OPTIMIZATION

All enterprises engaged in business activities with other enterprises are implicitly participating in the Value Net, whether they realize it or not. The critical business issue is whether they take advantage of the collective wisdom and resources of the Value Net to improve operational efficiency. The benefit of Value Net optimization is obvious, but it does require a collaborative effort to fully realize its worth. First, Value Net participants need to build the trust than encourages open, effective collaboration and the willingness to share selected, filtered business events and knowledge. More importantly, they need to participate in a collaborative decision- making process that is designed to optimize the performance of the Value Net. This requires a distributed decision support system that is built on top of the adaptive enterprise integration technology. It also requires the trust and discipline to become engaged at the right time. Below, we will discuss an adaptive, distributed decision support model that was first reported on in 2002 (Lin et al. 2002) called the "Sense and Respond Value Net Optimization" model.

This Sense and Respond Value Net Optimization model is a hybrid of the "Sense and Respond" organization change model described in Haeckel (Hacekel 1999) and the "Make & Sell" optimization model. Under the "Make and Sell" model, the enterprises optimized their businesses based on demand forecasts and production-planning (for example, a city bus system). Under the "Sense and Respond" model, organizations do not attempt to predict future demands, but rather seek to identify changing customer needs and new business challenges as they happen. This enables them to respond to them quickly and appropriately before they disappear or metamorphose into something else (for example, the taxi dispatching system, as opposed to the city bus system) (Haeckel 1999). The hybrid Value Net Optimization model we proposed expands the Sense and Respond model to Value Nets and utilizes supply chain technologies for capability and resource planning. It also used the adaptive organizational model for the flexible allocation of resources that would enable optimal decisions through collective efforts. Using the analogy of the bus and taxi system, it can therefore easily be seen that any city needs the economy and efficiency of a preplanned mass transit system and the dynamism of the taxi network. The collaborative decision support system enables the Value Net to utilize the best of both models.

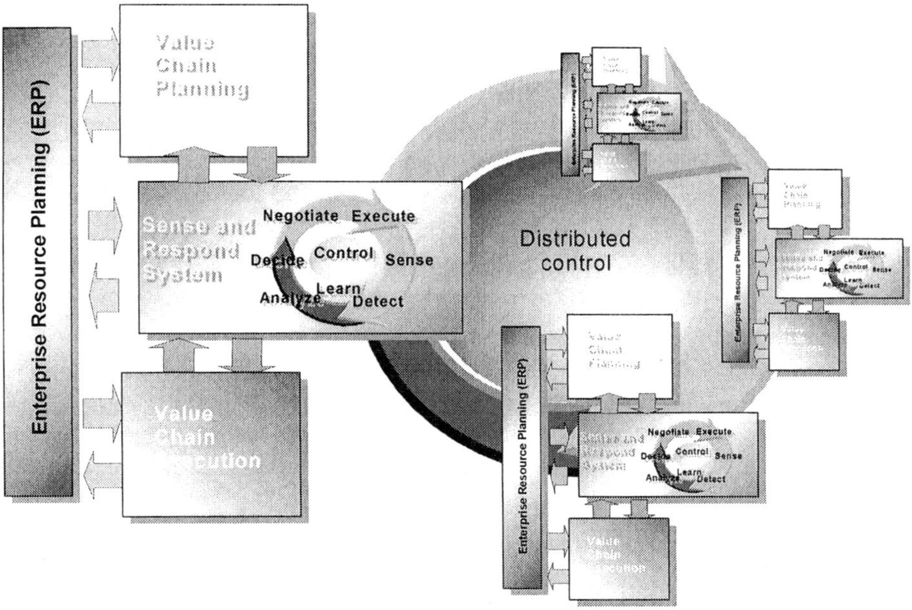

Figure 17-1. The Sense and Respond Value Net Optimization Control Model

As seen in figure 17-1, each Sense and Respond business unit is supported by one or more Sense and Respond control agents. These control agents interact with other agents in the Value Net (either inside or from outside their own enterprise) in a collaborative effort to optimize the performance of their own units, and of the whole Value Net.

These agents have the following capabilities:

- *Sense: utilize sensors embedded in the system or the Internet, to monitor the status of portfolios, skill deployments, inventory levels, supplier performance, sourcing, allocation, replenishment policies, etc.;*
- *Detect: analyze the raw data to recognize trends or events and to identify the causes of alerts;*
- *Analyze: perform risk assessment or a "what-if" analysis to understand risks and to explore alternatives and their potential impact;*
- *Decide: make strategic or operational decisions based on analysis and events. For example, the dynamic assignment of tasks, reallocation of resources and skills, procurement and order fulfillment;*
- *Negotiate: engage related partners in negotiation based upon decisions and the results of analysis. Utilize cross-enterprise Value Net optimization analytical tools to come to an agreement on universal optimal solutions;*
- *Execute: collaboratively carry out operations and measure results;*
- *Learn: Analyze the measured results and adjust the rules and policies as needed to improve the knowledge base and intelligence.*

It is important to note that Sense and Respond Value Net Optimization agents are not designed to replace humans in the decision-making process. Rather, they are the decision support tools that engage human as needed and provide critical information "just–in-time" to allow the decision makers to offer optimal solutions across the entire Value Net.

4. FLEXIBLE ENTERPRISE INTEGRATION

The implementation of the Sense and Respond Value Net Optimization system relies on the ability to link the data, application, infrastructure, processes, values and service to all enterprises involved in the Value Net. A foundation of this support is provided by Flexible Enterprise Integration technologies. Here we present an example that implements some of the critical functions needed to support the enterprise integration. This model can be used for both internal integration and cross-enterprise integration, so long as the interface points are well defined.

A flexible infrastructure is aimed at providing an integrated platform for the enterprises in the value net at the business and technical levels (Fowler, 2002). The flexible infrastructure can be logically categorized into several tiers as shown in following diagram (Figure 17-2):

a) *Business Platform* represents an organizational concept that groups individual component technologies into technical domains (layers), where business stands for Flexible Enterprise;
b) *Business Services* are infrastructure applications that shift responsibility for certain services out of the application domain into the infrastructure. Business services provide a set of physically shared components, such as credit card processing services and application hosting, which multiple applications can share;
c) *Business Commitments* are concepts that manifest the level of obligation between parties in the enterprise. An adaptive enterprise should allow business commitments to be externalized within specific parameters, so that they can be configured;
d) *Business Capability Patterns* are organized concepts that facilitate rapid mapping from client business requirements to the capabilities that can be provided via end-to-end infrastructure design. Business capability patterns structure the system component selection from many platform layers;
e) *Business Platform Adaptors* are organized concepts that enable the accessibility, usability and manageability of technology components within the platforms;

f) *Business Capability Adaptors* are organized concepts that enable the accessibility, usability and manageability of business patterns, services and commitments. The goal in using them is to provide capability interfaces to clients;
g) *Business Clients* are an organized concept that refers to either applications or users that access, use or manage the aforementioned artifacts.

We describe the above concepts in detail as follows.

Business Platform

There could be several ways of organizing the technologies for a business platform; (1) By technology similarity; (2) By architecture domain; (3) By corporate initiatives; (4) By business process; and (5) By support group. (See, Robertson 2002)

At least three approaches to build a Business Platform:

- *Physical. All components dealing with the tasks of physical connectivity, persistence and computation, including routers, mobile phones, laptops, desktops, servers, disks, etc;*
- *Functional. All components involved in data processing, logical storage, data exchanges, transformation and workflow. Examples include operating systems, relational databases, messaging bus, application servers, and integration servers;*
- *Interface. The components providing human-to-machine interaction, such as interaction voice responders, graphical user interfaces, or machine-to-machine interaction such as application programming interfaces.*

Business Capability Patterns

Business capability patterns and platforms are fundamental components of the overall adaptive infrastructure. Patterns are the information, insight, and experience; the "what is" and "how to" that are common to an existing class of application or business processes. This information is captured in a form that makes it easy to re-use with future applications of the same class. Re-usability is the key to making patterns effective. Patterns come in many forms: application framework, specifications, template and asset libraries. Pattern matching is necessary for developing the readiness of business capability pattern libraries and for their use.

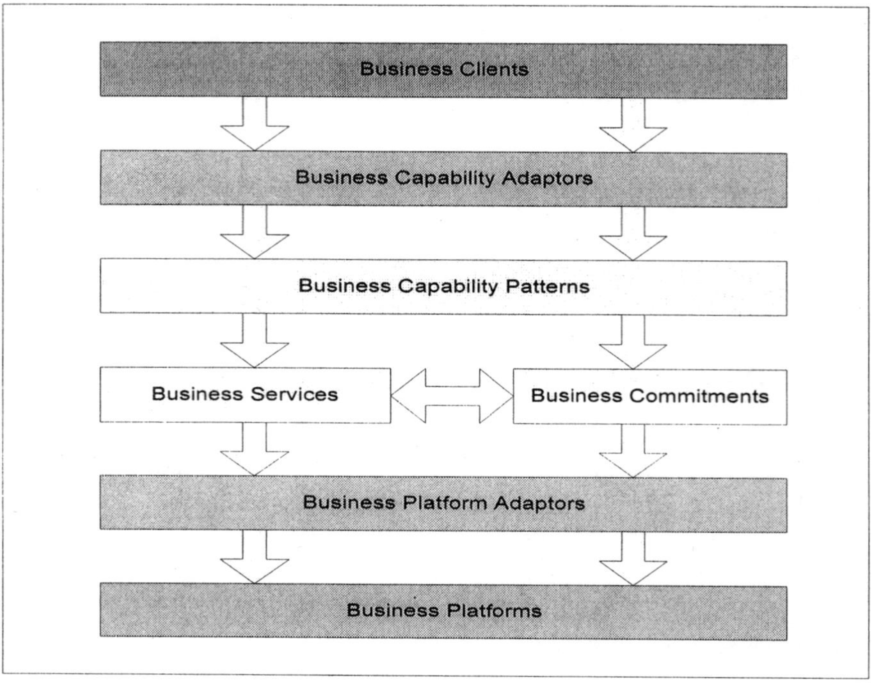

Figure 17-2. Business Integration Stack

Business Services

Business services are distinguished from the physical infrastructure by several features. They are:

1. Shared by the structures they support;
2. More static and permanent than the structure they support; and;
3. Under different ownership, they have their own life cycles (plan, build, run, modify, retire).

Examples of business services include network servers and credit card verification.

The most important work in developing an adaptive infrastructure is to ensure that key business services with a potential for re-use are shifted out of the application domain and into the infrastructure. To maximize the re-usability of the adaptive infrastructure, a high degree of de-coupling between the lifecycles of both business services and the applications at all stages of the life cycle is required, including design, build, deploy and upgrade. A notable example of de-coupling is in Web Services architecture, such as in the interface of Web Services hard codes with Simple Object Access

Protocol (SOAP) versus HTTP. This method involves very simple protocols. Universal Description, Discovery and Integration publish/find/bind will be a standard and is becoming a staple. Web Services Description Language (WSDL) de-couples the specification and implementation of business servers.

Business Commitment

A business commitment is an agreement or pledge to do something in the future. Business commitments are broadly defined as commitments related to issues. At the run-time, if the agreements or guarantees have been violated, one business party promises to take action, either to notify related parties or to remedy the violation. These parties can be internal agents within a business enterprise, or external ones like business partners, service requestors, or service providers. Business commitments exist in many forms and may be business contracts between partners, service level agreements (SLA) between service providers and service requestors, or an internal SLA specified by departments within an enterprise. A contract or SLA contains the terms and conditions to which all parties have mutually agreed. A business commitment is usually directional, meaning a commitment that has one initiator and one receiver. For example, in a shipping SLA, a carrier promises the service requestor to ship goods within X days. On the other hand, a service requestor promises the carrier to pay the bill within Y days. During the execution process, commitments may be violated for a variety of reasons, such as unforeseeable events or a change in the situation of one party or the other. It is therefore common for both parties to agree to take certain actions when mutual commitments are unfulfilled. Business commitments need to be monitored and violations have to be detected. Otherwise, commitments do not have the binding power to keep parties together.

A commitment consists of the following entities:

- *Actions* that it will perform and resources (data) that are required;
- *Resources* that are governed by the commitment;
- *Expectations* that the commitment will hold and recognition that each expectation is composed of S*ituations;*
- *Reponses* that bind actions with expectations;
- *Triggers* that initialize the evaluation of expectations.

To execute an action is to perform it; to execute an expectation is to assert it. A response will be formed in a commitment through the binding of expectations and actions. Examples of triggers include the arrival of events, timed events, managed resources status changes, data conditions etc. Some meta-actions can be defined to perform actions on the commitments themselves. One type of meta-action involves a commitment to an agent. This is usually a future commitment, but it can also be used to establish a past expectation. Another type of meta-action is to put in place an expectation relationship between agents. For example, an ATP agent *expects* an inventory agent to have an inventory level less than a certain quantified value. If the expectation is violated, action will be taken based upon the corresponding *response* that is defined by the ATP agent. A commitment can be applied to several agents, i.e., those agents *share* the same commitment. We can thus define the *locus of control* for the commitment "***C***" by defining a set containing the agents committing to "***C***" and the resources governed by "***C***". In an adaptive infrastructure, it is assumed that an agent will always communicate its commitments truthfully to other agents in response to queries and actions

Adaptors (both Business Capability and Platform)

Business capability adaptors provide connectivity from business clients to business services. Adaptors enable an application to connect to a component, perform transactions, and access services provided by that resource. A connection can be either transactional or non-transactional:

1. Platform event notifications, in which adaptors for specific managed resources pass to management beans, and;
2. Request processing, in which adaptors perform on behalf of management beans. The most promising Java-based connection architecture is the J2EE Connector Architecture (JCA) that defines the connectivity between an Application Server and Enterprise Information Systems (EISs). Application servers extend their containers and support connectivity to heterogeneous EISs based on JCA. On the EIS side, resource vendors provide a JCA-compliant managed resource connector for its EIS. A business connector can be plugged into an application server. The business connector and application server then collaborate to provide the underlying mechanisms -- transactions, security, and connection pooling. Adaptors are used within the same address space as their application server.

The adaptive enterprise integration technology is critical in supporting an enterprise's ability to rapidly enable the adaptive capabilities to change its operation models and rules at minimal cost,. It also allows business relationships to be quickly built through the linking of related processes to Value Net partners. The flexible enterprise integration technology is a necessary condition for value net optimization.

5. SUMMARY

Technologies for Fusion of Business and IT are integral to Value Net enterprises. They enable them to be agile and responsive. Both the Sense and Respond Value Net Optimization model and the Flexible Enterprise Integration technology are built on a number of advanced technologies. Some of these already exist; others are now emerging; and still others need further development. These are the "game changers" that will fundamentally change how enterprises will conduct business in the next two to 10 years (Lin et al. 2003). They are deserving of expanded studies.

Additionally, the Sense and Respond Value Net represents a significant change to the way in which business is conducted and values are managed. It is an emerging enterprise transformation that also deserves more studies, as it will take more than technology to achieve realization. It will take, at the very least, a new and more dynamic level of cooperation, collaboration and trust.

Regulatory bodies like national and multi-national governments, as well as other members of global economic and cultural environments, are extended Value Net participants. Their potential impact on the enterprises of the future needs to be explored in significantly more detail so that they will better understood and adequate preparations can be made to meet the challenges that are sure to come. It is a fundamental fact that we have a strong need to continue to explore these enterprises as partners in a greater business ecosystem that interacts, competes, co-exists and co-evolves into a true and trusted Value Net partnership.

ACKNOWLEDGEMENTS

The authors would like to acknowledge the contributions of the following colleagues who had either contributed to some of the work discussed in this paper or have helped to improve the presentation: Alexander Macaulay, Patrick Baker, Santhosh Kumaran, Roger Gung, Heng Cao, Markus Ettl, Shubir Kupoor, Steve Buckley, Henry Chang, Nathan Caswell, Anil Nigam, Patt Romero Cronin, and Jen-Yao Chung.

REFERENCES

1. April, C., BAM to Speed App Reports - Industry Heavyweights Target Business Process Performance, InfoWorld, 2002; 24: 44 (April): 27.
2. Fowler, M., *Patterns of Enterprise Application Architecture*, Addison-Wesley, 2002.

3. Haeckel S., and Slywotsky, A. J., *Adaptive Enterprise: Creating and Leading Sense-And-Respond Organizations*, Harvard Business School Press, 1999.
4. Havenstein, H., BAM bolsters data visibility. InfoWorld 2003; May (25)
5. Lin, G., Buckley, S., Cao, H., Caswell, N., Ettl, M., Kapoor, S., Koenig. L., Katircioglu, K., Nigam, A., Ramachandran, B., Wang, K. The new frontier: Sense and respond system for global value chain optimization OR MS Today, April, 2002.
6. Lin, G., et. al. The Adaptive Technologies Game Changers, IBM Enterprise of the Future white paper. 2003.
7. Robertson B., and Sribar, V. *The Adaptive Enterprise: IT Infrastructure Strategies to Manage Change and Enable Growth*, Addison-Wesley, 2002.
8. Strozniak, P. Exception management: let SCEM turn invisible supply chain problems into profit opportunities and improved service. Frontline Solutions 2002; 3: 8: 16.

Chapter 18

SUPPLY CHAIN MANAGEMENT WITH VIRTUAL MARKET IN ICT ENVIRONMENT

Tosshiya Kaihara
Graduate School of Science and Technolog,KOBE University

Abstract: Supply chain management (SCM) is recognized as one of the best means with ICT by which enterprises can make instant improvements to their business strategies and operations with ICT. This paper gives an overview of the virtual market approach with economic software agents and its applicability to SCM. Focusing on application areas this paper not only considers the theoretical research domain, but also proposes a new business model with a virtual market for SCM in an ICT environment. The paper clarifies that judicious constructions of the decision process, according to economic principles, can lead to an efficient distributed resource allocation in SCM, and also importantly the behavior of the system can be analyzed in economic terms.

Key words: Supply chain management, Multi-agent, Virtual market, Micro economics, B2B e-commerce

1. INTRODUCTION

A supply chain is a network of facilities and distribution options that performs the functions of procurement of materials, transformation of these materials into intermediate and finished products, and the distribution of these finished products to customers. Supply chains exist in both service and manufacturing organizations, although the complexity of the chain may vary greatly from industry to industry and firm to firm. Realistic supply chains have multiple end products with shared components, facilities and capacities. The flow of materials is not always along an arborescent network, various modes of transportation may be considered, and the bill of materials for the end items may be both deep and large.

The management of the physical flow of products amongst the nodes of the supply chain comes under the intensive study for effective operation in management of the supply chain management (SCM) (Fisher 1994). Since supply chains consist of several layers of business units within an ICT environment, resource allocation is quite an important operational criterion at a workshop level in SCM. As the number of potential business units in the supply chain increases, an effective management of product distribution (i.e. multi-echelon optimization) plays a more important role in the dynamic environment (Perrone et al. 1999).

Traditionally, marketing, distribution, planning, manufacturing, and the purchasing organizations along the supply chain have operated independently. These organizations have their own objectives that are often in conflict. Marketing's objectives of high customer service and maximum sales, conflict with manufacturing and distribution goals. Many manufacturing operations are designed to maximize throughput and lower costs with little consideration for the impact on inventory levels and distribution capabilities. Purchasing contracts are often negotiated with very little information beyond historical buying patterns. The result of these factors is that there is not a single integrated plan for the overall organization - there are as many plans as businesses. Clearly, there is a need for a mechanism through which these different functions can be integrated. SCM is a strategy through which such integration can be achieved.

SCM is typically viewed to lie between fully vertically integrated firms, where the entire material flow is owned by a single firm and those where each channel member operates independently. Therefore coordination between the various players in the chain is the key to its effective management. SCM, however, is conventionally based on the simple TOC (Theory of Constraints) concept (Goldratt and Cox 1992), and is not always concerned with optimal solutions for resource allocation in such complex supply flows within an ICT environment.

Currently the use of multi-agent systems for large-sized complex systems is increasing (Shen et al. 2001, Hannebauer 2002). The multi-agent paradigm has several characteristics, such as autonomy, pro-activeness, social ability, and emergence. In this paradigm, a global goal for the whole system is achieved as a negotiated aggregation of local objectives (Kaihara and Fujii 1998). In supply chain networks each business unit behaves independently and autonomously with simple goals for achieving local optima. The situation is quite similar to the distributed decision making mechanism in the multi-agent paradigm, so it is natural to model supply chain networks through multi-agent programming (Kaihara 1999). In such an environment each agent represents an independent business unit that has conflicting and competing individual requirements, and also may possess some localized information relevant to their utilities. To recognize this independence, we treat the business units as agents, allowing each of them to decide autonomously how to deploy the resources under their control to service their interests.

Within this model, a distributed SCM can be analyzed according to the following properties:

- Self-interest agents can make effective decisions with local information, without knowing the private information and strategies of other agents;
- The method requires a minimal communication overhead;
- Solutions do not waste resources. If there is some way to make some agent better off without harming the others, it should also be done. A solution that cannot be improved in this way is called the Pareto optimal.

Assuming that a resource allocation problem in SCM must be decentralized when considering a practical application, market concepts can provide several advantages:

a) Markets are naturally distributed and agents make their own decisions about how to bid based on prices and their own utilities for the goods;
b) Communication is limited to the exchange of bids and the process between agents as well as the market mechanism. In particular settings, it can be shown that price conventions minimize the dimensionality of messages required to determine the Pareto optimal allocation;
c) Since agents must back their representations with exchange offers, some mechanism, in some well-categorized situations, can elicit the information necessary to achieve Pareto and system optima.

Market-oriented programming is a multi-agent-based concept to facilitate distributed problem solving (Wellman 1996). In market-oriented programming we take the metaphor of an economy computing multi-agent behavior literally, and as a consequence directly implement distributed computation as a market price system. In the market-oriented programming approach to distributed problem solving, computing a competitive market for an artificial economy derives the resource allocation for a set of computational agents (Kaihara 2001).

First of all, we formulate a supply chain model as a discrete resource allocation problem with supply/demand agents, and demonstrate by simulation experiments the applicability of an economic analysis to this framework. Then we prove that the market mechanism can provide several advantages on resource allocation in SCM. Needless to say, the term 'resource allocation' in this chapter corresponds to 'product distribution' at shop floor level in practical SCM. In market-oriented programming we can also take the metaphor of an economy computing multi agent behavior literally, and directly implement the distributed computation as a market price system. A business application of our virtual market is given in the second half of this chapter. We propose a business model that includes a B2B e-Marketplace server with the virtual market that mediates amongst an unspecified range of relevant companies to demonstrate the applicability of an economic analysis to this framework (Kaihara 2000, Kaihara 2003). The proposed server is expected to facilitate a sophisticated e-Marketplace, which conducts Pareto optimal solutions for all the participating business units, for future ICT environments.

2. AGENT BASED VIRTUAL MARKET

2.1 Market Based Approach

In economics, the concept of a set of interrelated goods in balance is called general equilibrium. The general equilibrium theory guarantees a Pareto optimal solution at competitive equilibrium in a perfect competitive market (Layard and Walters 1978). The connection between computation and general equilibrium is not at all foreign to economists, who often appeal to the metaphor of market systems computing the activities of the agents involved (Shoven and Whalley 1992). The theory of general equilibrium provides the foundation for a general approach to the construction of a distributed planning system based on a price mechanism. In this approach the constituent planning agents are regarded as suppliers and demanders in an artificial economy. Their individual activities are defined both in terms of production and consumption of resources. Interactions amongst the agents are cast as exchanges, the terms of which are mediated by the underlying economic mechanism, or protocol.

2.2 Market-Oriented Programming

Market-oriented programming is the general approach for deriving solutions to distributed resource allocation problems by computing the competitive equilibrium of an artificial economy. It involves an iterative adjustment of prices based on the reactions of an agent in the market. The general concept of the negotiation mechanism in market-oriented programming is shown in figure 18-1, and for a detailed explanation of market-oriented programming refer to (Wellman 1996). Supply/demand functions represent an agent's willingness to sell/buy resources, respectively. They are defined as the relationship between price and quantity of the trading resource. Let *Pt(s)* be the price of resource *s* at time *t*. $\alpha_t ms$ and $\beta_t ns$ represent the supply function of supplier m on resource s at time t and the demand function of demander n on resource s at time *t*, respectively. The bidding mechanism computes an equilibrium price in each separate market. It involves an iterative adjustment of prices based on the reactions of agents in the market. Agents submit supply and demand functions ($\alpha_t ms$ and $\beta_t ns$) and the auction adjusts individual prices to clear, rather than adjusting the entire price vector by some increment. The mechanism associates an auction with each distinct resource. Agents act in the market by submitting bids to auctions. In this chapter bids specify a correspondence between prices and quantities of the resource that the agent offers to demand or supply as a basic study. Given bids from all interested agents, the auction derives a market-clearing price.

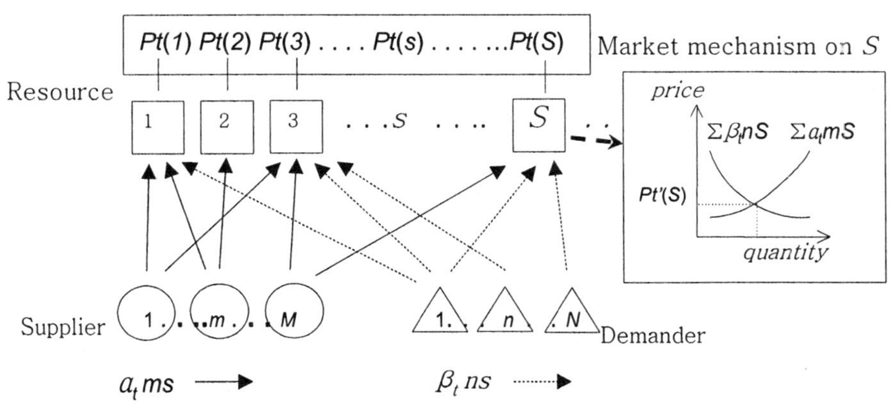

Figure 18-1. Negotiation Mechanism

The algorithm of the proposed market-oriented programming in SCM is shown as follows:

- *Step1*: A demand agent *n* sends bids to a market to indicate its willingness to buy the product *s* according to its current price *Pt(s)* in time *t*. The demand agent willingness is defined as a demand function in the bid message. The agent can send bids to the market within the limits of its domestic budget. Each product has its own market, and they construct a competitive market mechanism as a whole.
- *Step2*: A supply agent *m* sends bids to the market to indicate its willingness to sell the product *s* according to its current price *Pt(s)* in

time t. The supply agent willingness is defined as a supply function in the bid message. The agent can send bids to the market within the limits of its current inventory level.

- *Step3*: The market in product s sums up supply functions ($\Sigma \alpha_t ms$) and demand functions ($\Sigma \beta_t ns$), then revises balanced price $Pt'(s)$ of product s in time t, shown as "market mechanism on S" in figure 18-1. All the market must revise their balanced price via the same process.
- *Step4*: Check the balanced prices of all the products and if all the prices are fully converged, the acquired set of the prices is regarded as equilibrium price, then go to *Step5*. If not, go to *Step1*.
- *Step5*: If dealing time is up, then stop. And if not, t=t+1 and go to *Step1*.

Each agent maintains an agenda of bid tasks, specifying in which it must update its bid or compute a new one. The bidding process is highly distributed, in that each agent needs to communicate directly only with the auctions for the resources of interest. Each of these interactions concerns only a single resource; the auctions never coordinate with each other. Agents need not negotiate directly with other agents, nor even know of each other's existence. As new bids are received at the auctions, the previously computed clearing price becomes obsolete. Periodically, if any new or updated bids have been received each auction computes a new clearing price, and posts it on the tote board. When a price is updated, this may invalidate some of an agent's outstanding bids, since these were computed under the assumption that prices for remaining resources were fixed at a previous value. On finding out about a price change, an agent adjusts its task agenda to include the potentially affected bids. At all times, the market-oriented mechanism maintains a vector of going prices and quantities that would be exchanged at those prices. While the agents have nonempty bid agendas or the auctions new bids, some or all resources may be in disequilibrium. However, when all auctions clear and all agendas are exhausted the economy is in competitive equilibrium (Layard and Walters 1978, Shoven and Whalley 1992).

2.3 Market-Oriented Programming and SCM

An agent's activities, in terms of products required and supplied, are defined so as to reduce its decision problem for evaluating the tradeoffs of acquiring different products in market-oriented programming. These tradeoffs are represented in market prices, which define a common scale of value across the various products. The problem for designers of computational markets is to specify the strategy by which agent interactions determine prices (Kaihara 2003). Obviously the supply chain model is well structured for market-oriented programming, and that means the proposed concept takes advantage of the theory. A Pareto optimal solution, which is conducted by microeconomics, is attainable in the resource allocation problem in SCM. In the next section, we define several functions that formulate an agent's strategy for the resource allocation problem in SCM. The budget constraint of each agent is also considered for practical use in our definitions.

3. AGENT DEFINITIONS

3.1 Preliminaries

Several variables to formulate agent utilities in this paper are defined as follows:

x_{ki} : Input of resource i in agent k
y_{kj} : Yield of resource j in agent k
p_i : Purchase price of resource i per unit
P_i : Sales price of resource i per unit
E_k : Profit function of agent k
C_k : Cost of agent k
S_k : Sales of agent k
$\max C_k$: Budget of agent k
$l_{(i,j)}(=l)$: Index of production function from resource i to resource j
f_{kl} : lth production function in agent k
x_{kl} : Input resource amount into production function f_{kl}
y_{kl} : Output resource amount from production function f_{kl}
E_{kl} : Profit by production function f_{kl}
c_{kl} : Cost by production function f_{kl}
s_{kl} : Sales by production function f_{kl}

3.2 Production Function

Suppose supply agent k has a production function f_k described in equation (1).

$$Y_k = f_k(X_k) \tag{1}$$

Where, X_k and Y_k denote a set of input resources and yield resources in agent k described in equation (2) and (3), respectively.

$$X_k = \{x_{k1}, \cdots, x_{km}\} \tag{2}$$

$$Y_k = \{y_{k1}, \cdots, y_{kn}\} \tag{3}$$

In this paper we adopt the Cobb-Douglas function (Layard and Walters 1978) as a production function described by equation (4). Since the Cobb-Douglas function handles the economical scale in the market by an index constant b, in the range $0 < b < 1$, the production function is defined as a convex function. In other words, a diminishing returns function. If the production function is defined as convex, the market prices are established at a predictable level in the general equilibrium theory.

$$y = ax^b \qquad (Where,\ \ 0 < a,\ \ 0 < b < 1) \tag{4}$$

Then, the production function f_{kl} of agent k for input-output resource set $l(i, j) = l$ is given by:

$$y_{kl} = f_{kl}(x_{kl}) = a_{kl} x_{kl}^{\ b_{kl}} \tag{5}$$

Where, x_{kl} and y_{kl} denote the amount of input resource i for f_{kl} and the amount of yield resource j for f_{kl}, respectively. Then x_{ki} and y_{ki} are defined as:

$$x_{ki} = \sum_j x_{kl(i,j)} \tag{6}$$

$$y_{kj} = \sum_i y_{kl(i,j)} \tag{7}$$

3.3 Profit Function

Suppose a set of single unit purchase prices for a resource set $\{x_{k1}, \ldots, x_{km}\}$ is $\{p_1, \ldots, p_m\}$, and a set of single unit sales prices for a resource set $\{y_{k1},$

..., $y_{kn}\}$ is $\{P_1, \ldots, P_n\}$, then the total production cost of agent k is defined as:

$$c_{kl(i,j)} = p_i x_{kl} \tag{8}$$

$$C_k = \sum_l c_{kl} \tag{9}$$

and the total sales S_k of agent k is defined as:

$$s_{kl(i,j)} = P_j y_{kl} \tag{10}$$

$$S_k = \sum_l s_{kl} \tag{11}$$

Then the profit function E_k of agent k is finally acquired as:

$$E_{kl} = s_{kl} - c_{kl} \tag{12}$$

$$E_k = \sum_l E_{kl} \tag{13}$$

3.4 Profit Maximize Theorem under Budget Constraint

In this paper the budget constraint of each agent is considered so as to realize our market model. Suppose the maximum budget of agent k is $\max C_k$ then we have

$$C_k = \sum_l c_{kl} \leq_{\max} C_k \tag{14}$$

And, agent k should behave to maximize its profit E_k autonomously.

The basic principle of agents is to maximize their profits under the budget constraints. Their activities should follow the newly proposed theorem, named 'Profit Maximize Theorem', shown below.

[Theorem] Profit function E_k of agent k is maximized by minimized r_k, which satisfies the following conditions:

$$\forall l : \frac{\partial E_k}{\partial c_{kl}} = r_k \left(r_k \geq 0\right) \cap C_k \leq \max C_k \tag{15}$$

subject to,

- f_k is differentiable in any $x \in X_k$

- $$\forall l : \left.\frac{\partial f_{kl}}{\partial x_{kl}}\right|_{x_{kl}=x} > \left.\frac{\partial f_{kl}}{\partial x_{kl}}\right|_{x_{kl}=x+\Delta} \tag{16}$$

We have the following equation (17) by equation (8), (10), (12), (13).

$$\frac{\partial E_k}{\partial c_{kl(i,j)}} = \frac{\partial E_{kl}}{\partial c_{kl}} = \frac{\partial}{\partial c_{kl}}\left(s_{kl} - c_{kl}\right)$$

$$= \frac{\partial}{\partial c_{kl}}\left[P_j f_{kl}\left(c_{kl}/p_i\right) - c_{kl}\right]$$

$$= \frac{P_j}{p_i} f'_{kl}\left(c_{kl}/p_i\right) - 1 \tag{17}$$

The proof of the theorem is given in Appendix.

3.5 Demand / Supply Function Definitions

Since the Cobb-Douglass function shown in (4) is differentiable in which case,

$$\frac{\partial f_{kl}}{\partial x_{kl}} = a_{kl} b_{kl} x^{b_{kl}-1} > a_{kl} b_{kl} \left(x + \Delta\right)^{b_{kl}-1} > 0 \tag{18}$$

the proposed product function (5) perfectly satisfies the conditions (16). Demand function x_{kl}, which maximizes the agent's profit, is obtained by the Profit Maximize Theorem as follows:

$$\frac{\partial E_k}{\partial x_{kl(i,j)}} = \frac{\partial E_{kl}}{\partial x_{kl}} = \frac{\partial}{\partial x_{kl}}\left[P_j a_{kl}(c_{kl}/p_i)^{b_{kl}} - c_{kl}\right]$$

$$= a_{kl} b_{kl} P_j c_{kl}{}^{b_{kl}-1} p_i{}^{-b_{kl}} - 1$$

$$= a_{kl} b_{kl} P_j p_i{}^{-1} x_{kl}{}^{b_{kl}-1} - 1 = r_k \qquad (19)$$

Then we have,

$$x_{kl(i,j)} = \left[p_i(r_k+1)/a_{kl}b_{kl}P_j\right]^{1/b_{kl}-1} \qquad (20)$$

Supply function y_{kl}, which maximizes the profit, is also obtained by equation (5), (20) as follows:

$$y_{kl(i,j)} = a_{kl}\left[p_i(r_k+1)/a_{kl}b_{kl}P_j\right]^{b_{kl}/b_{kl}-1} \qquad (21)$$

We denote a concrete meaning for the Profit Maximize Theorem. It is obvious to maximize the production function f_{kl} at $r_k = 0$ by (19), because the function is defined as convex type. Then agent k has the maximum profit at $r_k = 0$, if it satisfies the budget constraint $C_k \leq \max C_k$ leads to maximize the profit function E_k.

3.6 Agent Utility: Price Elasticity

Generally the influence of demand factors into the demand is called 'elasticity' in economics (Layard and Walters 1978). Price elasticity, described in equation (22), is one of the major factors that control economic dynamics

$$\left|(dx/dp)\times(p/x)\right| \qquad (22)$$

In our market model, price elasticity, which characterizes the demand function, represents agent utility for purchasing resources.

$$(\text{Price Elasticity})_{kl} = \left|(dx_{kl}/dp_i)\times(p_i/x_{kl}) = 1/(b_{kl}-1)\right| \tag{23}$$

Agent demand utility depends on b_{kl} , and that means agent demand activity increases, as the price elasticity has a greater value in $0< b<1$ (refer (4)).

Suppose $R_k = r_k + 1$, then we have R_k elasticity as follows:

$$\left|(dx_{kl}/dR_k)\times(R_k/x_{kl})\right| = \left|1/(b_{kl}-1)\right| \tag{24}$$

By the comparison between (23) and (24), the reduction rate of input resource x_{kl} in the budget constraint depends on b_{kl}. Additionally budget change increases the amount of demand, as the value b_{kl} increases.

3.7 Market-Oriented Programming in SCM Model

In market-oriented programming we take the metaphor of an economy computing multi agent behavior literally, and directly implement the distributed computation as a market price system. The algorithm of the proposed market-oriented programming in SCM is shown as follows:

- *Step1*: Set initial price pi for all the resources.
- *Step2*: Agent k calculates x_{kl} by (20) assumed r_k=0, then computes C_k by (8), (9). If C_k > max C_k then go to Step 3, otherwise go to *Step4*.
- *Step3*: Modify rk followed by the Profit Maximize Theorem (Reduce r_k to satisfy C_k =max C_k).
- *Step4*: Define current demand / supply functions with r_k by (20), (21).
- *Step5*: Agent k sends the acquired demand / supply function as bids into the market to indicate its willingness to buy / sell resources.
- *Step6*: Market mechanism calculates the balanced prices of all resources in the competitive market.
- *Step7*: If all the balanced prices are sufficiently converged, then go to *Step 8*, otherwise go to *Step2*.
- *Step8*: Allocate all the resources under the acquired equilibrium prices.

4. EVALUATED DYNAMISM OF VIRTUAL MARKET

4.1 Experimental Simulation Model

A basic SCM simulation model shown in figure 18-2 is prepared to investigate the validity of the proposed approach by computer simulation. The model has a series of a three-layered market structure with two-layered agent groups. This model comprises the three types of agent in each layer and three types of goods. The interconnectedness of agents and goods defines the market configuration. Comparative analysis of the three market structures reveals distinct qualitative economic and computational behaviors realized by the proposed configurations.

Each agent has production functions to transform the resource from market ($M[i][j]$) to market ($M[i+1][j]$) , and the parameters are defined as $a[j]$, $b[j]$ in (4). The parameters in each agent group are described in table 18-1. The parameter b is set in common for each type of goods, because this parameter is deeply concerned with price elasticity of the goods, shown in (23). In this figure the outside demand function and the outside supply function, which are corresponding to sink and source in the experimental model, are defined respectively as,

$$x_i = \alpha_i {p_i}^{\beta_i} (\alpha_i > 0, \quad \beta_i < -1) \tag{25}$$

$$y_j = \alpha_j {P_j}^{\beta_j} (\alpha_j > 0, \quad \beta_j > 0) \tag{26}$$

and each parameter in (25), (26) is described in table 18-2.

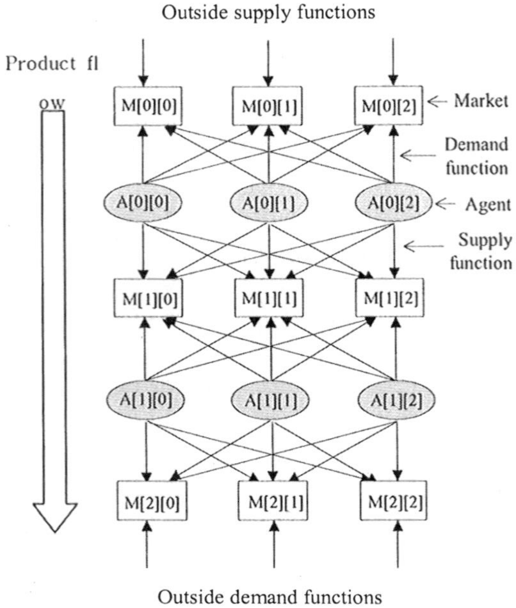

Figure 18-2. Experimental Basic SCM Model

Table 18-1. Production Function Parameters of Agents

———	A[0]	b[0]	a[1]	b[1]	a[2]	b[2]	Budget
Agent[0][0]	5	0.7	5	0.5	5	0.3	10
Agent[0][1]	3	0.7	8	0.5	3	0.3	10
Agent[0][2]	7	0.7	7	0.5	7	0.3	4
Agent[1][0]	10	0.3	10	0.6	10	0.4	50
Agent[1][1]	4	0.3	4	0.6	4	0.4	100
Agent[1][2]	8	0.3	4	0.6	4	0.4	6

Table 18-2. Outside Production Function Parameters

———	[0]	[0]	[1]	[1]	[2]	[2]
Supply function	100	1.5	100	1.5	100	1.5
Demand function	100	-2	100	-2	100	-2

4.2 Market Dynamism and Price Elasticity

Firstly, the dynamical changes in the amount of dealing goods for supply and demand, and secondly the prices of the goods, at each layer in the market structure are shown in figure 18-3 and figure 18-4, respectively.

Looking at these figures, it is obvious that both the amount of dealings and prices are converged into equilibrium. Since our methodology is perfectly endorsed by the 'general equilibrium theory' in a competitive market, we can get a Pareto optimal solution in the equilibrium. This means that the goods distribution policy followed by the acquired solutions, i.e. the amounts and the prices, are Pareto optimal in the entire market. An efficient SCM with a market mechanism is attainable by the proposed approach.

Furthermore it is observed that the number of iterations required to reach equilibrium appears to rise with the price elasticity. For example, Market [0][0] with 0.7 in price elasticity takes longer time to converge than Market [0][2] with 0.3 in price elasticity in figure 18-3(a). We attribute this to the natural characteristic of price elasticity, formulated in (24), and the experimental values completely agree with our prior estimations described in section 3.6.

Moreover it takes a longer time to converge into the equilibrium at the market in the middle layer, Market[1][..], compared with the other markets, Market [0][..] and Market [2][..]. This observation is explained by the following reason. We applied the functions described in (25) and (26) as outside functions in Market [2][..] and Market [0][..], and they are defined as static functions in the experimental model. On the other hand, Market [1][..] is operated by supply and demand agents with dynamic utility functions. As a result, Market [1][..] behaves dynamically and is more sensitive to trading situation in the competitive market.

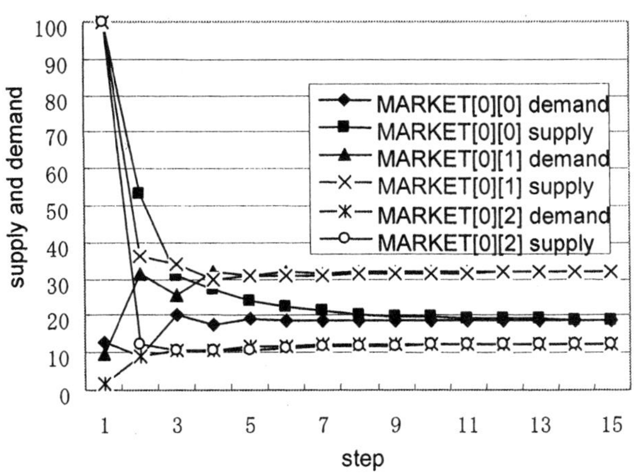

Figure 18-3 (a). Supply and Demand Oscillation (Market[0][...])

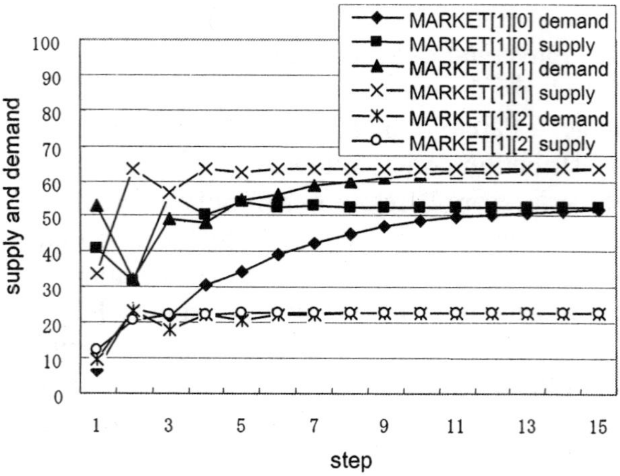

Figure 18-3 (b). Supply and Demand Oscillation (Market [1][...])

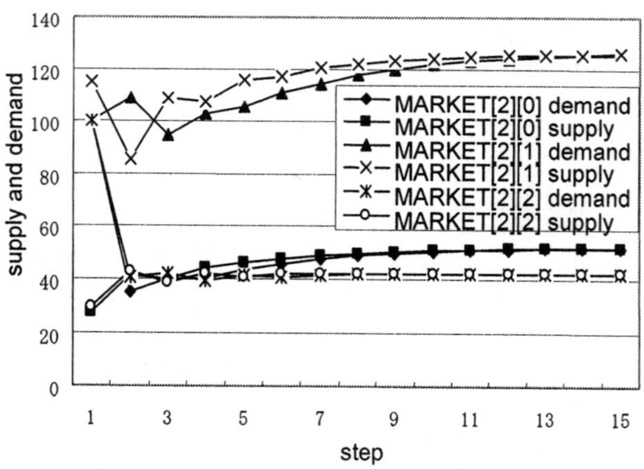

Figure 18-3 (c). Supply and Demand Oscillation (Market [2][...])

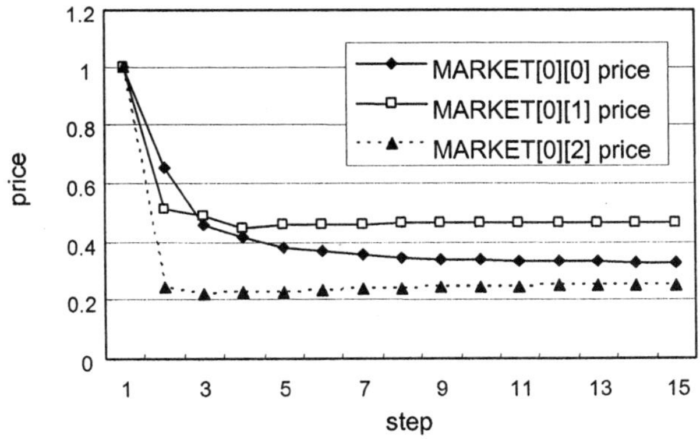

Figure 18-4 (a). Price Oscillation (Market [0][...])

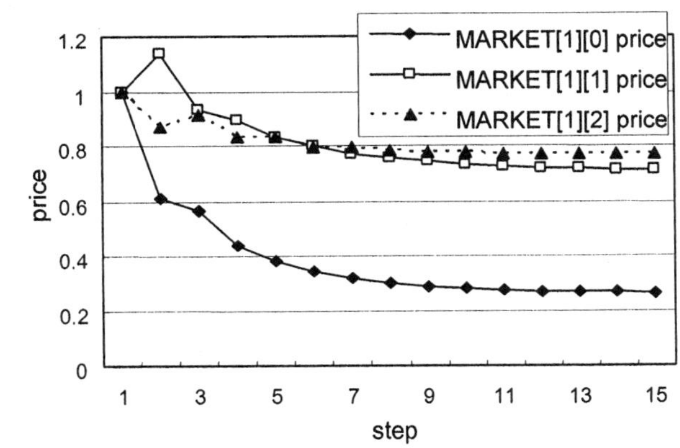

Figure 18-4 (b). Price Oscillation (Market [1][...])

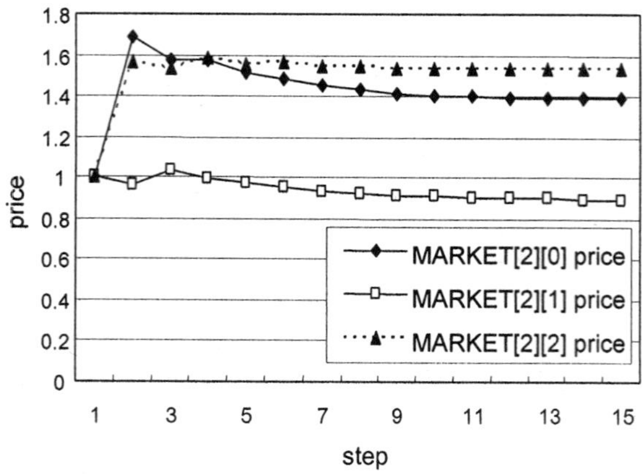

Figure 18-4 (c). Price Oscillation (Market [2][…])

4.3 Market Equilibrium

The market equilibrium dynamism should be explained by our market definition. A comparison between figure 18-3(a) and figure 18-4(a) illustrates that goods with the larger trading amount have a higher price in the equilibrium in Market [0][..]. That is because we defined the function shown in (26) as an outside supply function, which characterizes a positive correlation between the dealing amount and the price. On the other hand, goods with the larger trading amount have a lower price in the equilibrium in Market[2][..], see figure 18-3(c) and figure 18-4(c). The outside demand function defined in (25), which has negative correlation between the dealing amount and the price, influences the equilibrium. These experimental values are perfectly explicable by our market formulation. The middle layered market, Market [1][..], has a more complex dynamism in figure 18-3(b) and figure 18-4(b). The dynamism is emerged and explained by the agent utility parameters shown in table 1. In this case, Market [1][1] is high both on the dealing amount and on the price in the equilibrium. A set of Agent [0][..] has to offer a higher sales price to increase their profit, because b[1] in Agent [1][..] (=0.6) is the highest amongst b[..] in the second layer, but b[1] in Agent [0][..] (=0.5) is lower than b[0] in the first layer.

It is evident then that all the experimental values in the complex SCM model are perfectly explicable by our market formulation.

4.4 Experimental Summary

Several points to validate the proposed methodology have been enabled:

- A Pareto optimal solution is attainable by the equilibration process;
- The equilibration process scales with price elasticity of trading goods;
- Outside supply and demand function reduce oscillation in the equilibration process;
- The dynamism in the equilibrium highly depends on the utility functions of agents.

The experimental results successfully agree with the theoretical trends of perfect competitive market in microeconomics. It is obvious that each market is perfectly competitive and holds the market mechanism in general equilibrium.

5. INDUSTRIAL APPLICATIONS

5.1 Business Model

With the outset of the virtual enterprise, each business unit, which constructs the supply chain, generally uses the Internet for its procurement (Iwata and Fujii 1997, Laudon 1998). For the reason that the number of business units in a certain layer is countless in Internet world, a sophisticated matching mechanism is necessary to manage such a disordered environment (Padget et al. 2002). But since the matchmaking place is a kind of pure market in terms of its structure, the idea of SCM combined with the proposed virtual market, shown in figure 18-5, is a promising business model in practice.

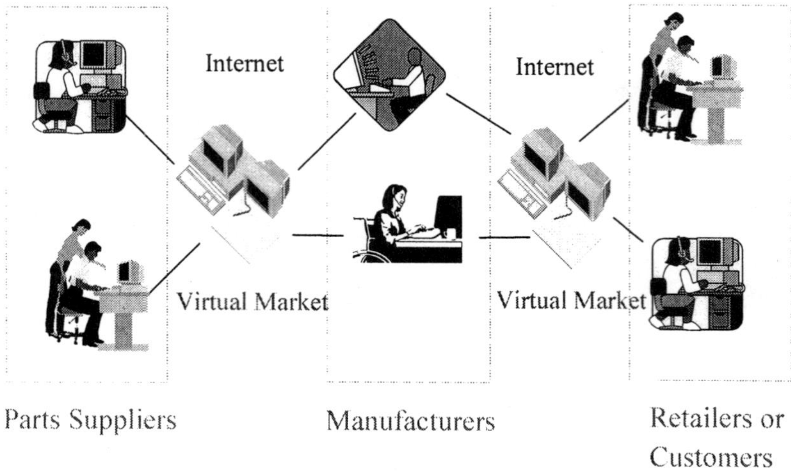

Figure 18-5. Virtual Market Based Business Model

Optimal partner searches in SCM based on their utilities are carried out automatically via the virtual market in each layer independently. Microeconomics theory endorses their Pareto optimality at a competitive equilibrium in the perfect competitive market, as described in the previous section.

5.2 e-Marketplace Server

It is quite difficult to implement some distribution algorithm into a large-scaled complex EC taking conventional approaches. Our approach is completely distributed and a Pareto optimal solution is attainable only by defining the supply/demand functions in every business unit, because the market mechanism equips a natural dealing protocol. We construct a prototype e-Marketplace server with a virtual market so as to validate the proposed business model.

We developed a prototype e-Marketplace server shown in figure 18-6. The server consists of i) Web server, ii) Relational database, and iii) Virtual market model:

- Web server : Microsoft Peer Web server (PWS);
- Server side script : Microsoft Active Server Pages (ASP);
- Relational Database : Microsoft ACCESS 2000;
- Virtual market model : Microsoft Visual C++.

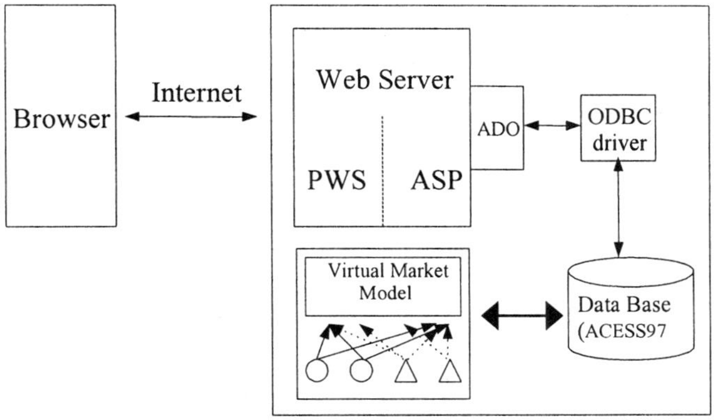

(ADO: ActiveX Data Object, ODBC: Open Database Connectivity)

Figure 18-6. e-Marketplace Server

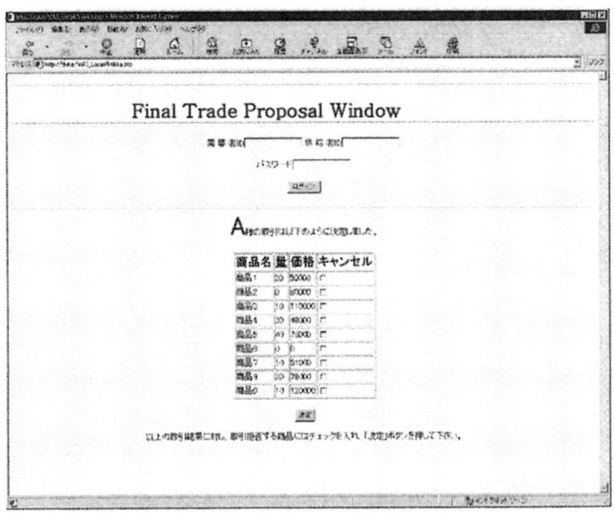

Figure 18-7. Final Trading Proposal Window

5.3 Transaction Flow

The product suppliers and demanders will access the e-Marketplace server via an Internet Browser. The transaction flow is as follows:

- *Step1*: All the suppliers and demanders create user accounts and register their personal information on the e-Marketplace server.
- *Step2*: Suppliers input their products with their preferable price and quantity.

- *Step3*: Demanders input their preferable price and quantity on the products to buy.
- *Step4*: The virtual market collects all the bids over a specified interval of time, then clears the market by computing a Pareto optimal solution at the expiration on the bidding interval.
- *Step5*: All the suppliers and demanders input their final decision on the proposed trade by the virtual market.
- *Step6*: The final trade is announced to all the participants.

To define the supply/demand function shown in equation (20) / (21), only a_{kl} needs to be acquired, because b_{kl} is unique to product s and it is attainable previously by the practical observation. When all the suppliers/buyers input only both the amount and the price for each product they want to sell/buy, then a_{is} is automatically calculated by equation (5). It is inevitable then, that the supply/demand function is defined by the system. A snap shot example of the final trading proposal window, calculated by the system, is shown in figure 18-7, for an example.

We have analyzed the performance of the proposed e-Marketplace server with the prototype system. The virtual market implemented in the server successfully produced a reasonable clearing of multiple bids with several business units.

6. CONCLUSIONS

In this chapter we proposed a SCM application with market economics. First of all, we formulated SCM as a distributed resource allocation system, based on general equilibrium theory and a competitive mechanism. The approach works by deriving the competitive equilibrium corresponding to a particular configuration of agents and markets. After defining production functions, we introduced a budget constraint for practical use and a newly proposed 'Profit Maximize Theorem' as an agent strategy. It has been confirmed by simulation experiments that the careful constructions of the decision process, according to economic principles, can lead to efficient resource allocations in SCM, and the behavior of the system can be analyzed in economic terms.

As an example of industrial applications with our virtual market, we described an auction server for the e-Marketplace with market-oriented programming that mediates amongst the various but unspecified trading companies, and demonstrated the applicability of the economic analysis to this framework. The proposed server is expected to facilitate a sophisticated e-Marketplace, which conducts a Pareto optimal solution for all the participating business units, in the upcoming agile VE era.

Since most efforts in ICT mainly focus on its technological side to develop better performance, or new functions, the symbiosis between ICT and value network is sometimes left out. The gap between technology and practice always exists, and this prevents the symbiosis from being established. Generally the software agent paradigm is quite effective to establish sophisticated interactions between human and software. It tries to elicit human preferences and requirements and ensure that they are represented in automated agent behavior. Therefore the proposed approach, based on the multi-agent paradigm, is placed at the intermediate layer between the technological layer and the application layer in an ICT environment, and it should play an important role to bridge the above-mentioned gap and produce real value for the ICT users.

There are two obvious challenges in this research. The first is to elaborate the negotiation protocol, possibly by exploiting several types of elaborated production functions. The second extension is to introduce bounded rationality into the agent. This extension makes the proposed model more practical and flexible in its implementation to a real e-Marketplace system.

ACKNOWLEDGEMENTS

This research was partly supported by the IMS (Intelligent Manufacturing System) Research Program - HUTOP Project, JAPAN (Grant No.0219), and the Grants-in-Aid for Scientific Research Kakenhi from Japan Society for the Promotion of Science (Contract No. (C) 12680456).

APPENDIX

[Theorem] Profit function E_k of agent k is maximized by minimized r_k, which satisfies the following conditions:

$$\forall l : \frac{\partial E_k}{\partial x_{kl}} = r_k \left(r_k \geq 0\right) \cap C_k \leq \max C_k \qquad \text{(A1)}$$

[Proof]
Note that

$$\frac{\partial f_{kl}}{\partial x_{kl}} = \frac{\partial f_{kl}}{\partial c_{kl}} \cdot \frac{\partial c_{kl}}{\partial x_{kl}} = p_i \frac{\partial f_{kl}}{\partial c_{kl}} \qquad \text{(A2)}$$

then

$$\left.\frac{\partial f_{kl}}{\partial c_{kl}}\right|_{c_{kl}=c} > \left.\frac{\partial f_{kl}}{\partial c_{kl}}\right|_{c_{kl}=c+\Delta} \quad \text{in any positive value } \Delta \qquad \text{(A3)}$$

Also, note that

$$E_{kl} = P_j f_{kl}(x_{kl}) - p_i x_{kl} = P_j f_{kl}(c_{kl}/p_i) - c_{kl} \qquad \text{(A4)}$$

then E_{kl} is regarded as convex by (A3). E_{kl} is maximized with the condition (A5).

$$\forall l : \frac{\partial E_k}{\partial c_{kl}} = 0 \qquad \text{(A5)}$$

Let C'_k denote total expense in (A4) and if $C_k \leq \max C_k$, then the maximum profit is given with (A5). Otherwise the maximum profit is not given with (A5) due to the budget constraint.

If $C_k > \max C_k$, agent k must consider to increase C_{kl} by Δc, and reduce $C_{kl'}$ by Δc shown in (A6) and (A7).

$$\left.\frac{\partial E_k}{\partial c_{kl}}\right|_{c_{kl}=c+\Delta c} = \frac{P_j}{p_i} f'_{kl}[(c_{kl} + \Delta c)] - 1 = r_{kl} \qquad \text{(A6)}$$

$$\left.\frac{\partial E_k}{\partial c_{kl'}}\right|_{c_{kl'}=c-\Delta c} = \frac{P_h}{p_g} f'_{kl'}[(c_{kl'} - \Delta c)] - 1 = r_{kl'} \qquad \text{(A7)}$$

then, we obtain that $r_{kl} < r_k < r_{kl'}$ from (A3).

Let ΔE_{kl} and $\Delta E_{kl'}$ denote the increased profit in f_{kl} and the diminished profit in $f_{kl'}$, respectively, then ,

$$\Delta E_{kl'} = \beta \Delta c\left(r_k < \beta < r_{kl'}\right) \qquad \Delta E_{kl} = \alpha \Delta c\left(r_{kl} < \alpha < r_k\right) \qquad \text{(A8)}$$

It is obtained that $\Delta E_{kl} < \Delta E_{kl'}$ from (A8), and that means the diminished profit is greater than the increased profit in any Δc. Therefore the profit function E_k of agent k is maximized with the condition

$$\forall l : \frac{\partial E_k}{\partial c_{kl}} = r_k \left(r_k \geq 0\right) \text{ in } C_k \leq_{\max} C_k \qquad \text{(A9)}$$

Finally, r_k is minimized with the condition (A9), since the profit function E_k is convex. This completes the proof.

REFERENCES

1. Fisher M. L. Making supply meet demand in uncertain world. Harvard Business Review 1994; May- June: 83-92.
2. Goldratt E. M., and Cox J., *The GOAL - 2nd edition*, Great Barrington. MA: North River Press, 1992.
3. Hannebauer M., *Autonomous Dynamic Reconfiguration in Multi-Agent Systems.* Berlin: Springer, 2002.
4. Iwata, K., Fujii S. New Manufacturing Era - Adaptation to Environment, Culture, Intelligence and Complexity, 29th CIRP International Seminar on Manufacturing Systems, Osaka University, Japan, March, 1997.
5. Kaihara T. "Supply Chain Management based on Market Mechanism in Virtual Enterprise." In *Infrastructures for Virtual Enterprises*, Camarinha-Matos, L. and Afsarmanesh, H., ed. Boston: Kluwer Academic Publishers, 1999, pp.399-408.
6. Kaihara T. Supply chain management with market economics. International Journal of Production Economics 2001; 73(1): 5-14.
7. Kaihara T. "B to B Electronic Commerce Server with Virtual Market." In *E-business; Key-issues, Applications and Technologies*, Stanford-Smith, B. and Kidd, P.T., ed. Amsterdam: IOS Press, 2000, pp501-507.
8. Kaihara T. A study on partnering mechanism in B to B EC Server for global supply chain management. The Japan Society of Mechanical Engineers (JSME) International Journal, Series C, 2003: 46(1): 3-8.
9. Kaihara T. Multi-agent based supply chain management with dynamic environment. International Journal of Production Economics 2003; 85(2): 263-269.
10. Kaihara T., Fujii S., "An Evolutionary Scheduling Paradigm using Coordinated Autonomous Agents." In *Globalization of Manufacturing in the Digital Communication*

Era of the 21st Century, Jacucci, G., Olling, G., Preiss, K. and Wozny, M., ed. Boston: Kluwer Academic Publishers, 1998, pp.553-563.

11. Laudon K, Management *Information Systems - New Approach to Organization and Technology*, Upper Saddle River, NJ: Prentice Hall, 1998.
12. Layard, P. R. G., Walters, A. A., *Microeconomics Theory*, London: McGraw-Hill, 1978.
13. Perrone G. Renna, P., Cantamessa, M..,Gualano, M., Bruccoleri, M., Lo Nigro, G., An Agent Based Architecture for Production Planning and Negotiation in Catalogue based e-Maeletplace, Proceedings of the 36th CIRP International Seminar on Manufacturing Systems, pp.47-54, 2003.
14. Shen W., Norrie, D.H. and Barthes, J.P., *Multi-agent Systems for Concurrent Intelligent Design and Manufacturing, London*: Taylor and Francis, 2001.
15. Shoven J. B., and Whalley J., *Applying General Equilibrium*, Cambridge: Cambridge University Press, 1992.
16. Padget, J., Shehory, O., Parkes, D., Sadeh, N. and Walsh, W., *Agent-Mediated Electronic Commerce IV*, Berlin: Springer, 2002.
17. Wellman M. P., A Market-Oriented Programming Environment and its Application to Distributed Multi- commodity Flow Problems, Proc. of ICMAS-96, 1996, pp385-392.

Chapter 19

INTEGRATED MANUFACTURING SYSTEM

Intelligent and Integrated Execution System to Provide 'Crystal' Shop Floor Information and Controlibility to Value Networks

Douglas Scott, Robert Pisa, BG Lee
Brooks Auotomation

Abstract: To adapt to new manufacturing paradigm and to improve visibility across value networks, factory level manufacturing execution system should be revised and integrated properly with other business information systems. In this respect, we introduce a new MES system called IMS for 300 mm wafer fabrication factory as an industry specific domain in this chapter. The concept of IMS can be expanded to other industries as an enabler for collaborative manufacturing execution.

Key words: Integrated manufacturing, Intelligent execution System

1. INTRODUCTION

As advanced 300mm factories (i.e. wafer fabrication) and foundries (i.e. customer order based wafer fabrication) move from the drawing boards to production, the financial stakes for the semiconductor industry have never been higher. With plants costing three billion dollars and up, and individual toolsets $20 million, why are we taking the leap to 300mm? We're banking on the savings to be realized. However, the savings are based on the assumption that a 300mm factory will operate more efficiently than a 200 mm one. And the efficiency should be expanded to the entire value chains beyond factory level

Getting there is not easy. Achieving high factory efficiency in 300mm plants requires material handling automation, and seamless integration between the various production systems and manufacturing tools, with less human decision-making at every point. This in turn requires a more intelligent and integrated automation solution, one that can balance the plant processes and business goals, making decisions efficiently in real time.

This chapter highlights the need for a new category of manufacturing system designed to meet this challenge: the Integrated Manufacturing System (IMS), an intelligent decision-making software platform that is open, flexible and integrated. This next-generation software will make it possible for 300mm factories and foundries to realize faster ramp up and improved production efficiencies, through a more highly automated plant. We describe what full automation under IMS looks like, and the capabilities you should look for in an IMS. And through utilizing IMS, the integration with other Enterprise level applications will be achieved in a more collaborative manner. The entire enterprise integration will be easier and more accurate than an existing factory level Automation system. The proposed idea could be used to build automation/tool in the other industries in order to provide better visibility and controllability on their shop floors thus will improve their value chain efficiencies. The integrated shop floor information will provide more accurate shop floor status (e.g. machine and resource status, WIP information, etc.) to other supply chain applications thus enabling better decision making in planning and execution.

1.1 The Challenges of 300mm Automation

Most semiconductor executives agree, at least in theory that 300mm factories need to be fully automated and integrated in order to provide accurate and integrated information to a higher level business information system to improve value network efficiency. The issue is when will this transition occur and how can companies best to manage this change. There is not yet a deep base of experience in how to achieve fully automated volume production in a 300mm semiconductor plant. In fact, not everyone agrees on the definition of full automation. For the purposes of this chapter, full automation means that key areas in semiconductor manufacturing, such as the process tools, recipe management, material handling, measurement and test, scheduling, dispatching, and Advanced Process Control (APC) are:

1. Automated in themselves;
2. Linked by automation, i.e., no continuous hands-on human intervention;
3. Controlled and coordinated dynamically by highly integrated and intelligent decision-making solutions;
4. Accountable to and controlled by factory managers and engineers, who are free to think in terms of business priorities, ship dates, and overall factory efficiency, balancing issues such as utilization, cycle time and throughput.

Most companies are taking a conservative course, starting with the basic forms of automation first, such as inter-bay Automated Material Handling Systems (AMHS) for moving materials, and Manufacturing Execution Systems (MES) for managing processes, and then planning to expand automation and intelligent decision systems as they move forward to achieve optimization of their supply chains. The question this chapter addresses is: "How will a 300mm factory improve efficiency?", and "Will existing automation and MES platforms be enough to get us there?"

New manufacturing concept: 300mm is not 200mm

300mm plants are not the same as 200 mm plants, and the software that runs today's factories may not be able to run tomorrow's. For example, let's look at just one component of an automated factory, AMHS. In most 200 mm factories, AMHS moves material automatically from bay to bay. Operators and handling equipment know from the MES that a given lot must be moved to stocker x, and where that stocker is located. The material is moved in advance to the correct stocker so that it is ready when needed. Since there is little or no intelligence employed, the AMHS system processes all the lot moves to their next location in FIFO order initiated upon lot completion of the current step.

In a fully automated 300mm plant, this approach will be replaced by tool-to-tool, just-in-time transport, with intelligent decision supported by integrated information. This is far more complicated, because now it is necessary to deal with multiple tools and their readiness/appropriateness for the work. "Is the tool available?" "Is a reticle available?" "If not, when?" "Should the lot be stored in the stocker?" "If the lot is hot (i.e. urgent), can it be moved with higher priority?" "How will all this affect other work in progress (WIP)?" "How the lot can meet the promised delivery date of sales order?" "When should the consumed chemical be replaced without effecting production schedule?"

In a fully automated factory, every WIP or tool decision becomes more critical than ever, rippling throughout the plant and potentially affecting other entities of supply chains, and the overall effectiveness of operations.

Complexity in process

Semiconductor manufacturing is inherently complex. As processes are linked in a fully automated operation, the complexities multiply. For example, one of the key requirements for 300mm plants is for wafer level tracking and process control. This means controlling each wafer through every process tool and chamber, and possibly modifying its process recipe based on APC. And the information of wafer should be exposed to enterprise level to make more accurate product delivery plans and meet end user's demand.

This wafer control capability in turn must be integrated with wafer sorter handling. Intelligent software can specify lot/FOUP (Front Opening Unified Pod called Carrier) splits and merges, and then, if a wafer sorter is attached to the stocker, the wafer transfers can be done while lots are waiting to be processed.

Factory owners may want to track a specific LCD panel on a larger substrate, for yield management, troubleshooting, or monitoring the current value of WIP.

The complexities of die/panel, wafer, lot, FOUP, and reticle control are even greater for foundries. Because 300mm wafers are so large, foundries will see average order sizes drop below 25 wafers, potentially below 10 wafers. This means that being able to efficiently combine and process multiple orders/lots in a single FOUP will be critical.

Another complexity is the factory's tool dedication policy, which has a major effect on factory performance. Tool dedication may be required for critical layers and devices, but not others. Factory systems must accommodate these process constraints, which can sometimes be quite complex.

So what does this tell us about the automation solution?

1.2 Three Key Requirements for 300mm Manufacturing

1.2.1 Integration

Integration may be one of the most important key requirements in manufacturing industry. As we have seen, the 300mm factory is a complex web of highly interdependent activities. Dispatching services must link to schedulers, material control systems, reticle management, equipment maintenance, reporting, and so on in an interconnecting web of activities, all supporting the same factory and business objectives. It is a balancing act of physical constraints and production goals, material handling and information handling, the visible and the invisible.

Integration goes far beyond interface standards, which merely bring order to how information is transferred between systems, but are silent on what those systems will do with the information. Even when integrating so-called "standard" products, wafer factories usually spend a lot of time, effort and money to get all the systems working together seamlessly.

1.2.2 Intelligent Decision-Making

Improving efficiency means evaluating trade-offs and making better decisions at the heart of the automated plant. As we have seen, the interconnectedness of the automated plant makes it impossible to rely on human decision-making at every critical juncture, simply because every juncture is potentially critical!

As the level of automation increases, there are fewer people in the factory. This reduces the risk of contamination, and creates a better work environment for many. It also has another powerful benefit: Instead of spending their time acting as porters, personnel can allocate more time to improving quality and overall performance.

This places a new burden on the software that is controlling the systems. Not only the actions, but also the decisions once performed by people need to be automated. We require both high levels of physical automation/integration, and high levels of data integration and information analysis. Data must be gathered, transformed into information, analyzed, and used for decision-making, both by the automated systems and by the people responsible for tool and factory performance.

In a word, we are talking about intelligence.

This intelligence must be robust enough to make informed, appropriate decisions not only for standard operating scenarios, but also for countless exception conditions, because abnormal cases may in fact be the "norm".

1.2.3 Open and Flexible Systems

Any automation solution must be open and flexible, because no company can be certain what the future will hold, especially with emerging 300mm technologies. A flexible platform allows adjustments to be made and new directions chosen as experience grows, technology advances, and business conditions change. Furthermore, automation won't happen all at once. Factories will evolve towards full automation in steps, with systems that are extensible and standards-based.

What's next?

Traditional MES solutions comprise only one component of a fully automated 300mm factory. The relentless drive for improved quality and performance requires systems that are more highly automated and better integrated. In Section 2, we discuss how IMS enables full automation and hence, greater productivity and higher profits across value networks.

2. FROM MES TO IMS

To manage a fully automated plant, a software platform is needed that encompasses MES and more. It must have the intelligence to make decisions based on information streaming in from all over the factory and enterprise level applications within supply chains. At the same time, it must provide a window on operations for the plant managers and supervisors who set the rules and objectives, and who make the decisions, which only human experts can make.

This software platform must be integrated with key factory systems, as it must supervise and control everything from scheduling to APC and to recipes to material handling. And it must be extensible, allowing factory owners to build on their systems and make adjustments as they learn from experience. To meet these requirements, a new category of software is needed: the Integrated Manufacturing System (Tables 19-1 and 19-2).

We define IMS as a set of integrated systems that improves overall factory performance by managing better the flow of information and materials through the factory. The normal functions of manufacturing – wafer processing, metrology and the transfer of WIP and durables – are managed without human intervention. IMS integrates the physical and logical systems in the factory, and cuts across many traditional disciplines, including factory economics, material handling systems, control systems, planning, scheduling, and execution. It includes integration of:

- Software with hardware, MES with logistics; AMHS with tools and load ports; tools with SMIF hardware and software
- The many different factors of production, the systems, equipment, materials, information, people, WIP, durables, procedures, facilities, recipes, processes, and other resources.

On a tool and WIP level, IMS makes decisions and routes material through the plant, making sure the right wafers are at the right tools, with the right recipes, on a just-in-time basis.

On a plant level, IMS makes sure that tools and processes are optimized, plant efficiency is high, and ship dates are met.

On an executive level, IMS allows a factory to focus on business objectives: costs, customer priorities, production and quality standards, and other key metrics that drive profitability.

Table19-1. A Comparison of MES and IMS (Scope)

MES Today	IMS vision
Interfaces systems	Integrates the factory
Helps manage equipment performance	Helps manage factory performance
Many decisions are based on local knowledge of tool or bay. Tries to optimize tool or bay performance. Decisions may be out-of-sync with plant-wide goals.	Decisions are based on local knowledge of tool or bay, + factory-wide status + factory-wide goals. Tries to optimize factory performance, with tool and bay decisions that are aligned with factory-wide goals.
In the normal course of events, processing and transport in the factory are largely people-driven	In the normal course, processing and transport in the factory are largely computer-driven, with authorized person override
Concerned with individual tools and lots	Concerned with overall performance, and factory-wide integration
Manages data flows	Manages information + material flows
Factory logistics	Factory, material handling and tool logistics
Focus on cost	Focus on profitability, and reducing risk
Fixed or manual operational scenario	Flexible operational scenarios, from manual through fully automated
Manages data	Transforms data into information, knowledge, and decisions
Manufacturing execution	Manufacturing excellence
Software + software service: manages logical systems	Software + hardware + full services, including factory productivity consulting: manages logical and physical systems.
User supplier relationship	Collaborative consultative relationship

Table19-2. A Comparison of MES and IMS (Functionality)

MES Today	IMS vision
Recommends next lot	Selects next lot
Recommends lot moves	Moves lots
Prepares dispatch list	Does plant-wide scheduling and dispatching
Independent production and maintenance	Integrated production and maintenance
Downloads recipes	Downloads recipes and fine-tunes them using APC
Specifies reticles	Moves reticles to scanners, and selects tools
Manages WIP	Manages WIP, reticles, carriers, and buffers: in-tool, on-tool (load ports), WIP and reticle stockers, whether in-bay, shared or central
Waits for "track out"	Anticipates "track out"
MES	MES + Decision Support (DSS) + AMHS + Recipe Management + APC + Equipment Engineering Systems (EES) + data mining and analysis + yield management
Execution focus	Performance focus – quality, speed, agility, variability, throughput, WIP levels, rework, cycle times, on-time delivery

MES Today	IMS Vision
Process steps tied to tool or tool types	Process steps linked to resources based on capability model
Relationships to resources and durables tightly linked	More general capability model implemented for all resources
Multiple factory and resource models	Shared factory and resource models
Loose coupling of metrology feedback	Tightly integrated metrology, for fast feedback
WIP Tracking centric	Job/Task centric; includes Material, Resource and Specification jobs
Material centric	Equipment and resource centric
Limited ability to define relationships between objects	Easy to expand and define new object relationships, which are extensible and flexible, to represent customer preferences
Hard coded or no workflow management	Flexible workflow management, integrated with other components
Data is distributed, duplicated and moved from one component to another	Data is integrated, shared, self-describing, and is naturally blended among different components of the system
Manual error recovery (abnormal situation handling)	Systematic error recovery

2.1 IMS and Full Automation

In today's factory environment, each system has its own models of the factory, its resources and its processes, with different resolutions, unaligned boundaries, and conflicting overlaps – creating communication headaches. In a fully automated, IMS-based factory, there will be a master set of high-fidelity models, a key step towards goal alignment and improved communication. Decisions made at a tool or bay, whether made by people or systems, will no longer be based on local knowledge and objectives alone, but will be in harmony with plant-wide goals.

Critical plant processes will be automated and integrated (Figure 19-1), using an integrated data model, with shared properties between applications:

- Manufacturing Execution Systems and Decision Support;
- Inter-bay and in-bay AMHS for wafers and reticles;
- Planning, scheduling and dispatching systems;
- Equipment Automation systems for process, metrology and test equipment;
- APC for run-to-run control and fault detection and classification;
- Equipment Engineering Systems for e-manufacturing, including remote diagnosis, maintenance and equipment upgrade systems;
- Data management and analysis systems;

– Enterprise applications.

The data model is a critical part of IMS. First, it allows all aspects of a factory activity to be integrated around a common database and software core that can make intelligent, real-time decisions on the plant floor. Second, it enables the IMS to report clearly to management on factory conditions, and makes it possible for plant managers to make better decisions about objectives, priority orders, change orders, tool constraints, maintenance schedules, and other aspects of daily plant operations.

At the plant level, this integration and unified data model enables detailed production control. At the wafer level, equipment and software systems support multiple lots per carrier, multiple devices in a lot, and even multiple device types on a wafer. Wafers are tracked and processed individually, and recipe control is wafer-specific. Real-time yield management systems automatically correlate test and probe results with process data, and assist in determining process problems, causes, and solutions.

2.2 The Benefits of IMS

The potential benefits of IMS and full automation are significant, affecting factory design, operations, economics and integration. Following are some of the ways they can be expected to improve factory performance.

2.2.1 Factory Design

With much of the integration puzzle solved, the times required to ramp volumes, yields and cycle times for new plants, technologies, processes, products and equipment will be drastically reduced. Factories will become efficient machines to build silicon machines. New plants will be highly automated, with a plant-wide command and control capability. Factories will be agile, able to respond quickly to changes in both customer demand and events in the factory and thereby achieve agile supply chains. Equipment and systems will be upgradeable, re-configurable and reusable to meet changed requirements.

2.2.2 Factory Operations

In the IMS-based factory, personnel will no longer have to spend long hours serving as the "system glue" in the factory, creating "work-arounds" for system and operational deficiencies. They can focus on improving operational performance. Misprocessing will be reduced, processing will be repeatable, and APC will enable fine-tuning of process parameters for individual batch runs or individual wafers. In-situ monitoring and integrated metrology will shorten the feedback time between wafer processing, and knowing the results of that processing, so process and equipment problems will be recognized and corrected quickly.

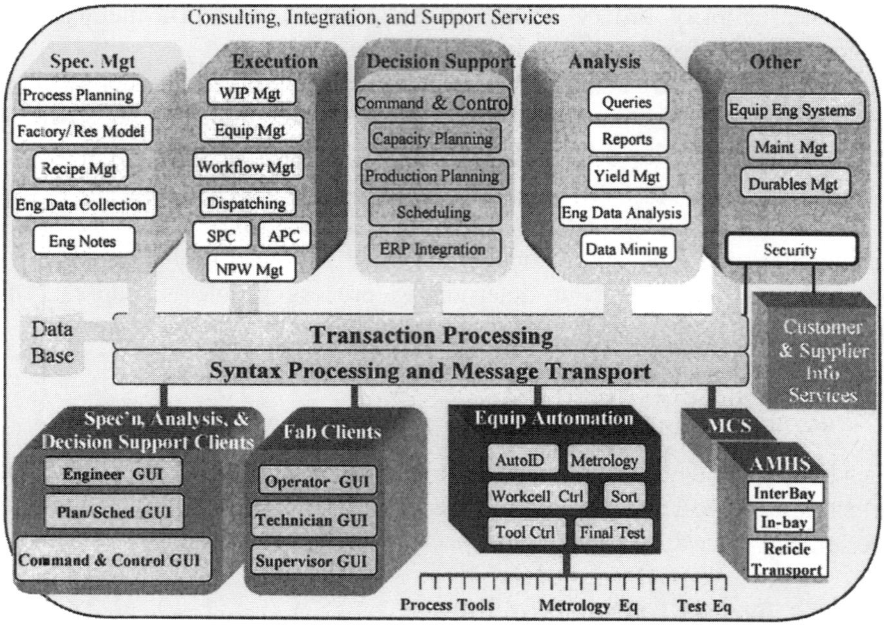

Figure 19-1. Integrated Manufacturing Solution

As experience grows, modeling and simulation tools and techniques will allow manufacturing personnel to predict the future outcomes of current decisions. In theory, factory operations will be predictable and almost mistake-proof. Variability in processing, equipment performance and cycle times will be low, improving on-time delivery. Production throughput per dollar of investment will be higher, and these results will be achieved with lower work-in-process. Opportunistic preventive maintenance policies will consider the state of each part of the overall manufacturing system, to increase equipment availability. Maintenance and production activities will be synchronized, and tool setups optimized.

2.2.3 Factory Economics

The economics of semiconductor manufacturing have been poorly understood to date, precisely because the processes involved are so complex. IMS, and the full automation it enables, will shed new light on the economic issues. Decision makers will know which practices increase revenues, which ones reduce costs, and which blend of choices will increase profitability. There will be a clear understanding of the various metrics (e.g. cost, agility, Economic Value Added, throughput, cycle time, capacity utilization, OEE, WIP inventory, turn rates, ramp-up efficiency), and the choices and trade-offs among them.

Factory owners will be able to dial up the mix of goals required to meet their specific plant, or value networks or corporate objectives, applying weighting factors for use by the plant's decision support systems, and by all personnel. Overall Factory Efficiency, or the metrics for individual goals, will be used to compare the performance of plant A with plant B, or to compare June with July, and drill down capability will allow production factory personnel to examine the factors contributing to each measured result. Factory design and factory operation decisions will be based on economics that are understandable, predictable, and measurable.

The preceding description of a fully automated factory and fully integrated supply chain may seem visionary, but we are getting closer to fulfilling that vision. In Section 3, we identify some critical requirements.

2.2.4 Information Integration

IMS is considered as a part of entire value network including ERP and SCM. Information supply from ERP to IMS and the information feedback from IMS to ERP and SCM should be considered as a part of IMS scope. It is very important requirement of IMS to operate the factory as dynamic execution based on real time information from factory floor and vendors. The rough information flow of IMS is described in figure 19-2. Even though the detailed implementation will be considered as an IMS solution, the information should flow by the standard message format such as XML/SOAP using a standard connection method such as JMS complaint message bus, The information flow will be the key for next generation system as IMS.

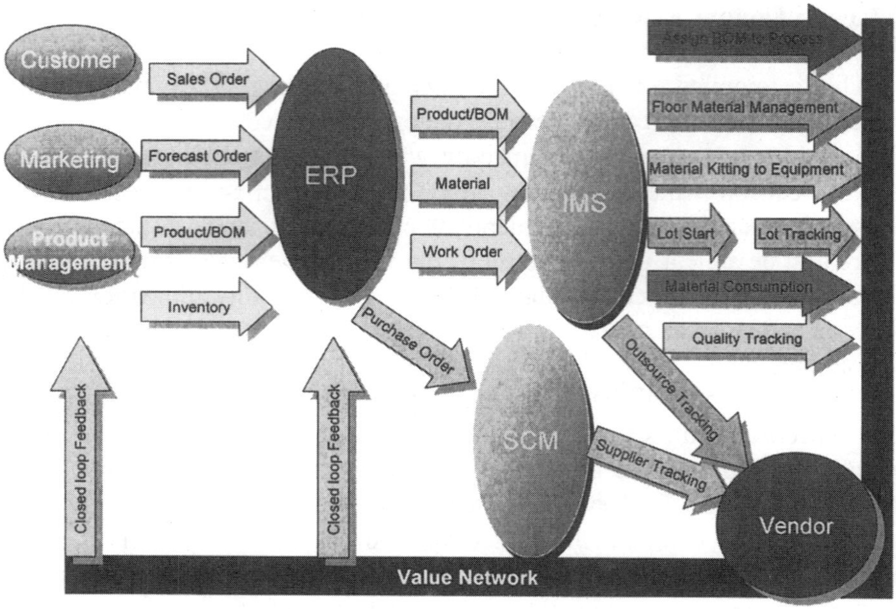

Figure 19-2. Integrated Information Flow

3. IMS REQUIREMENTS

IMS includes the typical functionality of MES, but adds many capabilities needed for 300mm automation. These include:

- Non product wafer management: preparation, qualification, alternate (downgrade) usage, usage count, source monitoring, cleaning/disposal cycle, dedicated return carrier, full identification, multiple monitoring purposes;
- 300mm carrier (FOUP) management: usage/cleaning count, automatic disposition, carrier exchange, carrier contamination control, multiple lot IDs, slot mapping, carrier hold, carrier history, manual/automatic sorter operation;
- Durable management: Reticles and other durables: life cycle management, availability, usage/cleaning count tracking, vendor and critical data management;
- R&D and engineering lot control: add/remove engineering steps/plans, recipe/ppid manual override, engineering release/skip, future actions (merge, split, hold, user defined), process look ahead;

- Multiple lots per carrier: carrier/lot tracking, multiple process jobs per carrier, selected slot processing, wafer-specific recipes;
- Multi-level equipment modeling: Chamber and load port configuration, configurable state model, user definable state rules;
- Wafer level tracking: Wafer level history, sub-wafer level identification, wafer level workflow, attributes, and data collection;
- Full automation support: Inter/in-bay transport support, load port state control, automatic workflow execution, multi-level rule-based dispatch for what next and where next, seamless interface with factory systems;
- Provide integrated information to enterprise level applications to make more accurate and intelligent decisions in a collaborative manner;
- Maintain information for supply chains from accurate shop floor information to enterprise applications;
- Bring the direct value of shop floor information to other enterprise level applications such as ERP, SCM and CRM.

3.1 Integration and Intelligence

In an IMS solution, MES applications must be tightly integrated with Command and Control Systems, Tool Automation, AMHS, Equipment Engineering Systems (EES), Reticle Management, Equipment Maintenance Management, Supply Chain System and Analysis/Reporting Systems. These applications must embody a high degree of software intelligence, which, as we have seen, is a critical part of a fully automated factory solution.

Key to this integration and intelligence is a comprehensive data model that is shared by all components of the system. In a non-integrated plant, each system component (e.g. MES, Material Control (MCS), and Maintenance) maintains its own duplicate database model of the factory, or even worse, it synchronously queries other systems for the simplest attribute values. Data is entered into each system, either manually or by import/export, introducing opportunities for error.

With true integration, each system component uses the same model. This means data is entered once and used everywhere, accessible by all modules in the system and by all users. Only with true integration is it possible to make intelligent decisions – whether by software or by humans – about dynamic processes such as the current state of process tools or the status of all WIP for a given customer order.

3.2 Openness and Extensibility

Since flexibility is a requirement of any 300mm automation solution, IMS requires an open architecture that can integrate multiple applications from multiple suppliers. This is compatible with the goal of most factory owners to implement automation in stages, with a predictable, achievable timetable. Semiconductor companies change over time, and their needs evolve. So IMS must be customizable, enabling business rule modification, to drive new behavior, and/or database extension, to allow for new objects or the extension of existing object properties.

Finally, there is a need to look for the support infrastructure and service capabilities. These are critical to an efficient ramp up, so you can reach first silicon quickly (a time to market factor), and expedite the plant profitability.

4. CONCLUSION

By its nature, as well as by its cost structure, the 300mm factory or foundry requires a significant evolution of automation software beyond what is currently called MES. Plant-wide activities will be integrated, from executive decision-making to material movement and tool allocation on the factory floor. The factory of the near future will be highly hands-off, operating as a finely tuned machine.

Accomplishing this will require systems that are more highly automated, better integrated, more robust and more functional than current MES solutions. Integrated Manufacturing Systems will give plant managers a much greater ability to run the plant more efficiently, and to monitor and control plant operations based on business goals, production realities and customer requirements.

It is very clear that the proposed system will reduce gaps between execution and planning by providing accurate shop floor information to enterprise level and hence value network level. The value of accurate and integrated shop floor data can provide a "fresh blood" to the entire IT organization. Through IMS, all applications of value networks can use the fresh information and data. With fresh data, order processing and planning systems for instance can achieve better decisions.

ACKNOWLEDGEMENTS

Many staff at Brooks collaborated on this chapter. The authors would like to acknowledge Steve West, Al Neil, and Tom Comstock for their valuable contributions.

Chapter 20

COLLABORATIVE SENSE-AND-RESPOND ICT FOR DEMAND-DRIVEN VALUE NETWORK MANAGEMENT

A Platform for Adaptive, Customer-Driven Value Network Planning and Fulfillment

Charalampos Makatsoris[1], Yoon S. Chang[2], Howard Richards[3]
[1]Orion Logic Ltd, [2]KAIST, [3]MAPS

Abstract: Consider the following scenario: "there are two products on sale, X and Y. Product X's demand is high while there is no demand for product Y. Orders for product Y are cancelled and a new large order for product X is placed automatically. The manufacturer and suppliers immediately know about this and have adapted operations by reallocating common subassemblies from Y to X and reset production to ramp up for X". In this scenario changes to the market needs have been sensed and the value network responded without relying on any forecasts or trends. This is on-demand business. New technologies can deliver the infrastructure for such new ways of work and deliver benefits, which previously would not be possible. Such technologies are relevant to businesses of all types and size, and can potentially offer significant advantages including the means and models to help companies reshape their value proposition and integrate their business with customers. A key area is that of collaborative, integrated planning and fulfillment using adaptive sense and respond capabilities across the value network. This chapter presents a commercial platform that integrates the demand chain with the supply chain and offers demand driven planning and fulfillment functionality across the value network.

Key words: Integrated supply chain, Supply chain management, Sense and respond ICT

1. INTRODUCTION

1.1 Business Challenges and Drivers for Change

Nowadays, the dynamic engagement of businesses in complex relationships with other businesses is a reality. Former vertically integrated companies are continuing to break down into smaller but more manageable agile business structures that are responsive to both the market and customer needs. Drives for greater effectiveness and efficiency in terms of customer perceptions and expectations have forced companies to cut to core competence and to work with other companies in closer knit networks aiming to provide astute customer solutions rather than continuing to push product.

Confronting the issues arising from working in closer partnerships with others yet maintaining the vigor of the company and its businesses is one of the most difficult challenges faced by industry today. There are many facets to the change process, but there is no one route that a company must take. Tightening up the supply chain, forming alliances with other companies, working in close harmony with specialist companies that help create new customer solutions are all part of the process of evolving to a better way of conducting business. In this chapter we focus on manufacturing networks that have a former history in loose-coupled working relationships with others as part of a supply chain. We also focus on companies intending to create a new network to serve customers in a particular market. Parts of the network may be different businesses within one company but geographically distributed. The other characteristic of these networks is that they are not formed just for one job but intend to have a semi permanent relationship. That is they intend to supply the market and customers for as long as possible and to create new opportunities for growth and profitability. The features looked for in these networks are,

- they must have an integrated competence that can sustain customer demand;
- they are efficient and effective;
- they are agile with respect to changes of the environment and customer demand;
- individual companies or businesses can retain some autonomy and can work in more that one network;
- they share information and knowledge to improve performance;
- there are good communication channels between all partners;

- there is mutual trust, shared risk and a perceived equitable distribution of profit.

To meet these objectives companies and businesses join in federal partnerships in which resources are virtually pooled and time-shared. All companies or businesses involved in the federation may not be involved in servicing a particular customer order but over a time period they serve other orders. The federation may be contracted or expanded by mutual agreement and on expansion other partners are sought. In the future this may be through eMarkets, as discussed in chapter 3.

This closely-knit federation has from a company's or business's perspective a sleeper engagement or a dynamic engagement at any time. The network over time is always active and has been described as an *adaptive value network* (AVN) that is:

> *An arrangement where companies form a web of close relationships and work together as a system that delivers the right customized product and expected service at the right quality in a coordinated manner and are responsive and adaptable to changes in the environment.*

Order taking by the network can happen at any node or by agreement by the partners. There is no overall management that is imposed by a hub company, although AVNs do not prevent this. It is entirely up to the partnership. This is an important flexibility and encourages a peer-peer approach. In the case that a hub company is responsible for the coordination of the AVN, they are responsible for material flow, information flow and most importantly customer delivery. But as has been stated above the role of the hub, can be interchangeable among peer companies in the network in response to market needs and also depending on how the order is structured. Jarillo (1995) also discussed a similar structure, which he called strategic network.

In this chapter we look at the role of ICT in helping to bring about the transformation process from supply chain to AVN. ICT is a key enabler for network partners and must serve them with fast, reliable and secure communication channels as well as a platform to do business with security, confidence and trust. In addition ICT must provide a much better facility to access information and knowledge and to use it effectively for better and faster decision making. When this is achieved benefits are gained for all, companies, stakeholders and customers. Technology can enable businesses to become responsive to demand changes and deal more effectively with coordination of their supply networks. In addition this transition will help cut costs and improve the way they manage large fluctuations in information and knowledge asymmetry across their supply networks.

As the industrial and market landscape changes ICT systems must respond to this and be able to evolve by use of new technology and new components. However the process of change is fraught with uncertainty. How to harness ICT with all its future potential benefits from technological innovation to the benefit of both the company and, their networks and customers is a key question. Flexibility must be secured, stakeholders assured and barriers, arising from any conflicting objectives, have to be overcome. A strategic plan is absolutely necessary to set out the aims and tactics that will be necessary. This will have profound implications on organization of the company and that of each network to which it belongs and the way customers are serviced in the future.

This unavoidably is a learning process for it is a fundamental change in business practices focusing upon the needs for customized products and all kinds of extra service. Businesses have yet to understand the complete requirements, apply their skills and knowledge and adapt as much as they can to a new mode of working in harmony with others external to their business. The challenges are severe:

- How do we belong to several supply networks at the same time, where some are managed by large corporations who force systems upon us and require information we do not have?;
- How do we handle multiple customer inquiries and orders to respond in real time to stay competitive?;
- How de we effectively and quickly deploy planning decisions in response to customer requests and demand changes across our networks and coordinate fulfillment to better manage capacity and profitability as well as ensure delivery on time?;
- How can we identify and match parts and work in progress to customer orders so we can track progress more effectively and also how can we use this information to immediately respond to unplanned events, new demand and demand changes?;
- How can we sense events as they occur and deal with those in the most effective manner with collaborative decision making across the value networks?;
- How can we respond in a collaborative and effective manner to meet end customer requests to ensure profitability for the companies involved and provision of service excellence and added value to the customers without eventual delivery being compromised?;
- How can we have a platform that allows us to do business and collaborate within the network but also with our customers and external suppliers but is also adaptable to the changing market environment?;

– How do we keep track of customer order flow and measure the value network performance?

In this chapter we present a software platform that addresses these challenges. Initially, a brief overview of the current perspectives for supply chain management is given, where we illustrate why many current systems are not always suitable for managing value networks and in particular why they cannot address some of the challenges listed above. Then we present the design features of our platform and discuss its functionality. Finally, we conclude with implementation issues and future developments.

2. OLD PERSPECTIVES FOR SUPPLY CHAIN MANAGEMENT

Traditional solutions have emerged from IT for single companies that may have several geographically dispersed units, factories, warehouses, design centers and so on. There have been two major approaches that are in current commercial usage. On one hand those delivered by company centric MRP-II type/ERP systems and on the other hand those delivered by traditional Advanced Planning and Scheduling systems often using the expensive and inflexible EDI connections to the supply chain.

The ERP system for manufacturing supply is not a real time system and is not intimately coupled with other companies in the network. It relies upon gathering information at intervals of time from the other companies or businesses in a supply chain, such as demand forecast quantities, product types in particular time windows, or confirmed capability to process a next stage of production in a product flow – again quantities, product types in particular time windows – and so on. The cycle ends with a static plan, which may be published in part to other partners or suppliers in the supply network. It is impossible to adapt to real time events as the inertia in the system is high and changes to plan difficult to make without wastage in all its manifestations. The company centric view of the planning process for the supply network is highly fragmented. Planning takes place independently and at different times and intervals. It is deterministic being primarily driven by fixed rules. Figure 20-1 illustrates a typical, basic planning process.

The process begins with an interactive but mostly manual planning process to determine which products should be made and where. It is a make to forecast plan not make to order so in most cases no attention is paid to customization. The result of the planning exercise is a time-phased product-plant matrix and it involves independent parts only. This is the master production plan, which is derived from market trends and forecasts and it uses only an available long term aggregate plant capacity. Typically this process takes place at long and fixed intervals of time. The master plan remains static regardless of the actual product mix or regardless to any changes to it.

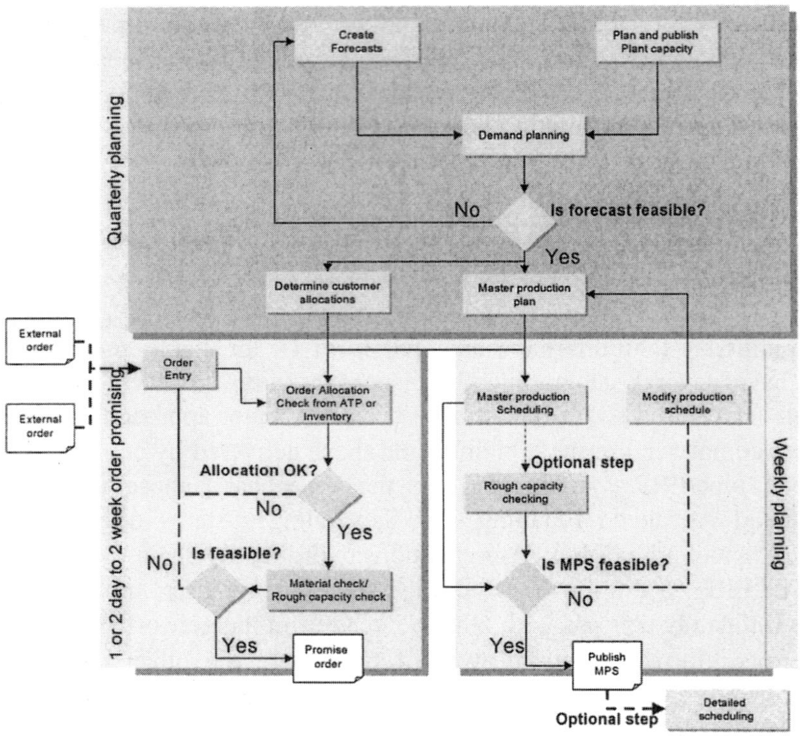

Figure 20-1. Typical MRP-II Based Process

Each manufacturing unit of the supply network may run its own MRP plans, typically once a week but based on the master production plan. The demands generated by the master plan for independent parts, generates manufacturing orders for every consumed part according to the independent parts' bill of material. Deterministic stocking policies and standard lead times generate requirements, thereby purchase orders are generated. An optional capacity requirements planning module may be used to calculate resource loadings. Any back orders and forecasts are aggregated and included within the Master schedule. Stocking policies are also considered when planning. It is very clear that this approach is not flexible but deterministic, predictive and rigid. This is far from the *"listen, serve and customize"* approach and retains the traits of *"make, sell and tell"*. Customer orders are not identified throughout the supply network due to a lack of pegging physical production to customer orders throughout the order lifecycle.

In summary the process used involves a batch allocation to production capacity approach and is based upon continually updated predictions (forecasts). This is the principle way of responding to the market and places considerable priority in getting predictions 'right'. On the plus side, if the manufacturing process takes an appreciable time, maybe weeks, time can be saved because inventory is 'available to promise' to customers either from finished or partially finished stocks or planned production. But predictions rarely meet with reality and as a result compromises customer service. Further to this, production to stock can be wasted, leading to reduced profitability and ultimately threaten the long term viability of the company.

To manage the order pipeline in this working environment, companies tend to give higher priority to allocate 'available to promise' quantities to top tier customers, whom account for a large percentage of the company's revenue, while compromising customer service to others. These planning failures to both the customer and the stakeholders precipitate a number of manual interventions to try to expedite an economic order flow. The real failure in the system is that it does not use real time information from either suppliers or customers, or even the production processes to be able to manage and control order requests and order processing in the real working and market environment.

Some Advanced Planning and Scheduling (APS) meet some of the failures in ERP with better algorithms and extra functionality, such as integrated supply chain and factory planning systems. They are increasingly used in industry as the core planning engines. Such systems process demands to determine optimal master plans and factory schedules. They are usually arranged hierarchically by planning sequentially and at different periods of time and frequency, meeting demands that comprise of primarily forecasts together with firm customer orders.

Although the planning process in early versions of APS systems is far more compact and efficient than the traditional MRP-II process, it remains a long way from the desired management and control system for an AVN. Produced plans probably show less variance to reality since capacity modeled constraints are taken into account. In addition, the integration of order promising and demand management function adds significant value. However, these systems fall well short of the responsiveness desired since they are still heterogeneous systems that employ fragmented data models and with some overlapping functionality.

Both planning practices used in industry today involve multiple steps and use a mixture of both automated and manual processes. These steps are early attempts to decompose a complex problem, involving managing, planning and fulfilling orders in a supply chain, into smaller and therefore more manageable sub problems. As Fox (Fox et al., 2000) argues, the decomposition into Master Production Scheduling, MRP, Rough Cut Capacity Planning, Detailed Scheduling and so on are primarily due to algorithmic limitations. Other limitations also accrue from the popular client-server architectural model. This decomposition arose out of organizational constraints, legacy systems, limitations on algorithms and limitations of IT architectures (Fox et. al. 2000).

A common issue for industry is getting a system that works seamlessly throughout a value network. Various constraints hinder this, but some of these constraints can be overcome in AVNs by working together and employing a common data model that can facilitate this collaboration. However, if manufacturing businesses belong to more than one value network at the same time what was a common data model may no longer be valid. Expensive ad hoc integration is only feasible today for large corporations, which try to overcome difficulties of data exchange arising from the many disparate data models and the logic they employ.

The concept of logical integration together with the use of a generic data model for an AVN management and control requires a complete rethink for the ICT tools and systems. This must be followed by a complete redesign and implementation methodology for responsive order taking, realistic planning and controlled fulfillment. A different decomposition and a new architecture are required to both physically integrate and logically integrate, as well as to coordinate the different relevant processes.

3. DESIGN REQUIREMENTS AND COMPONENTS OF SENSE AND RESPOND PLATFORM

3.1 Design Requirements

The platform presented in this chapter aims to solve operational planning and fulfillment problems that are fundamentally different to current approaches to supply chain management and is able to deliver sense and respond functionality.

By sense and respond we mean that:

- events that affect how resources are managed or the overall order flow throughout the orders' lifecycle are captured and continuously recorded in real time;
- mechanisms are in place to handle such events as they occur and update and repair plans in a structured manner to ensure that there is minimum impact from major changes or disruptions. Plans not yet released are modified accordingly so that completely updated plans can be released as late as possible;
- mechanisms are in place that deploy such decisions immediately across the value network.

These are the three key requirements for an operational sense and respond platform. In response to these requirements an ICT platform, code named, *Star Active*, has been implemented with the following design objectives:

- the system must be customer order oriented instead of product oriented. That is it should handle the complete lifecycle of orders instead of handling product families, which is the case of traditional systems. Special order constructs allow full trace-ability of orders across the value network with end-customer identification throughout an order's lifecycle. It also allows the pegging of customer orders to physical products and to work in progress;
- on the same platform seamlessly integrate the planning, fulfillment and order management processes using a common data model;
- provide mechanisms to capture new orders or capture changes to existing orders;
- match demand to supply by considering actual available capacity, work in progress, raw material and end product at each stage of the value network. It takes into account the different lot sizing constraints, such as manufacturing, procurement, transportation, Bill Of Material quantities, etc.;
- share planning information and requirements and coordinate plans with both external suppliers and value network partners by logical integration and physical integration of purchasing and order management business processes;
- provide a common data model to support collaboration and coordination and a flexible modeling framework for modeling of capacity for different types of physical processes and technologies. The latter allows for efficient handling of available geographically distributed capacities in the value network;

- the system should be "always on" to handle events and requests from anywhere at all times. This is in contrast to traditional systems, which handle batches of orders, load the models and provide a static solution;
- the system should be able to provide fast and reliable responses to events.

Star Active has been designed to meet these objectives. The platform was built upon research and development work performed in a three-year industrial led European project aimed at delivering multi-company proactive planning and reactive control tools to globally dispersed semiconductor manufacturing companies (Richards et. al, 1997, ESPRIT 20544, 1999) and their subcontractors and suppliers. Orion Logic Ltd bases the design presented in this chapter on the original system's commercial version.

3.2 Components of the Platform

To comply with the design objectives stated in the last section, Star Active has been implemented as a multi-agent system comprising static agents that is agents are not mobile from one host to another. This design approach was used for the following reasons:

- a more accurate modeling representation of resources in a value network can be achieved due to responsibility assignment of agents to a particular portion or domain of a value network. For example a particular warehouse or manufacturing line are modeled as distinct domains and hence an agent can be responsible for such a domain capable of solving a particular problem (e.g. resource planning or material sourcing);
- problem solving takes place in parallel with each autonomous agent responding to particular real time events relevant to an agent's capability and domain;
- the system can be both scaled up or down according to the complexity and needs of the value network (Figure 20-2);
- since an agent is attached to a particular domain, the agents are autonomous and static and are not mobile;
- since one agent creates a request for another peer agent and waits for the peer's response, the integration of the supply network is seamless since each agent exchanges on-line the right information with its peers in response to requests (Figure 20-2);
- the structure ensures complete control, autonomy, security and ability to customize to any entity in the supply network. Agents exchange abstract and largely non-confidential information. The handling of requests takes place internally by the relevant agents that use more detailed information,

such as particular policies, capacities, any necessary rules etc. That is information not visible or relevant to the requesting agent (Figure 20-2).

Agents are fully object oriented and agent to agent communication is handled by CORBA. Figure 20-2 illustrates a network of cooperating Star Active agents. Each unit in the network controls its own set of agents and hence data model, although agents are capable of negotiating with their peers by exchanging abstract and non-sensitive information.

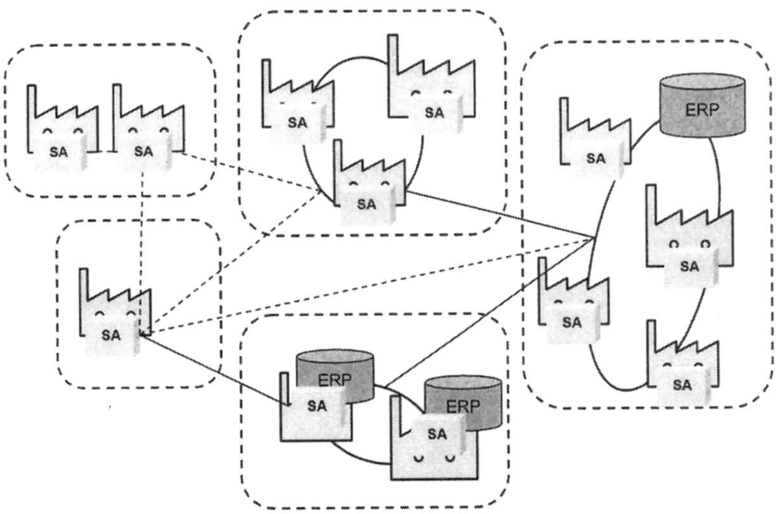

Figure 20-2. Network of Star Active (SA) Agents Deployed Across a Supply Network and Integrated to Other Systems

Star Active agents are of type service, resource management or user. The platform includes the following agents:

1. *Information manager (service type).* The information manager is responsible for:
- persistent storage of information;
- supply of structural and modeling information to other agents;
- management of dynamic information such as resource capacities, plans, orders etc.;
- integration to and exchange of data with external systems;
- collecting real time information from tracking systems;
- authenticating users who interact with the system through the user agent, based on user directory and user permissions and roles;

- one or more agents connect to the information manager to retrieve modeling information to carry out their particular tasks and also to store results in the data base.

  There is at least one information manager for a supply network node.

2. *Order management agent (service type).* The order management agent is primarily responsible for:
- capturing and processing external orders either from one or more user agents or from external systems such as CRM, e-commerce, customer's procurement system, ERP, mobile devices, call center etc.;
- processing customer or marketing forecasts, or internal orders;
- canceling or modifying existing orders;
- configuring products when capturing orders;
- creating 'Demand Orders' and submit those demands to subsequent agents for further processing (planning).

3. *Planning agent (resource management type).* When this agent receives a request, which includes a 'Demand Order' '*i*', which comes from a parent Order (e.g. customer order) j, for quantity *q* of product *k*, required by time *t*, it does the following:
- It searches for available planned production, work in progress or raw material (depending on the type of product *k*);
- It nets against q available quantity at time t or best time closest to *t* ;
- It creates a request for the capacity handling agent to create a new manufacturing order and plan it considering available resource capacity.

This agent employs a real time algorithm to search for available work in progress and together with the capacity handling agent identifies the best time closest to 'required by' time to plan the order. Both agents work in parallel and they are responsible for a particular resource domain. In a value network there will be several of these agents operating in parallel.

4. *Capacity handling agent (resource management type).* The capacity handling agent employs a comprehensive resource capacity model and uses fast non-linear algorithms to optimally allocate new manufacturing orders to resources with available capacity at particular time periods. The resource capacity models can be detailed or abstract depending on the particular modeling preferences and complexity of the resource domain in question.

5. *Material sourcing agent (resource management type).* The material sourcing agent employs a similar algorithm to that of the planning agent, which responds to 'Demand Orders'. Typically however, a material sourcing agent will model a warehouse and in case of SKU (stock keeping unit) shortages it will create a request for fulfillment sent to the upstream agent. The latter will typically be either a larger warehouse, a wholesaler or larger distributor (e.g. an upstream echelon in the network) or a manufacturing supplier.

6. *Planning exchange agent (resource management type).* For given a single demand order or a set of demand orders, the planning exchange agent:
- is based on the part ID and type in the demand order and the directory service of available agents;
- identifies which alternative agents can satisfy this demand order;
- creates requests to those agents to satisfy the demand orders;
- creates alternative demand orders containing permitted alternative parts in case the available agents cannot satisfy the initial demand orders in time;
- selects the best allocation from those proposed by the various agents in the network;
- determines transportation requirements, if applicable, and incorporates transportation constraints into the resulting plan.

7. *User agent (user type).* A user can interact with the system through this agent. This agent manages the user interface and specific user preferences and also provides all workflows relating to the functions the particular user is assigned to and has responsibility for (e.g. example order management, planning, performance reporting etc).

8. *Supplier coordination agent (resource management type).* This agent gets a request in the form of a demand order, then the agent:
- searches for available planned receipts from suppliers in the network;
- creates procurement requirements in case there are no planned receipts for material;
- issues those requirements to the relevant suppliers in the form of demand orders who receive them through the order manager agent, which triggers the processing and planning of those orders (Figure 20-2);
- responds in cooperation with the planning exchange to identify the most suitable supplier. In case the supplier does not have Star Active implemented, the requirements are sent to the supplier over the internet

or can be accessed through a standard web browser from which the supplier can respond until a timeout period expires.

9. System management agent (service type). The system management agent updates the status of the modeled resources and the status of the plan at regular intervals of time. For example,
   - it checks if timeouts are expired;
   - it changes the status of a plan to 'released' at the time specified;
   - it checks progress of plan to reality by comparing with real time data collected from an order tracking system;
   - it initiates order cancellation process on failed machines when triggered by production machine failures;
   - inventory changes are monitored and quantities are updated;
   - purchase order delays are logged and rescheduling is initiated by the scheduling agent

10. Scheduling agent (resource management type). This agent is responsible for one resource domain and performs the following functions:
    - creating machine timetable for released manufacturing orders;
    - applying production machine(s) sequencing algorithms for released orders;
    - calculating 'knock on' effect on released manufacturing orders in case of unplanned events and regenerates schedules and timetables;
    - updating production machine(s) loading profiles.

A solution delivered based on Star Active will comprise of one or more agent types and one or more agents of the same type (Figure 20-3). Hence is a solution for a network of such agents. A collection of such agents is assigned to a particular domain in the supply network. A domain is either a single manufacturing line, or a warehouse or a manufacturing unit etc. That is a domain is an area that contains a collection of resources that perform a particular process in the value network, for example silicon chip fabrication, thread dying, automotive final assembly, final product packaging and shipping etc.

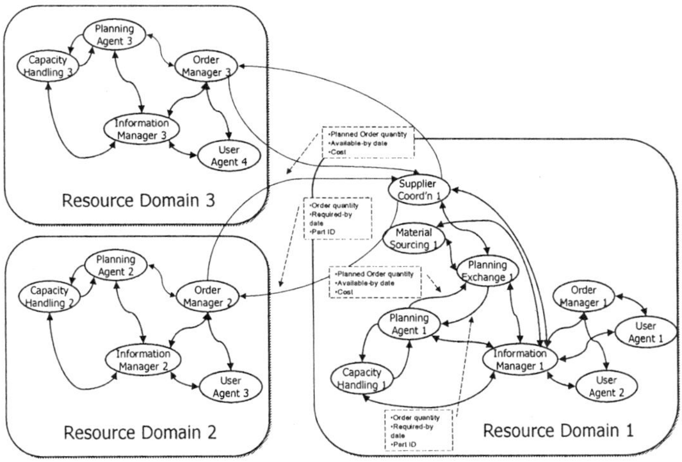

Figure 20-3. Agent Interactions across Resource Domains

4. SENSE AND RESPOND ICT PLATFORM FUNCTIONALITY

Star Active has been implemented as a platform, meaning that it can be specialized to cater for any industry requirement and operational business processes. Additional functionality may be added to meet particular needs. The basic functionality of Star Active is described next.

4.1 Collaboration and Generic Data Model

The platform delivers collaboration by the capability of the information manager to link bi-directionally to external systems. The various agents leverage this information through the information manager and the results are fed back to those systems. Collaboration is also delivered through cooperative problem solving of agents across resource domains and hence across participants in a value network. For example, the sales department is linked in real time to manufacturing likewise a particular value network company is linked to its suppliers and so on. Collaboration is assured through the use of coherent agent to agent communication protocols implemented by the requests and responses by those agents and also through the underlying coherent data model implemented by the information managers which is used by the various agents. The model includes the following constructs:

- Modeling constructs:
  - Part catalogues and directory, which includes pricing policies, discount schemes, substitute and alternative products, units of measure etc.
  - Customer information, including customer segmentation (tier), location, customized product catalogues, delivery preferences and others.
  - Bill-of-materials (BOM) information for each part including producedPartID, consumedPartID, consumedQuantity, processFlow.
  - Supplier information, including quality, minimum order quantities, parts catalogues etc.
  - Supply network structure, defined by location of supply network nodes such as customers, suppliers, warehouses, production units and transportation between each supply chain node
  - Factory information including ProcessFlows, ProcessOperations, Resources, etc.
  - User directory including user roles and access permissions
  - Agent directory
- Dynamic constructs:
  - Inventory availability at each node
  - Resource availability
  - Plans and schedules
  - Orders

Each node in the supply network or each resource domain is responsible for the maintenance of its model. This ensures autonomy of the partners in the supply network, easier and more efficient data collection as well as efficient data handling. This is in contrast to centralized solutions, which require the collection and maintenance of large amounts of data in a central database.

4.2 Sense and Respond Framework

Star Active is a dynamic system which has been designed to be available at all times to sense and respond to real time events. It employs an in-memory processing architecture meaning that supply chain models reside in memory and once events are sensed agents use this data for immediate problem solving. Star Active senses and responds to the following events (Table 20-1):

Table 20-1 (a). Sense and Respond Framework in Star Active.

Event	Example	How sensed	Agent involved for sensing	Agent involved for responding	Response
New demand capturing	New customer order, planning enquiry, planning scenario	Direct user input, B2B system, ERP system, customer's procurement or ERP system, Customer's Supply Coordination Agent	User agent, Order manager	Planning exchange, Material sourcing, Planning agent, Capacity handling	Insert new demand into plan, make order reservation, create planning scenario, exception message for inability to plan demand based on given parameters
Modification of existing demand	customer order change, customer forecast demand profile changed	Direct user input, order manager, external system (ERP or other B2B)	User agent, order manager	Planning exchange, Material sourcing, Planning agent, Capacity handling	Modification of existing demand, exception message for inability to modify existing demand
Material arrival	Timestamp of material arrival at a particular check point	External tracking system, manual user input	User agent, information manager	System management agent	Logged timestamp of arrival in information manager, compared to planned arrival time and exception raised accordingly

Table 20-1 (b). Sense and Respond Framework in Star Active (Continued).

Event	Example	How sensed	Agent involved for sensing	Agent involved for responding	Response
Early/tardy order	Late or early arrival of a planned manufacturing order or planned receipt	External tracking system, manual user input	User agent, information manager	System management agent, scheduling agent, capacity handling agent, material sourcing, planning agent	If early or tardy order exception is raised &issued to the relevant user agent, by the system agent then the resource management agents, propose alternative actions including reprioritization at equipment level, re-pegging of work in progress to urgent orders or change of due date (rescheduling)
Safety inventory level change	Material stock falls below a safety level or its JIT level	External tracking system, manual user input	User agent, information manager	System management agent, capacity handling agent, material sourcing, planning agent	System management agent compares sensed actual level against predefined level, issues early warning to relevant user agents and upon user approval, a stock replenishment order is inserted into the current plan

Table 20-1 (c). Sense and Respond Framework in Star Active (Continued).

Event	Example	How sensed	Agent involved for sensing	Agent involved for responding	Response
Resource breakdown	A machine cannot execute its schedule due to failure	External tracking system, machine interface, manual user input	User agent, information manager	System management agent, scheduling agent, capacity handling agent, material sourcing, planning agent	In case of machine breakdown, an exception is raised& issued to the relevant user agent, by the system agent then the resource management agents, propose alternative actions including reprioritization at equipment level, use of alternative equipment, re-pegging of WIP to urgent orders or change of due date (rescheduling)
Timeout expired	Order reservation timeout expired, supplier purchase order response time out expired	Information manager, system management agent by comparing current timestamp with timeout timestamp	System management agent, information manager	System management agent, information manager, user agent	Warning issued, reservation cancelled
Plan release time	Time to release plan for execution has arrived	Comparison of timestamp to plan start time and plan release rules	System management agent, information manager	System management agent, information manager, scheduling agent	Scheduling agent creates timetable and detailed resource schedule

This table illustrates Star Active's sense and respond framework. Note the table refers to a single supply network node or domain and that not all agents are involved in responding to particular evens. For example to respond to new demand there may be a case that only the material sourcing agent is involved. This depends on supply network complexity and solution configuration.

4.3 Order Management Framework

Star Active responds to new demands and changes to existing demands as they occur at any time and from anywhere. Demands are immediately processed and hence become visible to all partners. Demands can either be new firm customer orders, customer enquiries or forecasts, engineering demands for new design prototypes or planning requests generated by the planner for what if analysis etc. Once entered into the system, those demands have complete owner identification to provide the capability to trace the complete demand lifecycle throughout the supply network. Once a demand is entered and processed, the user has the following options:

- **make a commitment**, where this option updates the current demand profile and supply network plan, by including this new demand, but does not alter demands already planned or released;
- **make a reservation,** where this demand is tentatively planned but will be removed when a set timeout period is expired and no commitment has been made. If a commitment is made then the supply network plan is updated;
- **check only,** where in this case the system performs its complete cycle, considers constraints and also the current plans and demand profile but makes no reservations or commitments – it just returns a temporary plan for the particular demand. This option is primarily used for what-if analysis and fast feasibility checking when generating alternative scenarios.

An external demand relates to an independent part. Independent parts are those for which their demand does not directly depend on demand for higher level parts and hence they can be planned independently of other parts. Independent parts, include finished products or sub products that can be stocked at intermediate work in progress buffers and pulled into production when required in higher level subassemblies. Dependent parts are those in which their demand depends directly on the demand of other higher level, or parent, parts. This means that a single order for an independent part spawns a large number of 'sub-orders' for dependent parts and each has to be delivered on time so that the material flow is coordinated. Star Active employs a special hierarchical construct (data structure) to coordinate order flow, ensure end-customer identification and trace-ability and also to be able to peg physical parts to orders at all times, thus has the ability to track progress of specific customer orders at all times. This construct includes the components shown in table 20-2. Those components are of type demand, or supply or service. Each such order is specific to a supply network node's resource domain and these components are linked with their parents and children to ensure trace-ability across the supply network as well as end-customer identification.

Table 20-2. Components of Star Active's Order Management Framework

Component name	Type	Description	Created/Modified by agent
DemandOrder	Demand	is a request for a part p at a required-by time *t* to be fulfilled from new or existing ManufacutirngOrder, new or existing PurchaseOrder or available OnHandInventory quantity	Order Manager Planning Agent Planning Exchange Supplier coordination agent
ManufacturingOrder	Supply	is an order to manufacture quantity *q* of part *p* by time *j*	Planning agent Capacity Handling agent
PurchaseOrder	Supply	is an order to purchase quantity *k* of part *p*, where this part will be available at time *h*.	Supplier Coordination Agent
OnHandInventory	Supply	this is not an order but the available on hand inventory at location x at time *t* of part *p*	Material sourcing agent
TransferOrder	Service	This is an order to transfer from location a to location *b* (or from supply network node *a* to *b*) quantity of part *p* (or a certain number of lots) that relates to a demand order required by the node or customer	Planning exchange

Supply type orders are further linked to another data structure, which is the model of the plan. That is the steps required to complete the particular order. For example a manufacturing order for part *p* will require a number of time-phased steps to complete, when one considers its process flow (or manufacturing routing).

Figure 20-4 shows an example of an end customer order together with its sub orders to illustrate how the order management framework works. In this example there are 2 existing demand orders for independent parts already planned. These orders have created manufacturing orders and purchase orders. It can be seen that there is a direct referencing between demand and supply type orders. A third demand order which is urgent comes in at a later stage. The system re-pegs quantity from order Mfg0003/D0004 which is a customer forecast to the new urgent order and as a consequence all related orders and their quantities have been modified. Also the excess quantity for part P4_3.6GHz that arises from the supplier's minimum order quantity is pegged to the new order instead of ending up in inventory. Note also that specific orders contain a particular supply network node ID.

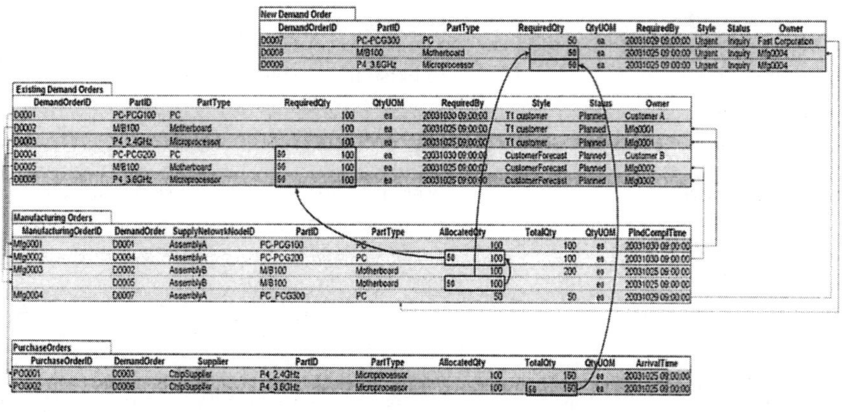

Figure 20-4. An Example of How the Order Management Framework Works.

Star Active's framework also supports dynamic re-pegging. That is a process where a quantity of a product in an existing planned order is re-pegged to a new order, which is of higher importance than the current planned order. This functionality is useful to respond to disruptive events and when planners would like to give higher priority to one customer against another. This functionality ensures that no wasted material or product occurs such as that which often ends up in inventory because of mismatches in lot sizing policies (see also chapter 4). Dynamic re-pegging priorities are determined by user-defined rules or policies depending on the type of demand and customer. Also note that manufacturing orders are assigned to particular AVN nodes, identified by the SupplyNetworkNodeID attribute. For simplicity, how quantities are pegged from inventory is not shown.

During its lifecycle, an order assumes a number of states including the following:
- Inquiry: an order is in this state if the owner of the parent order (i.e. the independent part) has been checked for feasibility but has not yet been confirmed;
- Planned: an order is planned once it has been confirmed and is therefore part of the master value network plan. In this case the system takes such orders into account when it processes new orders;
- Released: portions of the master supply network plan are released at predefined times for fulfillment, that is for production, or picking, and delivery;
- On hold: due to an unforeseen event, the order is on hold waiting a decision;
- Completed: all defined plan steps for a particular order have been completed.

The system management agent tracks each order throughout its lifecycle and updates an orders' status accordingly.

Star Active's order management framework supports the following order processing functions for independent parts. Those in turn drive requirements of dependent parts:
- full sales order: a standard sales order placed by a customer. This includes all components of the fulfillment process (planning, fulfillment, invoicing). It may include one or more order items each corresponding to a specific final product ID;
- full purchase order: standard purchase order created by requirements for supply to network suppliers and network partners (depending on the configuration of the value network). This order type may be linked to another demand order (e.g. sales order). It may include one or more purchase order items each corresponding to a specific part ID (i.e. subassembly, for assembly type manufacturing);
- quotation (RFQ-quote process): This represents a bid and not a sale. An order of this type can be converted to a standard order if the bid is accepted. It may include one or more order items each corresponding to a specific final product ID;
- stock movement: Represents stock movements between warehouses, or replenishment requirements, from customers in a Vendor Managed Inventory (VMI) scenario. This relates to a single part ID;

- direct back-to-back sales: Automatically creates an associated purchase order to be created. This order type provides for goods to be shipped from a network supplier, or network partner, directly to the customer. It may include one or more order items each corresponding to a specific part ID;
- forecast order: This demand type relates to a single part, or final product ID, and represents forecast time phased part requirements that may or may not be produced. This order does not relate to a real customer order and if produced will primarily be stocked until it is sold.

4.4 Product Modeling and Customization Framework

Star Active is an application in environments where customers are given options to customize their orders. Such customization is achieved by:

- Configuring product manufacturing: the platform allows process and manufacturing engineers to quickly combine existing manufacturing operations or specialize available abstract manufacturing operations and create new manufacturing process flows to make a customer specific product. For example in textile manufacturing it is very common for customer to select particular colors not previously used by the manufacturer. Another example is the production of a special semiconductor chip prototype, where the process engineer can build a process flow to make the prototype. This type of configuration is managed by the construct shown in figure 20-5. This customization can be done in a short space of time by an instance of the user agent, which is aimed at the process or manufacturing engineer, without any modifications of the underlying models and configuration of the system.
- Selecting product options: the platform allows multiple alternatives or substitute parts for each node in the bill of materials tree. The customer can then choose which option they require. The platform then tests out any compatibility constraints between the selected parts.

Due to lack of widely accepted product naming conventions and industry standards there are cases where various partners in the AVN have their own naming convention for the same parts. Star Active offers an intermediate layer that functions as a dictionary between the different product naming schemes across the AVN. This allows for identification, search, ordering and tracking of products across the AVN. It also allows the speedier product modeling since partners of the AVN do not have to keep changing the part numbers once they join an AVN.

4.5 Communication Framework

There are three types of communication supported by Star Active:

- Amongst agents which is handled by CORBA
- Between Star Active and external systems where a number of protocols and interfaces are supported including:
    - CORBA
    - XML and Web services
    - Data base to data base scripts
- Demand capturing channels including:
    - Web site
    - Customer's procurement system
    - Customer's ERP system
    - Direct user input through user interface
    - Through wireless device
    - Other order capturing system including CRM, ERP or e-marketplace

4.6 Resource Modeling Framework

The resource model is primarily used by the capacity handling agent and the scheduling agent. This model is a detailed representation of a production unit's capacity and includes process flows, process operations and resource loadings as shown in figure 20-5.

Capacity analysis and work allocation is performed by splitting the planning horizon into plan elements to allocate resource time available to each task. A plan element is a segment of the planning horizon where a particular resource has to carry out the tasks allocated to this resource. To identify earliest or latest resource availability, the agents search plan elements across the planning horizon as well as partners in the value network. During the search process, the part's process flow requirements, the current resource production rates and utilizations are taken into account. The agent considers alternative resources or alternative plan elements that can accommodate the new demand.

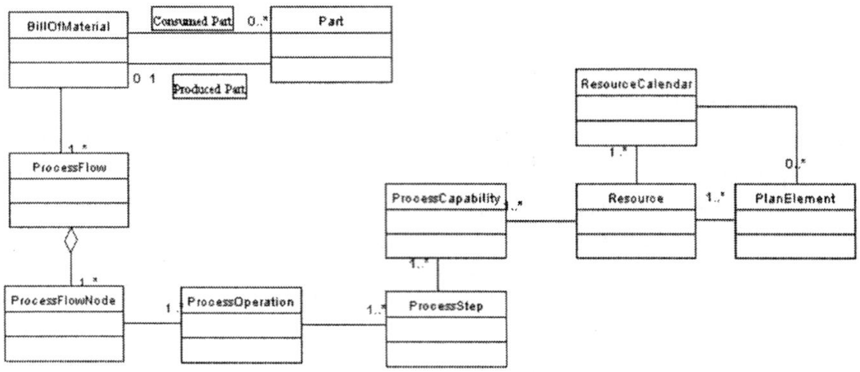

Figure 20-5. The Conceptual[1] Class Diagram of a Factory's Process Flows and Resources.

Star Active also provides the capability for more aggregate and less detailed capacity modeling. This is achieved if required by the use of static rate-based models of a resource domain that allows for fast modeling of simple resource domains. Such models are relevant to simple assembly processes or simple linear flow line based manufacturing processes. In addition such modeling is also used in cases where it is not possible to have the complete model of a component supplier.

4.7 Planning Cycle and Agent Cooperation

The system is continuously available ('live') for event and demand processing. Demand is captured from all communication channels, external systems, order acquisition devices, the internet, telephone, etc. through the order manager's relevant interface and then sequenced according to customer's segment, required date etc. The system processes demands at the time they are captured. The core processing functionality of the planning and fulfillment cycle is delivered from the system's planning core. The core contains the planning exchange and material sourcing, production planning and capacity handling agents. As mentioned earlier, those agents perform process requests in-memory where a copy of the data, relevant to their functionality is kept to avoid constant loading data from the database in order to process each request. Each external demand for fulfillment is processed initially by the planning core by breaking down the external independent-part demand into dependent-part demand orders, and identifies available material, planned production or receipts, or resource capacity. It then assigns a fulfillment time slot to each such demand order and the external, parent, and demand order. The cycle repeats for every level of the independent part BOM and supply network. This leads to an updated fulfillment plan and also to updated supply plans.

Planning is performed in layers. A layer is either a warehousing or a distribution echelon, a supplier tier or a production stage where material transformation takes place, i.e. the production of one parent part ID which consumes one or more child parts. Also a layer can have a combination of such functions. For example in the case where a part can be manufactured in-house, manufactured by another wholly owned production unit, or purchased from a supplier (not an AVN partner). The system creates a solution for one layer at a time, passes the requirements of the current layer to the next layer in the form of a derived or 'child' demand order and then if any constraints accrue from that layer, the solution on the former layer is adjusted to take into account the constraints of the latter. Typically to solve for a layer one would require the planning exchange and one or more resource management agents. In other words one can observe that the system propagates requirements backwards (an approach similar to backward planning) and coordinates by forward constraint propagation (an approach that resembles forward planning). The cycle begins with the input of one or more orders regardless of their particular type (e.g. full sales order, stock movement, forecast order etc.). Each resource management agent can solve for more than one order at a time that is more than one part IDs which clearly belong to a different layer. The planning exchange agents and the production planning, capacity handling and material sourcing agents continue in this manner until there are no unplanned demand orders.

5. SYSTEM IMPLEMENTATION

Star Active's agent based architecture provides the flexibility to model a supply network, deploy the required agents to support immediate user needs and then expand the solution by either adding more agents or expand it to other supply network resource domains, supplier, partners etc. without modifying the existing implementation.

Modeling is typically a one-off process for one or more resource domains since it requires the input of the data model into the information manager and the integration of the information manager to any existing systems (Table 20-3). Then the selection and configuration of relevant agents takes place that relate to the particular needs to be tackled in a project. Then specific business processes are modeled by configuring the user agent accordingly. That is the right screens are configured meaning that existing user agent components are used as is or extended. Then one or more resource management agents are deployed and connected to the information manager and have access to the resource domain model in order to retrieve required information to carry out their particular tasks. Additional agents can be added without any additional modeling effort and thus the system can be easily scaled to accommodate simple or very complex supply network models. Also there is the added advantage of extending the existing services and functionality of the basic agents and information manager to implement custom functionality and workflows with minimum effort.

Table 20-3. Sample Data Input from Other Systems to Star Active and Data Output from Star Active to External Systems

External System	Sample Input Data to Star Active	Sample Output Data from Star Active to systems
ERP/MRP-II	BOM, routings, Stock, Stock Locations, Sales Orders,	Available Inventory, ATP, Master Plan, Material Requirements, Purchase Requirements, Order Due dates
CRM	Customers, Orders, Channel partners	Order due dates, order reservations, order commitments, ATP
Manufacturing Execution System/Tracking System	Resources/equipment, equipment status, work in progress	Resource schedule, new repaired schedule, work in progress, expected planned receipts and deliveries, expected inventory

This architecture allows the complete or stepwise model of a value network with additional functionality delivered as necessary. It thus offers the ability to model a supply network and its integration with other systems together with added functionality without affecting the operation of the deployed system. One or more information managers can coexist. Each represents the model of a particular user-defined resource domain. This approach also allows much more controlled management of a project and lower solution implementation and maintenance costs. Figure 20-6 shows a screen capture of Star Active used for microelectronics industry.

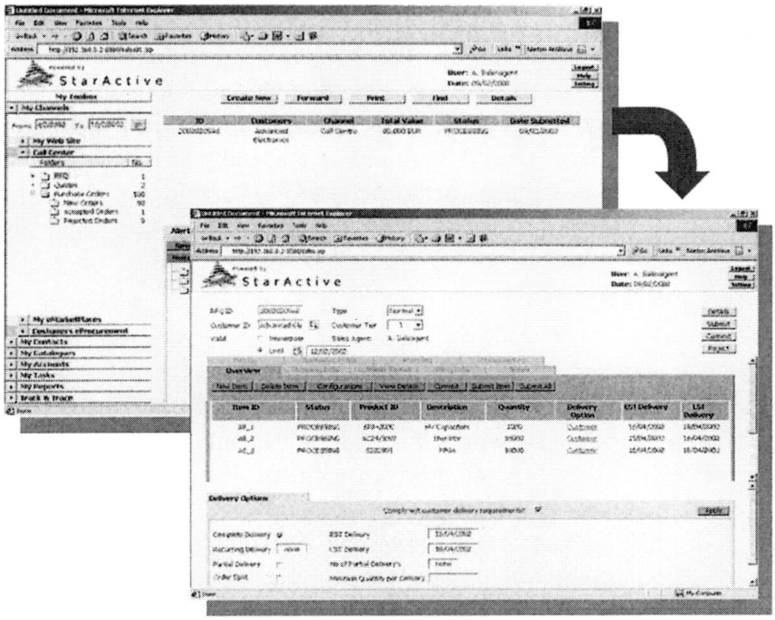

Figure 20-6. Sales Agent's Toolbox (Copyright: Orion Logic LTD, www.orionlogic.com)

5.1 Standard Work Flows

Based on the authors' many years' experience in supply chain consulting, it is very hard to change an operations manager's or planner's habits without introducing a convenient automated workflow based on the user's experiences and particular needs and practices. The ability to operate in a value network does imply a certain attitude change and the implementation requires education, careful planning and user training. Most software implementations fail because processes do not meet the real needs of the fluid organization or its people. Another reason for failure is when critical network nodes are omitted from the workflow. A flexible user interface, that allows the addition of functionality and customized workflow to manage information and the system's base functions, can provide the user with a virtual environment tailored to his or her need in the way he or she interfaces with it. In addition, various partners in the value network require a different view of information and also different handling of the information. This will be according to their roles, and achieved by adding either specific, or custom functions to the user interface agent. This helps to promote collaboration between the partners in a value network, by crossing organizational boundaries, and streamlines information dissemination to all parts of the value network an understanding of the complete planning and fulfillment workflow. For example the order taker, the planner and the factory managers and transport managers are all involved in planning and fulfillment processes, and in this respect they should manipulate common information, but in preference to do so each requires personalized GUI functionality.

6. CONCLUSION

The architecture and functionality, the system requirements and problem solving concepts of an adaptive, sense and respond planning and fulfillment system for adaptive value networks has been presented. We have shown how collaboration is delivered to support distributed operations and decision making in a value network as well as to enable individual companies to reinforce their position as partners in various other value networks.

Feeny (2001) defined five potential opportunities for improvements that business process automation can deliver: opportunity for automating administrative processes; new technology's ability to trigger a review of the business' primary infrastructure; supply chain reconfiguration and integration, use of technology to enable a virtual enterprise; intensified competitive procurement through electronic buying; increasing parenting value, which relates to performance improvement of individual business units through help from other "members of the network". The system presented in this chapter is intrinsically an adaptive, sense and respond platform that can leverage those opportunities and deliver benefits through the integration, into a single cycle, of planning with fulfillment execution within a value network. This is ensured through the provision of tools that allow fast response to disruptive or unplanned events that provide a systematic framework for effective and efficient control of fulfillment. The solution has been structured not only to deliver the state of the art in terms of an automated business process and decision support but also in terms of how benefits are delivered. The component architecture employed allows structured and controllable implementation in small steps, which reduce implementation risks and total cost of acquisition and ownership of the solution. In addition the solution presented is scalable both from the scope of functionality and the complexity of the value network, making it applicable to both large and small businesses alike. This ensures reduced barriers for use of new technologies to small businesses and also helps with improved collaboration between large and small businesses.

This 'networking' effect leads to responsive value networks, and reduction of costs and waste for all participants and ultimately to significant positive contributions to the bottom line for all partners in the network. Further to this, the networking structure of the platform allows the participation and handling of multiple constraints and events as they occur within the working and market environment, as sensed through multiple channels to markets and customers. For example how multiple requests arriving from multiple sales partners are handled and how capacity is effectively managed to best fulfill customer needs.

Companies seeking those opportunities invest heavily in purchasing a variety of state-of-the art systems in order to improve their supply chain management. But success implies a careful implementation. In order to get benefit from a planning and fulfillment system, it is very important for the vendor to thoroughly understand the clients' requirements and working environment. The client also has to understand what is being offered and needs to select a system that can address all the issues which include accurate event handling, flexible and easily customized user workflow and network collaboration. Then it is possible to truly optimize the business processes in the value network and to maximize the returns from the use of such a system. These characteristics were discussed in the context of the system presented in this chapter. This chapter also discussed how this system can bridge the gap between managing customer demands and managing supply in response to those demands. Finally, the concepts presented will illustrate the requirements and industrial needs when considering or designing real world adaptive, sense and respond supply network planning systems.

NOTES

[1]This diagram shows the conceptual relationship between classes to support the description of the base functionality. The actual implementation has certain differences for performance reasons and delivery of functionality, which is not part of this chapter.

REFERENCES

1. ESPRIT 20544, 1999, http://www.prosoma.lu/Showcase/Results/2717/HTML/descriptionsx.html
2. Feeny, D. Making business sense of the E-Opportunity. Sloan Management Review. 2001; Winter: 41-51.

3. Fox, M., Barbuceanu, M., Teigen, R. Agent-oriented supply-chain management, International Journal of Flexible Manufacturing Systems 2000;12:165-188
4. Jarillo J. C. *Strategic Networks, Creating the Borderless Organisation*, Butterworth-Heinemann, Reed-Elsevier PLC, 1993
5. ORION LOGIC LTD 2003, http://www.orionlogic.com
6. Richards, H.D., Dudenhausen, H.M., Makatsoris, C., Ridder D. L. Flow of orders through a virtual enterprise – their proactive planning, scheduling and reactive control. IEE Computing and Control Engineering Journal 1997; August: 173-179.6.

Epilogue

At the core of the dynamic industrial landscape is the organization and capabilities of enterprise. These organizational capabilities are the collective physical facilities, resources and human skills and knowledge that are time-shared and are brought together in a powerful mix through ICT to effectively compete in markets across the globe. It is only when these various characteristic elements are carefully coordinated and seamlessly integrated that economy of scale and scope manifest themselves with new innovative abilities, potency, resourcefulness and deftness. Traditional skills that were used to manage companies are no longer valid in inter-company networks and new skills have to be learnt and new knowledge acquired by understanding the mechanisms and joint behavior through which these new organizations will flourish. The critical task for management, as with single large companies, is to unify the organization such that its totality is more than equal to the sum of its component parts. And, management must also respect the aspirations of all partner businesses within the network.

The book has set out to provide the reader with a comprehensive understanding of the challenges faced by industry to achieve new levels of capitalism in which fluidity of organization is its strength and determinism its weakness. They describe the driving forces for change and offer new concepts, methodologies, design processes, management and control philosophies, and ICT tools and systems that benefit organization evolution and its strategic missions. Such missions take note of the all important customer choices and desires in the avid consumer society in which we are all accessories but at the same time they have to satisfy the stakeholder.

When dealing with the scale and scope of the compass of this book the reader should have sensed some of the difficulties of comprehension of the challenges and how to overcome them. This process of stimulating the imagination is paramount in the chaotic environment we have today. Thinking differently, utilizing experience from other fields as biology for instance enables a new way of working together to be developed. The suggestion that a view on the world of ants whereby their individual and collective behavior may offer model solutions for intelligent software agents as human proxies is a case in point. The characteristic that decentralized mechanisms can achieve a global coordination is also born out by economists that have long studied how local decisions can yield globally reasonable effects. These insights may or may not be catalytic ideas for new approaches to ICT but can lead to new research and development to be undertaken. This could eventually pay off in a number of ways, the example that autonomous intelligent agents with developed properties that allow them to work as useful human proxies in multi-agent communities to the benefit of an organized adaptive business network is one.

Our concluding remarks will take a journey through the landscape of component parts, both physical and virtual, that will lead to an adaptive value network (AVN) from conventional supply chains and reach out to understand how symbiosis with ICT can ensure their sustainability. This we will do by looking at networks, organization, connectivity and mechanisms to gain an insight on how they are combined in a united whole through intelligence and strong relationships. We will also take note of the responses to stimuli both within and without a business network. Finally we make comments of the value gained within AVNs and the change process necessary to achieve them.

It is far from an easy journey, we describe, to take us from where we are now in the modern industrial landscape to where we want to be in future. Interoperability standardization, culture and language differences, and market regulation currently limit connectivity. Moreover the right business processes to respond to the various stimuli external or internal have to be created and learnt and modified to meet the dynamic need. To assist with this, organs (e.g. skills of people and software tools) at a company's, or business's, disposal have to be adapted or changed to meet the wider needs of the business network. In the context of the business network environment, real time mechanisms of planning, tendering, procurement, network execution processes and so on have also to change to meet the new supra-organizational goals. Moreover, the difficulties of local autonomous demands living together with those of the supra-organizational network have to be resolved. Strategies for companies have to reflect this, as do any inter-company or inter-network collaborative agreements. Belonging to more than one network and being able also to have free action in the market place seem to be conflicting objectives but can be resolved with the right overall strategies and actions.

NETWORKS

Networks are the oldest social invention. But ICT technology has created a new ability to form networks, to form the connections and develop functionality for constructive mutual interdependence between companies or businesses. These networks may be transient affairs and are often created to tackle one specific task, such as a construction project or a tendering task with after support services. But they may also have longer lifetimes as manufacturing and service networks for meeting sustained customer demand. But before any networks can be formed there has to be a willingness to reify by a group of consenting companies who wish to work in some kind of alliance; or there is some catalyst that acts as a stimulus for companies to seek out others with different capabilities and capacities to join together in a stronger grouping, as with tendering for instance; or it may be a forced evolution from a supply chain with loose couplings to a much stronger network that will meet demand and stakeholder concerns more effectively. Once formed elements of mutual trust, greater cooperation needs, sharing needs, learning together, teaching each other are exposed and have to be solved by answering the question 'how to achieve ambitions for greater collaboration and consolidation', which is demonstrated eventually through becoming an AVN.

Networks, as constellations of companies or businesses containing their suppliers, are not constant over time. Parts of the network may be inactive that is they are not actively connected to respond to either outside or inside stimuli. Over time connections switch off and on at different times and different connections are built. Moreover, active parts of a network respond to different stimuli at the same time. Such outside stimuli may be end customer requests, or orders, and inside stimuli, workplace events, business strategy and mission. Over time, all parts of a network should be used. If there is not enough work for the entire network then the network may be contracted, or even disbanded; but if the reverse is true then it can be expanded and as a consequence other partners are sought.

Large multi-national corporations (MNCs) have aimed for global power and have a geographic dispersion to attack competitors in their home market. But even MNCs have realized network solutions, organized by alliances with others, to gain competitive advantages by supplying value added innovative solutions over the traditional product-based market strategy of an MNC that does not have the full competence to provide it or indeed provide it in time. (" 'Best practice' manufacturing capabilities (are) no longer good enough to face the challenge of future competition and value creation" Chapter 14). MNC virtual networks (MNCVNs) may grow from a more innovative idea through exploitation towards a more dynamic and diversified business reality. Chapter 14 spells out an evolutionary framework in which this might happen. The importance of traditional manufacturing strategy is highlighted but this alone is not good enough, a higher value creation, astute market positioning, and dynamic mobility are new added game rules to accommodate. "In a manufacturing system, taking advantage of and leveraging other companies' existing resources seems to be more important than owning these resources – this implies that various types of inter-firm collaborations are therefore the preferred architecture."

SMEs are already acting in local regional clusters to give them more strength in business. In the high-tech electronics or software industrial sector they have shown to reach out to the global arena by cross clustering. They are orientated to high levels of innovation but the network although mutually supportive does not intend to have long-term relationships since they come together for one purpose in mind to deliver integrated services and product. This can now be a complementary business approach for all SMEs in high-tech, or indeed other sectors if opportunities arise.

ORGANIZATION

Organization is the social and work structure that unites all the organs available to it. There is much diversity, elasticity and strength in AVNs, they pool and time-share resources, share information and knowledge, are customer supportive and offer a superior service excellence with careful and astute responses to demand as well as are agile in action. How this is achieved in practice is partially through organization as well as other factors already mentioned. Organization must be defined more carefully to understand the implications.

Organization is an orderly structure, defining where decisions are made and having the necessary skills, knowledge and authorizations in place to make them. That is being furnished with the proper organs for management and control within and has an intelligent considered response to outside stimuli. Such organs are physical resources, virtual resources (e.g. manufacturing processes) and software proxies. In practice it is a powerful combination of all, as people are very important in how new information is responded to and for difficult decision making. But they need tools and an organized system to help them to carry it through. Individual companies and businesses, which may have their own underlying network structure, have an organization as well as the supra-organization of the AVN to contend with. These organizations need to fit closely together in the parts that count to meet the mission of the AVN, but at the same time protecting the individual companies involved that have their own strategies and missions to ensure longer life and stakeholder expectancy. Belonging to an AVN though provides for more opportunities, and makes better economic sense, in the highly competitive and more assertive world of today.

People are the most powerful resource and have to be freed up from mundane tasks that are more suitable to automation. Examples of this are easily found on a manufacturing shop floor where automation of processes has been rising since the availability of low cost high-powered microprocessors. Some high-tech manufacturers are increasing the amount of automation and are providing integrated manufacturing systems (IMS), which are more robust and have more functionality embedded within it than traditional manufacturing and execution systems (MES). It gives "(local) plant managers a much greater ability to run the plant more efficiently, and to monitor and control plant operations based on business goals, production realities and customer requirements." The customer requirements are those of the supra-organization of the AVN, or other networks to which the local-plant belongs or via a local enterprise resource management (ERM) system, or however the interface is designed. The IMS provides superior management and control of local affairs to reduce both wastes in time and in capacity thus increasing the capacity available on demand to stimuli from for instance the AVN. This capacity may be potentially increased in a number of ways. "Opportunistic preventive maintenance policies will consider the state of each part of the overall manufacturing system, to increase equipment availability. Maintenance and production activities will be synchronized, and tool setups optimized.........On a plant level, IMS makes sure that tools and processes are optimized, plant efficiency is high, and ship dates are met." Chapter 19 sees manufacturing becoming a 'hands off' process and "The factory of the near future will be highly hands-off, operating as a finely tuned machine."

In future the local manufacturing plant will have other enhanced facilities to improve the quality of information it provides and also the action it takes to control and manage production. Such features as Auto-ID for identifying unique parts that will "In the near future, every single object will be connected to the Internet through a wireless address and unique identifier." Having unique part identification enables better customer order decisions, or engineering decisions for new design prototypes, to be made more precisely through scheduling, tracking and expediting carried out at both the local plant level and the global AVN level (Chapter 1 and Chapter 15). Today most orders in WIP are not precisely identified in the management and control system but only by production batch number, or order line item, and may contain many customer order parts. Tracing of unique order parts by Auto-ID is an important health and safety control feature for food supply networks. Another future feature for a better automated system is the establishment of much more intelligent local sensor networks than are available today. This for instance would be to monitor plant facilities and equipment with more finesse to indicate when maintenance should be carried out and so as to increase the amount of useful uptime for capacity availability. Furthermore, "Ambient Intelligence (AMI), with sensors and actuators embedded in the environment capable of identifying individuals entering the environment, and not only delivering personalized services with pre-emptive capabilities, but also adapting chips to the particular needs of the moment (adaptive computing)." Chapter 2 adds a new dimension to the future of plants and businesses.

Local plants or businesses, other than manufacturing, when seeking new opportunities want to advertise their competencies and capabilities to others that may be known or unknown to them but with a potential willingness to join with them in a joint enterprise. An example of this may be taking action for a public tender for hospital supplies, another example may be to provide a service to a customer in which both manufacturing and logistics need to be solved. In the latter example a company seeks a local manufacturer and a local logistics provider to serve a customer in their locality. The company that is seeking partners may be located on the other side of the planet. The degree of difficulty increases by geographical removal from a customer and also when the companies' characteristics and track record are unknown. This limits opportunity taking with current technology. In future it will be possible through a single electronic market, as proposed for Europe, in which a database that stores data on important characteristics of companies and uniquely identified to them will exist (Chapter 3). Then companies seeking partners may securely access the database known as the 'Registry'. This is another example of a virtual organ. A more fascinating organ is one that acts in the virtual world on behalf of people as a personal assistant, or may be for a business as a business assistant. (Agent systems may correspond 1-to-1 with the individuals (firms or divisions) in the system being modeled, and the behaviors are analogs of real behaviors...... agents are software objects that proactively operate on behalf of their human masters in pursuing delegated goals." Chapter 1)

Personal assistants, or proxy agents have been under development for some time but using them in the context of AVNs and electronic markets is a complex challenge, not only resolving the legal issues that can arise as a consequence of carrying our their tasks in the virtual world, but also the shear complexity and control of the task assignment. There is no doubt that autonomous agents have great potential and will help considerably in the dynamic world of enterprise. This providing that a virtual world with agents can be modeled well enough and that there are good enough tools available to design them, that their roles are clearly defined, and that a suitable infrastructure can be provided for them to operate within the dynamic world of either an AVN or eMarket. Research and development programs continue to answer these provisos. The question remains when will this happen and can we harness their power well enough to make economic gains and create new opportunities.

People themselves are the most important organ for a company and an AVN. They may be located within one business node or operate as mobile workers, for instance at customer sites. They can make or break the system if responses are not given quick enough, the right decisions made, or that they do not understand the technology available to them or comprehend the organization or supra-organization they are dealing with. It is important that people are educated and provided with sufficient skills for the job to be done and that learning is a continuous process. They must also be empowered to make decisions for the tasks they manage and control and be provided with tools for the job through the other organs which serve them as well as others, but people need to be 'tuned in'. This is not to say they cannot think independently for this should be encouraged, as innovation will be created through them. And companies will only survive through innovation.

There are numerous organs available to companies today including some virtual ones not mentioned above, CRM, ERM, KM, mobile services, etc appearing throughout the book. But two powerful technologies will help the integration of these in the context of AVNs, they are the Semantic Web, "the semantic web is an extension of the current web in which information is given a well-defined meaning better enabling computers and people to work in co-operation."; and Web services, "..these promise to revolutionize the way in which organizations build applications, simplify integration projects and create a connection between business partners." These technologies are available, but work on ontologies still needs to be further developed.

Organizations demand high reliability and place extraordinary demand on the organs within them and their effectiveness is the key to collaborative work with others. We have briefly given above some of the component parts to be found within this book both in terms of the local node organization in a network with others and those virtual parts that are switched on or off as part of the supra-organization. We have discussed them as entities but we need now to turn our attention of how they are switched on and interconnect to provide the higher value required in an AVN.

CONNECTIVITY

Connectivity is the way the various organs within an organization are virtually coupled with other nodes in the constellation network of companies or businesses. This coupling is not continuous but spike like bursts of data and information flowing through the Internet or other channels and these are not always between the same business nodes within the network. Coupling may also be made to end customers, other networks and the eMarket as favored by a particular business node, or may be even the single market of the future. Coupling can take place through a number of different communication channels.

An important point has been raised in the book concerning the nature of the coupling. In traditional supply chains companies or businesses may cooperate without any formal agreement for joint collaboration. In other words they share no joint goal or mission and focus only upon their own individual interest. Whereas in creating a constellation of companies or businesses that are willing to do joint collaborative work such a formal agreement and a commitment exists, which specifies mission and goals. Fundamentally the agreement is to pool resources (this may be anything from an acceptable minimum amount of resources by the network to up to 100%) and time-share them when the occasion of mutually identified need arises, such as order processing. These constellations of business nodes can morph into an AVN if the collaborative coupling, local organs, relationships and mechanisms are capable of supporting it. So, the very important parts of this coupling are new business processes that were not in place in traditional supply chain operations.

The last paragraph was aimed at former supply chains merging into AVNs. But if creation of a Network has to start from scratch, using the eMarket perhaps to discover prospective partners, then a new mission is in the mind of at least one business. A process of discovery in future will perhaps start with the process of coupling with an eMarket say to gain maximum visibility of company electronic registries. But until such an eMarket is up and running business partners will most likely be found in the region of the prime business and coupling may be via more traditional means. Transient network constellations are already found in practice they have made agreements together perhaps to find a solution to answer a 'call for tenders' on a public authority website.

However, the mindset for a local business at one of the business nodes in a network constellation needs to adjust to new market demand. If we take for example a manufacturing business, they were traditionally focused on product manufacture, local productivity, and quality excellence. They know how to produce, how to split the work down into manageable parts, and how to control it. But when such a manufacturing business is part of a network moving toward an AVN then products are no longer the focus but customer orders are, which may be a customized service package (this may contain new innovative products with embedded services). And the strategies behind the agreements between the business and the network, such as a time-share of its production capacity and the way orders are moved in and out of the business and any special services that have to be provided establish new working practices. Thus different business processes have to be devised and learnt, and involve both synchronous and asynchronous information and knowledge flows between business nodes in the network, as well as with the ubiquitous mobile workforce.

Building the right business processes is a key for good collaborative coupling between players in the network. Such that when an external stimuli from the environment, as a customer order request, occurs parts of the network constellation are automatically switched on and each coupling as it occurs for the duration of the order request processing is like a well trodden path. This has to be well tried and faultless. To help with this process proxy agents or personal assistants may in future be used as part of the coupling process to seek out capacities, to warn of pending tasks, to supply information for decision making and so on, in other cases such as maintenance tasks mobile services may be used. They work together with the active mechanisms of the system and assigned personnel. The tasks are highly dynamic and couplings have also to take care of important internal events at each business node and for new external stimuli. Such internal events can arise because of time-shared capacities with other networks, loss of capacity at a crucial time and so on. The couplings within the network have to be carefully coordinated to maximize the returns for a business node as well as providing an excellent customer service from the business network by eventually delivering their high quality personalized order at promised time and date. The coupling to an external customer is also not straightforward as it depends upon customer preference and choice provided. Contact with people is still a preferred choice and must be accommodated but how to get the right communication channel mix is still a challenge today. Moreover, none of the available channels should be considered separately when responding to external stimuli except for any financial benefit that might be offered as a customer incentive.

There are a large number of business processes that must be dealt with and others will evolve. Some of the categories are procurement, order taking/processing, distributed design, distributed maintenance, performance enhancing, handing capacity usage, establishing new competencies, tendering, and seeking new partners. This leads on to another important issue relationships.

RELATIONSHIPS

Relationships between partner members of business networks intent upon becoming eventually an AVN have to be strong. This strength must be developed through a mutual trust and individual business node performance. One of the dominant themes with any relationship analysis is how people perform, so it is essential that all personnel be provided with the right education and continued counseling on how work is being done. Special multi-media tools will be of benefit to this the educational process as well as the close involvement of experts with novices to ensure knowledge transfer. In the future perhaps seeking centers of excellence, or educational experts via an electronic market registry may boost this process, "not only will SEEM facilitate the interaction student/teacher, but it will also increase the quality of the education" (Chapter 3). People are not expected to be part of any cybernetic system except through new input. But they are expected to use their intellect and knowledge to find new ways to analyze, synthesize and solve problems by interacting astutely with the information, distributed knowledge sources and the personalized tools that they have available to them. They, as a consequence of their discoveries, which may be helped by their personal assistants in future, may request changes to the system to plug gaps and improve the smooth running of the management and control system. This may be by another automated link, or decision loop. Perhaps they can see ways of how to get more capacity by fine-tuning, establishing new balances, and refine maintenance procedures and practices. In other ways people may interact with other people, such as customers. Positive but doable responses should always be given if the system is to work well. Maximum returns, in future, for productivity increases are now likely to come from people's actions and innovation where the maximum waste resides, "with the start of the new millennium, there has been a recognition that most of the productivity increases will come from interpersonal productivity". There will be a complete range of people in the workplace with different levels of experience and acquired knowledge. Their jobs will be different and their activities exposed to different events and stimuli. Some have to learn that ICT is a servant to them and not vice versa, they are the problem solvers not the machine. But the machine can work better for them if they continuously tune and modify its ability to carry out its tasks. It is crucial that this process is carried out to achieve symbiosis.

There are two basic relationships, static and dynamic. The former includes agreements, commitments, contracts, and tender responses. The later includes working together for order processing, responding to tenders, planning and scheduling, performance enhancement, product/total package design and customization, tracking unique order items, and so on. There are local tasks too that impact on the dynamic relationship that is handling and improving capacity, improving competence, and local marketing. There are also two types of relationship human-human and human-machine. The former can have challenges brought about by different language and culture. The later may be via a proxy agent or electronic personal assistant.

An elaborate social structure will develop over time as progression toward an AVN evolves. Learning to deal with both the external and internal environment by morphogenesis proves an AVNs viability. The AVNs specialized receptors are capable of responding to both a wide range and a fine gradation of stimuli that provides it with a networkability and differentiation to compete in the global market place.

Lastly but not least is the management of the business network. This may be led by the prime contractor to the end customer, but it may also be through an intermediary. The way AVNs will be managed is still an evolving scenario.

MECHANISMS

There are a number of virtual mechanisms available to any network, as the reader will see in the book, that respond to stimuli both external or internal and deal with information transfer, explicit knowledge formulation knowledge transfer, distributed planning and scheduling, order processing item tracking, financial tracking, decision assistance, and negotiation and so on. These mechanisms are assisted by the automated physical tools and sensor systems all acting in concert in the ideal real time world. Moving from systems as they are today to this integrated ideal in which every actor in the network understands all is quite a challenge and involves a number of development routes including new standards, multi-agent systems and web services among others. These mechanisms hold in virtual model form representations of the physical world. They also include algorithms that help to solve particular problems and seek to optimize in an automated way but they are at the same time responsive to real time events and not solely dependent upon deterministic equations based upon historical data or any data that is predicted. Prediction and history are no guarantee of success as any holder of stocks or shares will know too readily.

Although CAD tools and knowledge management and transfer tools are quite complex in distributed design, distributed order processing is equally complex, if not more complex, as it is subject to receive more random events. Cybernetic systems are available with the necessary feedback loops to ensure capacity control, item process timing control, and responses to changes in orders, or breakdowns in material flow. They have the ability to act upon particular stimuli to activate parts of the constellation into connecting pathways and procedures by utilizing the distributed organs available to optimize for instance order flow requirements. These depend in part on tailoring the actual manufacturing and order models from a core model for discrete manufacturing but also depend upon intervention of people at critical decision points, as order taking, order flow adjustment when reacting to change, and dispatching stages. New systems will evolve using intelligent autonomous multi-agents with the addition of Auto-ID technology and sensor networks to add a new dimension for controllability.

Rather than the conventional algorithms used today these may be enhanced or replaced by more advanced algorithms, models and techniques, such as game theory (a good introduction to this is given in Chapter 1). This is particularly the case for business node capacity management when working in multi-networks, playing the market and also for identifying most profitable production options. For example traditional capacity management algorithms based on MRP-II type logic have been proved in practice and also by academic research that they are not relevant to dealing with such issues and needs.

INTELLIGENCE

We have seen that organs available are escalating for business networks. They must be used to their fullest extent and the information they provide is a key to the evolution of the system. Imperfect systems provide large amounts of information. Smart organizations utilize this information by recognizing patterns and oddities, discovering order out of disorder. People should be encouraged to have imagination to have the ability to formulate a problem from what they see and not sit back and be run by rules and practices. Conveying this knowledge of a problem in its right form to others can result in innovative new solutions to a whole range of things to the benefit of the business and AVN. Creativity is a subtle process and must be encouraged if AVNs are to be achieved.

In the near future ambient intelligence (AMI) will have the potential in the workplace to help the synergy between people the system and the environment. How it will be used in practice is still too early to say but it will probably involve sensor networks, mobile agents, and adaptive personalized chips in some fashion to help people to become one with the working environment and be sensitive to its changes. New research and demonstrator projects over the next few years building upon previous research experiences will open up more opportunities for the AVN to capitalize upon.

The level of automation will increase to tackle the immediacy of real time control and cut out unnecessary loops where manual intervention currently exists to cobble different working process systems together. This has been illustrated in Chapter 19 for a manufacturing plant producing integrated circuit wafer arrays for silicon chips. Automation will also increase to provide a rapid service for information flow aiding the buyer and seller alike. Such automation of necessity has to have new levels of security mechanisms embedded within it protecting safe areas of information but maintaining flexibility, checking credit worthiness of the buyer, and carefully monitoring and controlling request authentication. Trusted and secure information flow of this kind will improve RFQs and eventual order processing. But order processing cannot be constructed from automating what we have now, people must be involved using their intellectual capacity to improve the coupling between buyer and supplier. No longer can we tolerate a passive approach by selling what is available but to understand the customer, what they want and how they perceive value, and for an AVN this is the end customer. "The future goal must be to build the business working back from the consumer by adopting radically new thinking which translates into radically new solutions. Solutions which are constructed around the changing drivers of customer value are unlikely to be just the products and services that have been offered in the past" (Chapter 8). This example applies to all the business processes.

ENVIRONMENT

There are three types of environment: that within the business node; that within the bounds of the supra-organization and its linkages; and that external to the network. Environments are dynamic and more highly so in some industries than others. The manufacturing world is highly dynamic, particularly that of high-tech industries. Stimuli arrive from customers seeking to purchase packaged services, or changing orders, changes to market regulations, changes to employment laws, changes to health and safety regulations and so on, but by far the most is from customers or prospective customers. Supra-organizational stimuli can be failed competence, seeking new competencies, order flow events including failures, sensor inputs, capacity change inputs, plan changes and so on. Internal business node stimuli include organ activity, sensors signals, event messaging, action messaging. A business node that handles multiple networks is one of the most difficult challenges but is a typical case.

SYMBIOSIS

Symbiosis can mean living together in harmony or fused in some way. It should promote the best of the combined potential of the capacity of the ICT and the intellectual capability of the people using it. It is a living system in the virtual world and tuned by people to mutual advantage, if it is to work well. But how can we achieve this symbiosis? We have a plethora of ideas elaborated throughout this book but sifting through them finding the keys to unlock the special routes to take is difficult and will be discussed more fully in the change process below. However what is important is that the people using ICT have to feel comfortable with it in all its respects of mechanisms and communication channels. Partly this can be resolved through personalized interfaces or personal assistants (proxy agents) that steer them and help them with searching, digesting and decision making. And partly this can be resolved through ensuring a seamless ICT with different organs, mechanisms and connectivity. But importantly people must use their knowledge and skills and have the right education programs. The road to achieve full symbiosis will be hard and may never be achieved to the satisfaction of the AVN or the people operating within it.

VALUE GAINED

Adaptive Value Networks as the name implies have value or create value and are adaptive to the environment. Adapting does not mean just adjusting to the environment but the process by which it adjusts and the connections that need to be made are important. It is a process of orientation and balance taking account of pliancy and responsiveness. How these processes begin depends largely upon 'external' stimuli. Outside here is meant outside of the AVN or before this is created outside of a latent value network that is ready to form at short notice from a constellation of companies or businesses that have some agreement to work in partnership if the right occasion arises. On the other hand 'internal' has several meanings in this context. It can be inside one of the companies, which may have its own wholly owned network of businesses, or the inside of an adaptable value network or a latent network. The response to an outside stimulus depends in all cases on the connectivity, relationships, mechanisms, intelligence and the symbiosis. The latter, symbiosis is a fusion of all and its strength depends upon the bonds and cohesive forces. This may be termed the 'networkability'. Value comes from the belonging to the network and through its transaction processes. It also takes into account customer perceptions of cost to them and value to them.

NEW GENERATIONS OF ICT

There is still a need to beware of hype and over selling of particular technology. The 'Third Wave' of e-commerce is a case in point. Technology is created with certain new benefits but how these benefits are converted into practical solutions for modern business depends upon us all. The days of buying a package are long gone. Now all must be aware of what is it we are trying to achieve and out of all that is available can it be harnessed in the right way to serve our need today and use it to build upon to help serve the need of tomorrow.

Like all revolutions there are more innovations than we can usefully employ. The trick is to see their potential for further improvement. There will be those with an original idea and there will be followers. But all must have a vision of the future for their business to be able to use ideas or by pioneering innovations. The book has provided the reader with a number of new technologies, new concepts for use, and new components that are being developed. We will briefly revisit some of these to see when and how they can be deployed for AVNs, how they may improve the infrastructure, the available organs and the mechanisms.

In the near future ambient intelligence will transform the workplace. Smart sensors integrated with the local IMS systems will contribute to improvements in resource utilization through smarter maintenance and also through agents that handle capacity much more effectively. Consequently a business will become a smarter player in an AVN. Every order component part will have an ID tag and together with ambient intelligence assigned to monitoring the order flow, better reactive control will be made possible. The smart play will be how to handle the increased power for control through autonomous agents in a multi-agent domain of an AVN whereby capacity for laggards to plan is continually negotiated. The virtual control environment is a very fluid balance between resource and an order part. People remain extremely important to continuously tune the system through analysis and synthesis of its historic results utilizing refined information processing, possibly using grid computing. Astute mathematical techniques will be used to refine the control process.

Management too can be improved through recognizing that the AVN is fluid and not fixed. Use may be made of self organizing data networks, the semantic web, distributed agents, and SEM and other new technology as it emerges. Management however will evolve through the use of an ICT system that models business processes in a mutable digital form, reflecting reality possibly through process calculus models of the third wave.

Management and control of an AVN in the future will be performed based on facts and actual 'wants' and needs of customers rather than anticipated market trends. Hence more emphasis and priority will be given to execution processes and to how such customer demands will be met more effectively and efficiently. Emphasis will also be given on how to sense and respond to events as they occur in a manner that is cost effective and also acceptable by the customer. This will not only involve managing local disturbances within a business but will also involve how an AVN is configured and who the partners are to deliver the right products at the right time. It will also involve how decisions can be made collaboratively, deployed in real time and implemented accurately across the AVN partners. The selection and implementation of the right ICT infrastructure will therefore become of strategic importance for a business since it will potentially position a business as a key, strategic player in an AVN. The result is that this partner will be the one who always gets more business.

INDUSTRIAL INERTIA

Generalizations cannot be made about the way companies are reacting to the new challenges for business. Today there are companies that are slow on the uptake, there are those that have been burnt with poor experiences in particular with software or supply chain evolution, there are those that are helped by local government initiatives and there are those with enough money and shareholder support to experiment and so on. In Europe, in 2002 it was revealed that only 10.4% of enterprises state that eBusiness constitutes a significant part of the way they operate today. Moreover only 6.8% believe that eBusiness has significantly changed customer-supplier relationships. However, statistics in the UK show that on average of the microbusinesses and small and medium sized businesses 65% have Internet access. And the larger the company and the higher the turnover this increases to 92%. The use of Broadband is increasing and is now at 27% for all users. These figures may not reflect what is happening worldwide in absolute terms but the trend to more and more take up does. So the main inertia seems to be utilizing the Internet in the right manner. This to some extent is not surprising since early forms of e commerce reflected the normal world of selling by providing a virtual shop and simple transactions only were done over the Internet. There were some successes but in general frustration and disappointment was felt. Much more was wanted to provide for more customer opportunities and the ability to deliver what the customer really wanted not the standard products they had to sell.

On the whole large companies and corporations have been fully engaged with getting their supply chains into a better shaped cooperative structure and have heavily invested in ICT. This is not surprising since they have more to lose. Consultants and ICT providers have given them support and perhaps influenced the direction of activities. The key drivers have been cost and operational efficiency. But now many boards and corporate heads are taking charge of ebusiness initiatives and removing the responsibilities from IT departments. Value to stakeholders must be assured. "More fundamentally, the core challenge may be not only the profitability issues but more crucially a life-warning signal that indicates a business course or a rule-of-the-game has been changed. If the company fails to sense the change and still focuses on its product order-winning criteria, its innovation power will be damaged and sustainability will be jeopardized" (Chapter 14). Companies and their supply chains are trying to meet the challenge of obtaining appropriate good quality information to and from the supply network. Today information does not flow without restriction in a supply chain meeting all needs in a timely manner from the start of the product concept phase in engineering then through production, logistics and sales to product end of life maintaining good customer support at all times. The larger company wields its influence through standards organizations, government research programs, and through academia and the ICT providers and wherever else to ensure that it meets its objectives.

For the small and medium enterprise (SMEs), the situation is somewhat different. SMEs are generally less informed than the larger company about ICT and its probable benefits. Importantly, "SMEs and micro companies represent the 95% of the European companies and employ more than 70% of the manpower…they are the solid platform of the European economy" (Chapter 5). And there are 35million SMEs worldwide. But considerable numbers of SMEs and micro-businesses lack the necessary education and need help. Research initiatives that are designed by local associations or authorities have proved to be good enablers in Europe and a number of case studies are given within the book. These research projects have given rise to a large number of isolated eMarkets unable to interact with others to increase both the scale and scope. Examples of temporary and more permanent networks of companies are given and a case where global joint design project in textiles is made possible. SMEs however largely start from their home area base learning to get strength from others in more formal arrangements. A factor that limits them is affordability of ICT and the ease with which they may seek partners wider field in joint enterprise. New possibilities such as SEEM and grid computing may help in the future but for now help by associations and larger companies wishing to establish value networks can help with the learning process of working with others in joint enterprise. Although the very large company and the SME may be engaged in activities to work toward AVNs there is little evidence that the medium sized company is doing so and as a consequence they may be falling by the wayside.

THE CHANGE PROCESS

This book is mainly about the change process opportunities. It sets out the state-of–the-art, new concepts, possibilities of new technologies and research programs. It is a journey through a wealth of knowledge. We have seen how SMEs are reacting, what currently are the limitations for stimulating them into action, what can be done about removing some of these constraints, examples of early explorations in working together and the possibilities for the future both in terms of the types of activity to be encouraged and ways in which they can maximize their potential. We have also seen the other end of the spectrum of the large company growing into value networks in a stepwise fashion and more revolutionary approaches by large multi-national corporations with their strategic alliances. A realignment of industry is taking place to take account of developing markets in Asia and other underdeveloped areas; and in the western world to concentrate more upon the added value of pre-production and post-production stages. In the next years there will be more 'off-shore' workplaces and industrial moves for a better world balance in economic competence.

The change process toward AVNs from conventional supply chains will be long and hard. Mentioned throughout the book are a number of key dates when new technology is ready and stretches well out beyond 2010. In part I the scale and scope we see that the knowledge revolution just like the industrial revolution will take a number of years to complete, but unlike the industrial revolution which started in one country the knowledge revolution involves the whole planet at its outset. It is an exciting time for this century and what will be achieved will depend on how well we employ our imagination. The case studies presented within the book give some indication of how far we have got to progress. So many companies are still steeped in tradition but nevertheless so much is changing. Unlocking the full potential of companies is 'what is needed', bringing their potential to the marketplace or in collaboration with others will bring about further gains to the world economy.

The book has provided the reader with some of the conflicting objectives at play. It also should provide the reader with a greater understanding of the types of AVN and how they may be modeled and equipped with ICT to their mutual benefit, how interaction with their customers may be improved, how they can manage their day to day commitments and obligations. It also provides the reader with a view of the research and development activities and the emerging technologies that could be useful instruments for transformation. The inputs to companies and business networks are many and varied and the pace of change they are experiencing in their environment is fast moving. There are trail-blazers and there are laggards but those which adapt to the new environment and the surge of new ideas in the right way at the right time will reap the major benefit. It calls for all to be mentally sharp and to spot the opportunities as they arise. There is no panacea for change or any prescriptive solution, judgements are necessary, and investments have to be weighed against the future benefits. Relationships with others have to be worked at and the workforce educated to fit in with new expectations. Both AVNs and companies need strategies to carry them through and these need to be continually modified to match the conditions prevailing. Hard graft by people motivated in their work using their intellect will pay off with the right differentiation in the marketplace to compete and grow.

ICT providers and consultants to date have provided much of the impetus for industrial change. They must now help to provide affordable solutions for all. Industry has to fully engage with strategic plans. To be part of an AVN demands full dynamic engagement. Research and development, improved education standards, emerging technologies, new ICT standard agreements, and governmental policies to reduce regulation and improve legislation for electronic commerce will go a long way to promote the process of change. Early trail blazers for AVNs will demonstrate the feasibility of different new solutions. Lastly the change process is in the hands of the company and its stakeholders and the people employed. It calls for leadership and imagination of the highest order and a determination to carry through to the objective.

Contributing Authors

Shoumen Datta is the Executive Director of the MIT Forum for Supply Chain Innovation at MIT. He can be contacted at <http://supplychain.mit.edu/innovation/shoumen.htm>

Bob Betts is a president of Timogen systems, a supply chain software company, USA. He can be contacted at <http://timogen.com/bob_pr.htm>.

Mark Dinning is in the Supply Chain Engineering Group at Dell Inc. He can be contacted at <Mark_Dinning@Dell.com>

Feryal Erhun is an Assistant Professor in the department of Management Science and Engineering, Stanford University. She can be contacted at <http://ww.stanford.edu/~ferhun/>

Tom Gibbs is a director of industry marketing at Intel.

Pinar Keskinocak is a Coca-Cola Chair and Assistant Professor in the school of Industrial and Systems Engineering, Georgia Institute of Technology. She can be contacted at <http://www.isye.gatech.edu/people/faculty/Pinar_Keskinocak/>

Hui Li is a PhD Student, Dept of Aeronautical Engineering and Astrophysics, MIT. She can be contacted at <huili@mit.edu>

Mike Li is a Research Scientist, Dept of Civil and Environmental Engineering, MIT. He can be contacted at <mikeli@mit.edu>

Micah Samuels has been working in Operations at Amazon.com since 2002. He can be contacted at <micahsam@amazon.com>

Florent Frederix is a Scientific Officer - Communication Networks, Security and Software, Applications Unit D5 in European Commission. He can be contacted at <Florent.FREDERIX@cec.eu.int>

Isidro Laso is a Scientific Officer of the New Working environment Unit in European Commission. He can be contacted at <Isidro.LASO@cec.eu.int>

Juno Chang is a Professor of Media Technology at SangMyung University, Seoul, Korea. He can be contacted at <jchang@smu.ac.kr>

Min-Hyung Kang is a doctoral candidate at KAIST Graduate School of Management, Korea. He can be contacted at <woodstck@kgsm.kaist.ac.kr>

Flavio BONFATTI is an associated professor of Software Engineering, University of Modena and Reggio Emilia. He can be contacted at <bonfatti@unimo.it>

Paola Daniela MONARI is a CEO of SATA. She can be contacted at <p.monari@satanet.it>

Thomas E. Callarman is the Director of the Institute for Manufacturing Enterprise Systems and Associate Professor of Operations in the College of Business at Arizona State University. He can be contacted at <thomas.callarman@asu.edu>

John W. Fowler is a Professor in the Industrial Engineering Department at Arizona State University. He can be contacted at <john.fowler@asu.edu>

Esma S. Gel is an Assistant Professor of Industrial Engineering at Arizona State University. She can be contacted at <esma.gel@asu.edu>

Michele E. Pfund is a Manager of Decision Analysis and Portfolio Optimization for Johnson & Johnson Pharmaceutical Research & Development, L.L.C. She can be contacted at <Michele.fund@asu.edu>

Dan L. Shunk is a Professor of Industrial Engineering at Arizona State University. He can be contacted at <dan.shunk@asu.edu>

Guido Grau works at ITV Denkendorf in the Management Research Department. He can be contacted at <geg@itv-denkendorf.de>

Marcus Winkler works at ITV Denkendorf (department Management Research). He can be contacted at <marcus.winkler@itv-denkendorf.de>

Chris Lawer is a doctoral candidate at the Cranfield School of Management in the UK and is also the founder of The OMC Group (http://www.theomcgroup.com). He can be contacted at <Christopher.lawer.phd.02@cranfield.ac.uk>

Simon Knox is a Professor of Brand Marketing at the Cranfield School of Management in the UK. He can be contacted at <s.knox@cranfield.ac.uk>

Rainer Breite is a lecturer in Industrial Management and Engineering of Tampere University of Technology, Finland. He can be contacted at <breite@pori.tut.fi>.

Hannu Vanharanta is a professor in Industrial Management and Engineering in Tampere University of Technology, Finland. He can be contacted at <hannu.vanharanta@pori.tut.fi>.

António Lucas Soares is an Assistant Professor at the Faculty of Engineering of the University of Porto and a researcher at INESC Porto R&D Institute. He can be contacted at <asoares@inescporto.pt>

Jorge Pinho de Sousa is an Associate Professor at the Faculty of Engineering of the University of Porto and area leader at INESC Porto R&D Institute. He can be contacted at <jsousa@inescporto.pt>

Patrizia Fariselli is a head of the Unit Nomisma-NET (http://www.nomisma.it), Italy. She can be contacted at <farisellip@nomisma.it>

Malcolm McDonald is a Professor of Marketing and Deputy Director Cranfield School of Management with special responsibility for E-Business. He is also a Fellow of Royal Society of Art (FRSA). He can be contacted at <m.mcdonald@cranfield.ac.uk>

Hugh N. Wilson is a Visiting Fellow and Director of the Center for e-Marketing at Cranfield School of Management. He can be contacted at <hwilson@vcmeasom.u-net.com>

Dr Yongjiang Shi is a senior research fellow and research director of the Centre for International Manufacturing in Cambridge University. He can be contacted at <ys@eng.cam.ac.uk>

Yoon S. Chang is with Korea Advanced Institute of Science and Technology (KAIST), Korea. He can be contacted at <yoonchang@kaist.ac.kr>

Duncan McFarlane is a Senior Lecturer in Manufacturing Engineering in the Cambridge University Engineering Department. He can be contacted at <dcm@eng.cam.ac.uk>

Christian Dannegger is a Vice President of Applications & Solutions, Whitestein Technologies. He can be contacted at <Christian.Dannegger@living-systems.com>

Klaus Doere is a Senior Researcher at Whitestein Technologies. He can be contacted at <Klaus.Dorer@living-systems.de>

Grace Y. Lin is an Associate Partner in Business Consulting Services, IBM Global Services, USA and an elected member of the IBM Academy of Technology. She can be contacted at <gracelin@us.ibm.com>

Jun-Jang Jeng is with IBM Research, USA. He can be contacted at <jjjeng@us.ibm.com>

Ko-Yang Wang is an IBM Distinguished Engineer and Research and Innovation Executive in Business Transformation, IBM Global Services (IGS), USA. He can be contacted at <kyw@us.ibm.com>

Toshiya Kaihara is an associate professor at Kobe University Japan. He can be contacted at <kaihara@cs.kobe-u.ac.jp>

Douglas Scott is a pioneer in developing Manufacturing Execution Systems (MES).

Robert Pisa is Director of MES Product Management at Brooks Automation. He can be contacted at <robert.pisa@brooks.com>

BG Lee is the Application Marketing Director in the MES business unit of Brooks Automation. He can be contacted at <Byunggon.Lee@brooks.com>

Harris Charalampos Makatsoris is a founder and managing director of Orion Logic Ltd., a UK based, software and services vendor of adaptive supply chain optimization and execution software. He can be contacted at <h.makatsoris@orionlogic.com>

Howard D Richards has been associated with European research and development since the early 1980s, working on manufacturing science and technology, new forms of planning and control software, software for extended and virtual enterprises, and e-markets. He can be contacted at <richards@rmplc.co.uk>

Index